Databook of Adhesion Promoters

2nd Edition

Anna Wypych

ChemTec Publishing

Toronto 2023

Published by ChemTec Publishing
38 Earswick Drive, Toronto, Ontario M1E 1C6, Canada

© ChemTec Publishing, 2018, 2023
ISBN 978-1-77467-012-5 (hardcover), ISBN 978-1-77467-013-2 (epub)

Cover design: Anita Wypych

Library and Archives Canada Cataloguing in Publication

Title: Databook of adhesion promoters / Anna Wypych.
Names: Wypych, Anna, author.
Description: 2nd edition. | Includes bibliographical references and index.
Identifiers: Canadiana (print) 2022026953X | Canadiana (ebook)
20220269556 | ISBN 9781774670125
   (hardcover) | ISBN 9781774670132 (PDF)
Subjects: LCSH: Adhesives. | LCSH: Adhesion.
Classification: LCC TA455.A34 W97 2023 | DDC 620.1/99—dc23

Table of Contents

1 Introduction	**1**
2 Information on data fields	**3**
3 Adhesion Promoters	**17**
3.1 Acids	**17**
Methacrylic acid	17
3.2 Acrylates	**19**
Bomar BR-541S	19
Bomar BR-543MB	20
Bomar BR-741	21
Bomar BR-742S	22
Bomar ER-744BT	23
Bomar BR-941	24
Bomar BR-970BT	25
Bomar BR-3641AJ	26
Bomar BR-7432GB	27
Bomar BR-7432GI30	28
Bomar BR-14320S	29
Bomar XR-741MS	30
Elvaloy AC 2103	31
Sarbox SB401	34
Sarbox SB510M35	35
Sarbox SB520E35	36
Sarbox SB520M35	37
Sarbox SR9050	38
Sarbox SR9051	39
Sarbox SR9053	40
Sarbox SR9054	41
3.3 Amines, amides, amidoamines	**42**
Ancamide 261A	42
Ancamide 2353	43
Ancamide 2396	44
Ancamide 2424	45
Ancamide 2482	46
Ancamine K54	47
Cohedur H 30	49
Jaylink JL-103M	51
Jaylink JL-106E	52

Markoba HMT 53

Nourybond 276 54

Priamine 1073-LQ-(GD) 55

Priamine 1074 57

3.4 Benzene derivatives **59**

Ekaland BQD 30 DS 59

Ekaland PPDN 30 DT 60

Ekaland PPDN 30 DX 61

Ekaland PPDN 50 HU 62

3.5 Carbamic resin **63**

Alnovol UF 410 RPC 63

3.6 Chlorinated polyolefins **64**

Eastman Advantis 510W 64

Eastman AP 550-1 (25% in Aromatic 100) 67

Eastman AP 550-1 (25% in xylene) 71

Eastman CP 153-2 (25% solids in xylene) 74

Eastman CP 310W 77

Eastman CP 343-1 (25% solids in xylene) 80

Eastman CP 343-1 (50% solids in xylene) 84

Eastman CP 343-1 (100% solids) 88

Eastman CP 343-3 (25% solids in xylene) 91

Eastman CP 347W 95

Eastman CP 515-2 (40% solids in aromatic 100) 98

Eastman CP 515-2 (40% solids in toluene) 102

Eastman CP 515-2 (40% solids in xylene) 105

Eastman CP 730-1 (20% solids in xylene) 109

Eastman CP 730-1 (100% solids) 113

3.7 Crosslinkers **116**

Visiomer TMPTMA 116

Visiomer TRGDMA 117

3.8 Epoxides **118**

Isopropyl glycidyl ether 118

Phenyl glycidyl ether 120

3.9 Esters **122**

Phthalate diethylene glycol diacrylate 122

Radcure ODA 123

Uniplex 260 124

3.10 Inorganic compounds   **126**

Markoba CB20   126

Markoba CB-S   127

Markoba CB23   128

3.11 Ionomers   **129**

Loxanol MI 6721   129

Lupasol SC 61 B   131

Lupasol SK   133

3.12 Isocyanates   **135**

Desmodur BL 2078/2   135

Desmodur BL 3175A   137

Desmodur RFE   139

Nourybond 289   142

Nourybond 290   143

3.13 Isocyanurates   **144**

Dynasylan VPS 7163   144

Vulcabond MDX   146

Vulcabond VP   147

3.14 Lignin   **149**

Lignin   149

3.15 Maleic anhydride modified polymers   **150**

Amplify, TY 1451B   150

Amplify, TY 4817   151

Bynel™21E4817   152

Bynel™41E1352   154

Bynel™41E3351B   156

Eastman G-3003   157

Eastman G-3015   160

Fusabond A560   163

Fusabond E100   165

Fusabond E158   167

Fusabond E204   169

Fusabond E205   171

Fusabond N216   173

Fusabond E226   175

Fusabond E265   177

Fusabond E528   178

Fusabond E564   179

Fusabond M603   180

Ricobond® 1731 HS   181

Xibond 120   182

Xibond 140   183

Xibond 160   184

Xibond 180   185

Xibond 220   186

Xibond 230   187

Xibond 240   188

Xibond 250   189

Xibond 255   190

Xibond 260   192

Xibond 280   193

Xibond 285   194

Xibond 315   195

Xibond 330   196

Xibond 370   197

Xibond 375   198

Xibond 830   199

3.16 Melamine   **200**

Actmix HMMM-50GE F140   200

Cohedur A 200   202

Cohedur A 250   203

Cohedur RDL   204

Cohedur RK   205

Cyrez 963   206

Cyrez 964   207

Cyrez CRA-100   208

Markoba HMMM   209

3.17 Metal-organic complexes (non-silicon)   **210**

Chartwell B-505.1   210

Chartwell B-515.1/2H   212

Chartwell B-515.1W   214

Chartwell B-515.71W   216

Chartwell B-525.1   218

Chartwell C-515.71/1.5H   220

Chartwell C-515.71HR   222

Chartwell C-515.72HRW   224

Chartwell C-515.72HRX   225

Chartwell C-545.1   227

Chartwell C-600   228

Chartwell D-535.1   230

3.18 Metal-organic complexes (non-silicon)+silica   232

Chartsil B-515.1/2H   232

Chartsil C-505.1/2H   234

Chartsil C-523/2H   236

3.19 Monomers   238

Bis(2-methacryloxyethyl) phosphate   238

Fancryl FA-512AS   240

4-Methacryloxyethyl trimellitic anhydride   241

3.20 Oligomers   242

Sarbox SB400   242

UA-1605N   243

3.21 Phenol novolac resins   244

Alnovol PN 760   244

3.22 Phosphoric acid esters   245

Sipomer PAM 100   245

Sipomer PAM 200   247

Sipomer PAM 4000   248

Sipomer PAM 5000   249

3.23 Polymers and copolymers   250

Bomar BR-3741AJ   250

Cecabase RT 2N1   251

Cecabase RT 945   252

Cecabase RT BIO 10   254

Cleartack W130   256

Escorex 2173   257

Poly DNB   258

Polytex E-100   259

Sulfonex M-80   260

3.24 Polyols   262

Hypomer FX-2460AF   262

Hypomer FX-2860A   263

Hypomer FX-4365AF   264

Polypol 610   265

Polypol 615   266

Polypol 653   267

Polypol 663   268

Polypol 676   269

Polypol 693   270

Priplast 1837   271

Terrin 168	273
Terrin 168G	274
Terrin 170	275
3.25 Resorcinol	**276**
Cofill 11 GR	276
Cohedur RS	277
Cohedur VP KA 9197	278
Markoba RSC	279
3.26 Rosin	**281**
Eastman ester gum 8D resin	281
Foral 85-E CG hydrogenated rosinate	284
Foral 105-E CG hydrogenated rosinate	287
Foralyn 5020-F CG hydrogenated rosinate	290
Pexalyn 9085	293
Pexalyn 9100	296
Pexalyn Ester 10	299
Pexalyn SR	302
Pexalyn T100	304
Regalite R1090 hydrocarbon resin	307
Regalite R1100 CG hydrocarbon resin	310
3.27 Silanes	**312**
Carbo NXT	312
CoatOSil 1770 Silane	314
CoatOSil 2287 Silane	316
CoatOSil DRI Waterborne Silicone	318
CoatOSil MP 200 Silane	319
Dowsil AZ-720 Silane	321
Dowsil Primer C OS	323
Dowsil SZ-6030 Silane	326
Dowsil Z-6026 Silane	328
Dowsil Z-6062 Silane	331
Dowsil Z-6094 Silane	333
Dowsil Z-6119 Silane	336
Dowsil Z-6120 Silane	339
Dowsil Z-6121 Silane	342
Dowsil Z-6124 Silane	346
Dowsil Z-6137 Silane	349
Dowsil Z-6269 Silane	352
Dowsil Z-6341 Silane	356

Dowsil Z-6376 Silane   359
Dowsil Z-6675 Silane   362
Dowsil Z-6883 Silane   365
Dowsil Z-8090 Silane   368
Dowsil Z-9805 Silane   370
Dynasylan 1122   373
Dynasylan 1124   376
Dynasylan 1146   379
Dynasylan 1161 EQ   382
Dynasylan 1175   385
Dynasylan 1189   388
Dynasylan 1401   391
Dynasylan 1505   392
Dynasylan 2201 EQ   394
Dynasylan 4150   397
Dynasylan 6490   398
Dynasylan 6498   400
Dynasylan 6598   402
Dynasylan 9116   404
Dynasylan 9896   405
Dynasylan AMEO   406
Dynasylan AMEO-T   409
Dynasylan AMMO   411
Dynasylan BDAC   414
Dynasylan BTSE   416
Dynasylan DAMO   417
Dynasylan DAMO-T   420
Dynasylan GLYEO   422
Dynasylan GLYMO   425
Dynasylan Hydrosil 1151   429
Dynasylan Hydrosil 2627   432
Dynasylan Hydrosil 2776   435
Dynasylan Hydrosil 2907   438
Dynasylan Hydrosil 2909   441
Dynasylan Hydrosil 2926   444
Dynasylan MEMO   446
Dynasylan MTES   449
Dynasylan MTMO   450
Dynasylan MTMS   453
Dynasylan OCTEO   454

Dynasylan OCTMO	455
Dynasylan PTEO	456
Dynasylan SIVO 110	457
Dynasylan SIVO 121	458
Dynasylan SIVO 202	459
Dynasylan SIVO 210	461
Dynasylan SIVO 214	463
Dynasylan SIVO 408	465
Dynasylan SIVO 560	466
Dynasylan SIVO Clear Ec	467
Dynasylan TRIAMO	468
Dynasylan VTEO	471
Dynasylan VTMO	474
Dynasylan VTMOEO	477
Geniosil GF 31	480
Geniosil GF 56	483
Geniosil GF 60	486
Geniosil GF 62	489
Geniosil GF 69	492
Geniosil GF 80	494
Geniosil GF 82	497
Geniosil GF 93	499
Geniosil GF 94	502
Geniosil GF 95	504
Geniosil GF 96	507
Geniosil GF 98	510
Geniosil XL 10	513
Geniosil XL 12	516
Geniosil XL 32	519
Geniosil XL 33	521
Geniosil XL 65	523
NXT	526
NXT Z 52S	528
Si 69®	530
Si 75®	531
Si 264™	533
Si 266®	535
Silcat 17 Industrial Silane	536
Silcat RHE Silane	538

Silox 23	540
Silquest A-137	542
Silquest A-151NT	544
Silquest A-171	547
Silquest A-172NT	550
Silquest A-174NT	553
Silquest A-178	556
Silquest A-186	558
Silquest A-187	560
Silquest A-189	563
Silquest A-1100	566
Silquest A-1102	569
Silquest A-1106	572
Silquest A-1110	574
Silquest A-1120	576
Silquest A-1128	580
Silquest A-1130	582
Silquest A-1160	585
Silquest A-1170	587
Silquest A-1387	590
Silquest A-1524	592
Silquest A-1871	594
Silquest A-1891	596
Silquest A-2120	598
Silquest A-2387	601
Silquest A-Link 15	603
Silquest A-Link 25	606
Silquest A-Link 35	609
Silquest A-Link 597	612
Silquest A-Link 599	615
Silquest G-170	617
Silquest PA-1	619
Silquest RC-1	621
Silquest VS-142	624
Silquest VX-225	625
Silquest Y-9627	627
Silquest Y-9669	629
Silquest Y-9936	632
Silquest Y-11699	634

Silquest Y-15744   636
Silquest Y-15866   637
Silquest Wetlink 78   639
SiSiB® PC1100   641
SiSiB® PC1120   642
SiSiB® PC1200   643
SiSiB® PC1220   645
SiSiB® PC1710   647
SiSiB® PC1711   648
SiSiB® PC1800   649
SiSiB® PC2000   650
SiSiB® PC2200   652
SiSiB® PC2300   654
SiSiB® PC2310   655
SiSiB® PC2320   656
SiSiB® PC2521   657
SiSiB® PC2640   658
SiSiB® PC2720   659
SiSiB® PC3100   660
SiSiB® PC3200   661
SiSiB® PC3300   662
SiSiB® PC4100   663
SiSiB® PC6110   665
SiSiB® PC6120   666
SiSiB® PC6130   667
Wacker® Adhesion Promoter AMS 60   668
Wacker® Adhesion Promoter AMS 70   670
Xiameter™ OFS-6011 Silane   672
Xiameter™ OFS-6030 Silane   674
Xiameter™ OFS-6032 Silane   676
Xiameter™ OFS-6040 Silane   680
Xiameter™ OFS-6062 Silane   683
Xiameter™ OFS-6075 Silane   686
Xiameter™ OFS-6076 Silane   689
Xiameter™ OFS-6094 Silane   692
Xiameter™ OFS-6106 Silane   696
Xiameter™ OFS-6224 Silane   700
Xiameter™ OFS-6610 Silane   704
Xiameter™ OFS-6697 Silane   706

Xiameter™ OFS-6920 Silane 709
Xiameter™ OFS-6925 Silane 712
Xiameter™ OFS-6940 Silane 715
Xiameter™ OFS-6945 Silane 718
3.28 Silane+silica **720**
Coupsil 6109 720
Coupsil 8113/8113 GR 721
Dynasylan SIVO 160 722
3.29 Silane+silicate **725**
Geniosil CS 2 725
3.30 Silane+titanate **728**
Dowsil P5200 728
Dowsil PR-2260 732
3.31 Silicate+silica **735**
Dynasylan 40 735
3.32 Silicic acid ester **736**
Dynasylan A 736
Dynasylan A SQ 737
Dynasylan AR 738
Dynasylan M 740
Dynasylan MKS 741
Dynasylan P 742
3.33 Silicic acid ester+SiO$_2$ **743**
Dynasylan XAR 743
3.34 Sucrose acetate isobutyrate **744**
Subcrose Acetate Isobutyrate 90 744
Sustane SAIB 746
Sustane SAIB MTC 748
3.35 Sulfur compounds **750**
Duralink HTS 750
3.36 Titanates **752**
Tytan AP20 752
Tytan AP100 754
Tytan AP110 756
Tytan AP120 758
Tytan AP130D 760
Tytan AP310 762
Tytan TAA 764
Tyzor AA-65 766
Tyzor AA-105 768

Tyzor GBA   769
Tyzor GBO   772
Tyzor IAM   774
Tyzor TPT   776
3.37 Zirconates   **778**
Tyzor® 212   778
Tyzor® 217   780
Tytan APZ900   782
Zirconate Zr-402   784

1 Introduction

Adhesion promoters form a very important group of additives, without which many industrial products (29 product groups made out of 30 different polymeric materials) cannot perform according to requirements. The previous publication on this subject was mostly related to silanes, which formed the most widely used group of these additives. The information on silanes was based on the book that was published at the beginning of the 1980s. Since then, many new additives were introduced into the market. Many of these new additives are not based on silanes but on one of 37 chemical groups of chemical compounds needed for a variety of products in which silanes do not function, are too expensive, or better performance can be achieved with these new additives, some of which are environmentally friendly and come from renewable resources.

Databook of Adhesion Promoters contains extensive data on the most important products in use today. Two groups of data are included: data for some chemical compounds used for the manufacture of adhesion promoters (data included come from many available sources) and commercial products (data from a single supplier of material).

The information on each adhesion promoter included in the **Databook of Adhesion Promoters** is divided into five sections: General information, Physical properties, Health and safety, Ecological properties, and Use & performance. The data belong to almost 150 data fields, which accommodate a variety of data available in the source publications. The description of each section below gives more detail on the composition of information.

In the **General** information section, the following data are displayed: name, CAS #, EC #, Acronym, Active matter, Amine number, Amine value, Chemical class, Chemical structure, Common synonym, Complexed organics, Composition, Empirical formula, Functional organic group, Functionality (average), General description, IUPAC name, Metal content, Mixture, Moisture content, Molecular formula, Molecular mass, Molecular structure, Moisture contents, Name, Number of metals, Organoreactive group, Purity, RTECS number, and Solids content.

The **Physical-chemical** properties section contains data on State, Odor, Color, Color (Gardener), Color (platinum-cobalt scale), Acid number, Acidity, Ash content, Boiling point, Freezing point, Glass transition temperature, Melting point, Pour point, Chloride content, Cloud point, Density, Evaporation rate, Hildebrand solubility parameter, Hydrolysable chloride, Hydrolysis half-time, Hydroxyl number, Kinematic viscosity, Melt flow rate, Neutralizing agent, Nitrogen content, pH, Refractive index, Saponification value, Solubility in water and solvents, Solubility (diluents), Specific gravity, Specific heat, Sulfur content, Surface tension, Thermal decomposition products, Thermal stability (TGA), Vapor density, Vapor pressure, Viscosity, Viscosity SUS, and Volatility.

Health and safety section contains data on Autoignition temperature, Flash point, Flash point method, Explosive LEL, Explosive UEL, NFPA Health, NFPA Flammability, NFPA Reactivity, HMIS Health, HMIS Fire, HMIS Reactivity, Hazardous combustion products, Agency ratings, Hazard ingredients labeling, Hazardous hydrolysis products, UN number, UN Risk Phrases, R, UN Safety Phrases, S, DOT Class, ICAO/IATA class, IMDG class, TDG class, UN/NA hazard class, UN risk and safety phrases, ICAO/IATA Class, IMDG Class, OSHA hazard class, Rat oral LD50, Mouse oral LD50, Guinea pig dermal LD50, Rabbit dermal LD50, Inhalation rat LC50, Skin irritation, Eye irritation, Ingestion, Inhalation, Effect of repeated or overexposure, First aid: eyes, skin, and inhalation, Exposure personal protection, ACIGH, NIOSH, and OSHA exposure limits, Carcinogenicity by ACGIH, IARC, NTP, and OSHA, Mutagenicity, Teratogenicity, and Specific target organ.

Ecological properties section contains data on Bioaccumulative potential, Bioconcentration factor, Biodegradation probability, Aquatic toxicity LC50 (Green algae, *Rainbow trout, Bluegill sunfish, Fathead minnow, Zebra fish,* and *Daphnia magna*), Biological, Chemical, and Theoretical Oxygen Demand, BOD/COD ratio, and Partition coefficient (log K_{oc} and log K_{ow}), and Stability in water half-life.

Use & performance section contains information on Manufacturer, Outstanding properties, Potential substitutes, Recommended for polymers, Recommended for products, Recommended applications, Processing methods, Concentration used, Guidelines for use, Food approvals, Alternative products, and Conditions to avoid.

The above data are given, whenever available, for more than 400 most important adhesion promoters produced and used today. This book is best used together with the **Handbook of Adhesion Promoters**, which contains an analysis of scientific and patent literature available on the subject today. Both books are complementary and form the most comprehensive, actual, and accurate information on the subject of adhesion promoters.

2 Information on data fields

The fields used in the databook are listed in alphabetical order. The information on data in a particular field includes a glossary of the term used, the unit of measurement, typical methods used to measure quantities available in the databook.

ACID NUMBER
Twenty-five grams of an adhesion promoter is placed in 125 ml Erlenmeyer flask, and 50 ml alcohol is added to dissolve the sample. If the sample is not completely soluble, 50 ml of equal amounts of alcohol and acetone are used. This sample is titrated with 0.01N NaOH or KOH in the presence of bromothymol blue used as an indicator. ISO standard uses a method of titration similar to just described ASTM standard, but phenolphthalein is used as an indicator, and 0.1 N NaOH is used as a titrating agent. The results are expressed in mg KOH per 1 g of sample.

ACRONYM
Abbreviations of adhesion promoter names can be found in the standard terminology (ASTM D1600-14 Standard Terminology for Abbreviated Terms Relating to Plastics). ISO standard has a separate section for symbols used for some additives (ISO 1043-2:2011 Plastics -- Symbols and abbreviated terms -- Part 2: Fillers and reinforcing material). Abbreviations of some adhesion promoters used in rubber may be given in a separate standard (ISO 6472:2010 Rubber compounding ingredients -- Symbols and abbreviated terms). In most cases, abbreviations are created for convenience by manufacturers and users.

ACTIVE MATTER
Active matter gives the percentage of adhesion promoter in the total weight of the commercial additive.

AMINE NUMBER
Equivalent weight is a calculated value determined from the amine number (also known as the amine value or amine alkalinity). The amine number is determined by titration of the amine acetate ion by a dilute, typically 1N HCl solution. For pure material, the amine number can be calculated using the molecular weights of the pure compound and KOH (56.1 g/mol).

ANIMAL TESTING, ACUTE TOXICITY
Acute toxicity describes the adverse effects resulting from a single exposure to a substance. Typical methods of measurement include LD50, which is the amount of a solid or liquid material that is required to kill 50% of test animals in one dose.

AQUATIC TOXICITY
Aquatic toxicity is an adverse effect on marine life (aquatic organisms) that results from exposure to a toxic substance. Aquatic toxicity can be expressed as the lethal concentration of a chemical substance in milligrams per liter that caused death to 50% population of aquatic species (LC50) during the time of the experiment (usually 24, 48, or 96 hours).

ASH CONTENT
The ash content equals the weight of the ash divided by the weight of the original sample multiplied by 100%.

ATMOSPHERIC LIFETIME

The atmospheric lifetime of a species measures the time required to restore equilibrium in the atmosphere following a sudden increase or decrease in the concentration of the species in question in the atmosphere. The longer the lifetime, the higher the global warming potential of species.

AUTOIGNITION TEMPERATURE

The lowest temperature at which a material will ignite and sustain combustion in the absence of a spark or flame. The properties of liquid can be tested by standardized methods (ASTM E659-15 Standard Test Method for Autoignition Temperature of Chemicals). The ignition temperature is the temperature in degrees Centigrade at which the substance shows spontaneous combustion when touching hot bodies.

AVERAGE PARTICLE SIZE

Average particle dimension (usually diameter).

BIOCONCENTRATION FACTOR

It is the ratio of a substance's concentration in tissue of an aquatic organism to its concentration in the ambient water in situations where the organism is exposed through the water only and the ratio does not change substantially over time.

BIODEGRADATION PROBABILITY

The probability that an organic substance will biodegrade under aerobic conditions. This field includes the general statements regarding the biodegradation probability of adhesion promoters. Experimental data are included in the following fields: Biological Oxygen Demand, Chemical Oxygen Demand, and Theoretical Oxygen Demand.

BIOLOGICAL OXYGEN DEMAND

The biological oxygen demand, BOD, is the mass concentration of dissolved oxygen consumed under specific conditions in a given time (e.g., BOD-5 stands for 5 days test) by the aerobic biological oxidation of a chemical or organic matter in water. BOD is an empirical test, which evaluates the ultimate aerobic biodegradability of organic compounds in water. The following ISO standards can be applied: ISO 10708:1997 Water quality -- Evaluation in an aqueous medium of the ultimate aerobic biodegradability of organic compounds -- Determination of biochemical oxygen demand in a two-phase closed bottle test and ISO 5815-1:2003 Water quality -- Determination of biochemical oxygen demand after n days, BODn.

BOILING POINT

The boiling point is the temperature in degrees Centigrade at which the substance undergoes a transition from the liquid into the gaseous phase under normal pressure. Tapped density is an increased bulk density attained after mechanically tapping a container containing the powder sample.

BLUEGILL SUNFISH

In order to assess bioaccumulation of substance, a test organism such as *Bluegill sunfish, Daphnia magna, Fathead minnow, Rainbow trout,* or other is selected. The test substance is administered as a suspension directly into the water. A test organism is observed by a specified number of hours. If deleterious effects due either to toxicity or pathogenicity are observed, sequentially lower doses should be tested. The experiment should establish a an

LC50 value for a particular substance and a test organism (LC50 is a dose required to kill 50 percent of the test organisms).

CARCINOGENICITY

A carcinogenic material is one that is known to cause cancer. The process of forming cancer cells from normal cells or carcinomas is called carcinogenesis. A summary of findings included in this field is based on general principles of a material assessment, which includes: a – There is limited evidence of carcinogenicity from studies in humans. A cause and effect interpretation is credible, but that alternative explanations such as chance, bias, other variables, etc. cannot be ruled out. Again, science can never prove a hypothesis, only disprove one. Scientific "facts" are established only when a preponderance of the evidence supports a hypothesis, and there is 1) no evidence to disprove it and 2) no equally viable alternative hypotheses. b – There is sufficient evidence of carcinogenicity from studies in experimental animals, which indicates there is an increased incidence of malignant and/or a combination of malignant and benign tumors (1) in multiple species or at multiple tissue sites, or (2) by multiple routes of exposure, or (3) to an unusual degree with regard to incidence, site, or type of tumor, or age at onset. c – There is less than sufficient evidence of carcinogenicity in humans or laboratory animals; however, the substance is structurally related to other materials that are either human carcinogens or reasonably anticipated to be human carcinogens. d – There is convincing relevant information that the material acts through mechanisms that are likely to cause cancer in humans. Carcinogenicity lists are maintained by ACGIH, NTP, IARC, and OSHA.

CAS NUMBER

A number assigned by the Chemical Abstracts Service that uniquely identifies a chemical substance.

CHEMICAL CLASS

Non-systematic classification of adhesion promoters based on their main component. Proprietary and masterbatch entries have their contents usually unknown.

CHEMICAL OXYGEN DEMAND

The amount of oxygen required for the chemical oxidation or decomposition of compounds in water.

CHLORIDE CONTENT

The weight percentage of chloride in the adhesion promoter

CLOUD POINT, MMAP

Measured in 1:2 mixture of methylcyclohexane and aniline (a measure of compatibility)

COEFFICIENT OF THERMAL EXPANSION

The coefficient of thermal expansion is a fractional increase in volume per unit rise in temperature.

COLOR

This field gives a description of color that is typical of a commercial product or pure specimen.

Color, Gardner scale
The Gardner scale is a visual scale (described in ASTM D1544, Standard Test Method for Color of. Transparent Liquids)

Color, Platinum-cobalt scale
The Platinum-Cobalt Scale is a color scale that was introduced in 1892 by chemist Allen Hazen (also known as Hazen or APHA scale). ASTM D1209, Standard Test Method for Color of Clear Liquids (Platinum-Cobalt Scale), contains a precise description of measurement.

Common name
Many adhesion promoters have commercial names (see under label – Name). This frequently does not permit recognition the chemical composition of the adhesion promoter. The common name helps in recognition of the chemical nature of the material.

Common synonym
A common synonym has a similar application as the Common name. It is used to help in the identification of the chemical nature of the product.

Complexed organics
The field contains information on the percentage of organic matter reacted with metal (e.g., silicon) to form adhesion promoter (e.g., silane)

Composition
Components of formulated products are given if available.

Daphnia magna
In order to assess bioaccumulation of substance, a test organism such as *Bluegill sunfish, Daphnia magna, Fathead minnow, Rainbow trout,* or other is selected. The test substance is administered as a suspension directly into the water. A test organism is observed by a specified number of hours. If deleterious effects due either to toxicity or pathogenicity are observed, sequentially lower doses should be tested. The experiment should establish an LC50 value for a particular substance and a test organism (LC50 is a dose required to kill 50 percent of the test organisms).

Decomposition temperature
Decomposition temperature is determined using thermogravimetric analysis. Typically either beginning or temperature range is given.

Density
Several methods are used to determine the density of the adhesion promoters, such as hydrometer, digital density meter, displacement, and pycnometer methods.

Density temperature
The temperature at which density was determined.

DOT CLASS

Transportation instructions require DOT Hazard Class in which materials are divided into the following classes: 1 – Explosives, 2 – Gases, 3 – Flammable and combustible liquids, 4 – Flammable spontaneously combustible solids, 5 – Oxidizers & organic peroxides, 6 – Poisonous & infectious materials, 7 – Radioactive materials, 8 – Corrosives, 9 – Miscellaneous. In addition, UN number and packaging group is also given.

EC NUMBER

The number assigned by the EU commission to a substance (previously EINECS and ELINCS) is used to identify a compound.

EMPIRICAL FORMULA

The empirical formula is the molecular formula of a chemical compound. The order of atoms follows the Hill system, which is utilized by the Chemical Abstracts Services and by the Beilstein Institute. Within the empirical formula, C is the first element symbol, H is the second, the other element symbols are added in alphabetical order. The empirical formula does not take into account any crystal water content. Using this field for searches requires that the above guidelines are strictly followed.

ENTHALPY OF VAPORIZATION

Enthalpy is a thermodynamic function of a system, equivalent to the sum of the internal energy of the system plus the product of its volume multiplied by the pressure exerted on it by its surroundings. Conversion into vapor requires the absorption of the enthalpy of vaporization.

EVAPORATION RATE

An evaporation rate is a rate at which a material will vaporize (evaporate, change from liquid to vapor) compared to the rate of vaporization of a specific known material (e.g., butyl acetate)

EXPLOSION LIMIT, LOWER, LEL

The explosion limits are the lower and the upper border concentration in volume percent of flammable gas or vapor mixed with air between which the mixture can be made to explode by heating or by a spark.

EXPLOSION LIMIT, UPPER, UEL

The explosion limits are the lower and the upper border concentration in volume percent of flammable gas or vapor mixed with air between which the mixture can be made to explode by heating or by a spark.

EYE IRRITATION

The information included in this field comes from existing human experience, animal observations related to the potential irritation of the human eye, and *in vitro* studies.

FATHEAD MINNOW

In order to assess bioaccumulation of substance, a test organism such as *Bluegill sunfish, Daphnia magna, Fathead minnow, Rainbow trout*, or other is selected. The test substance is administered as a suspension directly into the water. A test organism is observed by a specified number of hours. If deleterious effects due either to toxicity or pathogenicity are observed, sequentially lower doses should be tested. The experiment should establish an LC50 value for a particular substance and a test organism (LC50 is a dose required to kill 50 percent of the test organisms).

FIRST AID
The emergency treatment to be administered to an injured or sick person before professional medical care is available.

FLASH POINT
The flash point is the lowest temperature in degrees Centigrade at which so much vapor develops under normal pressure that it results in a flammable mixture together with the air over the liquid level. Different methods are used in the test, with the Cleveland cup being the most suitable method for testing adhesion promoters. Cleveland open cup is used to determine flash and fire points of liquids with a flash point above 79°C and below 400°C, such as promoters (ASTM D92-12b Standard Test Method for Flash and Fire Points by Cleveland Open Cup Tester). Standard gives the methods of determination using manual and automatic Cleveland open cup apparatus. About 70 ml of test liquid is heated first rapidly, then slowly close to approaching an expected flash point. Test flame is applied to the surface to ignite vapors. Test flame is a natural or bottled gas flame (full description included in the standard). Test flame is applied first when the temperature is 28°C below the expected flash point and then in 2°C intervals. The flash point is the lowest temperature at which vapors are ignited by the test flame.

FLASH POINT METHOD
The following abbreviations are used to describe the method that was used for the data included in the field "Flash point": CC – closed cup, CCTC – closed cup tag closed, COC – Cleveland open cup, OC – open cup, PMCC - Pensky-Martens closed cup, TCC- tag closed cup, TOC – tag open cup.

FREEZING POINT
The temperature at which the liquid and solid phases of a substance of specified composition are in equilibrium at atmospheric pressure.

FUNCTIONAL ORGANIC GROUP
The field contains the name of the functional group which can react with the polymer site of the joint.

FUNCTIONALITY, AVERAGE
The average number of functional groups in the adhesion promoter

GENERAL DESCRIPTION
The field contains information on the product's structure.

GLASS TRANSITION TEMPERATURE
The temperature region at which polymer transitions from hard, glassy material to a soft, rubbery material

GLOBAL WARMING POTENTIAL

The Global Warming Potential (GWP) allows the comparison of the global warming impacts of different gasses. It is a measure of how much energy the emissions of 1 ton of a gas will absorb over a given period of time, relative to the emissions of 1 ton of carbon dioxide (CO_2). The larger the GWP, the more that a given gas warms the Earth compared to CO_2 over that time period.

HEAT OF COMBUSTION

It is the quantity of heat liberated per unitary weight when a substance undergoes complete oxidation. It is expressed in MJ per kg.

HEAT OF FUSION

It is the change in enthalpy resulting from heating a given quantity of a substance to change its state from a solid to a liquid.

HENRY'S LAW CONSTANT

The solubility of a gas in a liquid is proportional to the pressure of the gas over the solution. Henry's law constant is a proportionality factor of this relationship.

HILDEBRAND SOLUBILITY PARAMETER

The Hildebrand solubility parameter gives a numerical estimate of the degree of interaction between materials. It helps to evaluate solubility, particularly for nonpolar materials.

HMIS CLASSIFICATION

A rating system (HMIS - Hazardous Materials Identification System) has been devised by The National Paint Coatings Association to assist emergency responders. The following are simple explanations of numerical symbols. Health: 0 – Like the ordinary material, 1 – Slightly hazardous, 2 – Hazardous – use breathing apparatus, 3 – Extremely dangerous – use full protective clothing, 4 – Too dangerous to enter – vapor or liquid. Flammability: 0 – Will not burn, 1 – Must be preheated to burn, 2 – Ignites when moderately heated, 3 – Ignites at normal temperature, 4 - Extremely flammable. Reactivity: 0 – Normally stable, 1 – Unstable if heated – use normal precaution, 2 – Violent chemical change possible – use the hose from a distance, 3 – Strong shock or heat may detonate – use monitors, 4 – May detonate – evacuate the area if materials are exposed to fire.

ICAO/IATA CLASS

In classification for transport by cargo aircraft, the statements are based on the Dangerous Goods Regulation issued by IATA (International Air Transport Association). The class and the packaging group are stated. No account is taken of special regulations stipulated by individual countries or airlines.

IMDG CLASS

This field contains the hazard classification for transport by sea.

INGESTION

Ingestion is the act of taking something (food, medicine, liquid, poison, etc.) into a body through the mouth. Synonyms include "swallowing," "taking internally," or "eating."

INHALATION
Inhalation is the drawing of air or other substances (fumes, mists, vapor, dust, etc.) into the lungs (the respiratory system). A common synonym is "breathing in."

IUPAC NAME
International Union of Pure and Applied Chemistry, IUPAC, standardized names of organic compounds and created a systematic naming system which is given here.

KINEMATIC VISCOSITY
Kinematic viscosity is the ratio of absolute (or dynamic) viscosity to density usually expressed in centiStokes.

LAMBDA VALUE
See K-factor. Both are the same, but Lambda value is used in Europe.

LC50
Lethal Concentration 50, LC50, is the concentration of a chemical that kills 50% test animal population. This measure is generally used when test animals are exposed to a test chemical in the form of gas or mist.

LD50
Lethal Dose 50, LD50, is the dose of a chemical that kills 50% of a sample population. In full reporting, the dose, treatment, and observation period should be given. LD50, LC50, ED50, and similar figures are only comparable when the age, sex, and nutritional state of the animals are specified.

MANUFACTURER
The name of the manufacturer is given for commercial adhesion promoter having a particular brand name. For generic compounds, the name of the manufacturer(s) is also frequently suggested.

MASTERBATCH
Masterbatch is a solid or liquid additive for plastics used for imparting special properties to the product.

MAXIMUM GAS YIELD
It is a temperature or its range in which the maximum decomposition takes place as determined by thermogravimetric analysis.

MELT FLOW INDEX
Melt Flow Index (MFI) is a measure of how many grams of a polymer flow through the die in ten minutes.

MELTING POINT
The temperature at which a given solid will melt.

METAL CONTENT
Metal content (e.g., silicon) in the commercial adhesion promoter by weight.

MIXTURE
A statement whether adhesion promoter is a mixture of compounds or pure product.

MOISTURE CONTENT
Moisture concentration in a commercial adhesion promoter as declared by its manufacturer.

MOLECULAR MASS
The sum of the atomic weights of all the atoms in a molecule. The weight of a molecule of any gas or vapor as compared with the hydrogen atom as a standard.

MUTAGENICITY
The capacity of a chemical or physical agent to cause permanent alteration of the genetic material within living cells. Tests of chemical substances and physical agents for mutagenic potential include microbial, insect, mammalian cell, and whole animal tests.

NAME
A proper name is used, which, in the case of a commercial product, is a brand name given by the manufacturer and the most commonly used name in the case of generic compounds.

NEUTRALIZING AGENT
A chemical compound that was used for neutralization.

NITROGEN CONTENT
The weight percentage of nitrogen in the sample of adhesion promoter.

NFPA CLASSIFICATION
A rating system has been devised by the National Fire Protection Association, NFPA, to assist emergency responders. The following are simple explanations of numerical symbols. Health: 0 – Like the ordinary material, 1 – Slightly hazardous, 2 – Hazardous – use breathing apparatus, 3 – Extremely dangerous – use full protective clothing, 4 – Too dangerous to enter – vapor or liquid. Flammability: 0 – Will not burn, 1 – Must be preheated to burn, 2 – Ignites when moderately heated, 3 – Ignites at normal temperature, 4 - Extremely flammable. Reactivity: 0 – Normally stable, 1 – Unstable if heated – use normal precaution, 2 – Violent chemical change possible – use the hose from a distance, 3 – Strong shock or heat may detonate – use monitors, 4 – May detonate – evacuate the area if materials are exposed to fire.

NIOSH REL
National Institute for Occupational Safety and Health, NIOSH, recommended exposure limits (RELs) will be based on risk evaluations using human or animal health effects data and on an assessment of what levels can be feasibly achieved by engineering controls and measured by analytical techniques. The RELs are given in either mg/m^3 or ppm.

ODOR
Any property detected by the olfactory system. In the case of chemical materials and particularly adhesion promoters, it helps to distinguish different materials and select non-intrusive materials for the application.

ORGANOREACTIVE GROUP
The chemical structure of an organic group of adhesion promoter.

OSHA PEL
A Permissible Exposure Limit, PEL, is the maximum amount or concentration of a chemical that a worker may be exposed to under the U.S. Occupational Health and Safety Administration, OSHA, regulations. The PELs are given mg/m^3 or ppm.

OUTSTANDING PROPERTIES
The most important properties, which may help in selection for the application.

OZONE DEPLETION POTENTIAL
The Ozone Depletion Potential (ODP) is a number that refers to the amount of ozone depletion caused by a substance. The ODP is the ratio of the impact on ozone of a chemical compared to the impact of a similar mass of CFC-11. Thus, the ODP of CFC-11 is defined to be 1.0.

PARTITION COEFFICIENT
The organic carbon adsorption coefficient, K_{oc}, is only applicable to individual substances. The K_{oc} can generally be calculated from the octanol/water partition coefficient, K_{ow}. The following equation can be used: $\log K_{oc} = 0.937 (\log K_{ow}) - 0.006$. The $\log K_{ow}$ is determined in a laboratory without the use of organisms. It is a measure of how polar the substance is by determining whether the substance partitions primarily to water or to octanol. Substances that partition primarily to octanol are likely to bioaccumulate in the fat of organisms.

pH
The pH value is the negative decadic logarithm of the concentration of hydrogen ions (dimensionless).

POUR POINT
The pour point is a temperature at which a substance still remains liquid.

PROCESSING METHODS
Manufacturers recommendations are given.

PURITY
The weight percentage of pure compounds in the adhesion promoter.

RAINBOW TROUT
In order to assess bioaccumulation of substance, a test organism such as *Bluegill sunfish, Daphnia magna, Fathead minnow, Rainbow trout,* or other is selected. The test substance is administered as a suspension directly into the water. A test organism is observed by a specified number of hours. If deleterious effects due either to toxicity or pathogenicity are observed, sequentially lower doses should be tested. The experiment should establish an LC50 value for a particular substance and a test organism (LC50 is a dose required to kill 50 percent of the test organisms).

RECOMMENDED FOR PRODUCTS
Manufacturers' recommendations are given as to the product application for manufactured goods.

RECOMMENDED FOR POLYMERS
Manufacturers' recommendations are given in this field.

REFRACTIVE INDEX
The ratio of the velocity of propagation of an electromagnetic wave in a vacuum to its velocity in the medium. It is a parameter that helps to identify chemical compounds and means of their selection for use in compositions having predesigned optical properties. Refractive index has been measured at two temperatures, 20 and 25°C, unless otherwise indicated.

REL
PEL and REL are acronyms used by the safety industry to define Permissible Exposure Limits (OSHA term) and Recommended Exposure Limits (NIOSH term).

RELATIVE PERMITTIVITY
The dielectric constant is a measure of the behavior of the substance when introduced into an electric field (it indicates the multiple to which the capacity of a condenser increases if the substance is between the plates instead of vacuum).

RTECS NUMBER
The Registry of Toxic Effects of Chemical Substances (RTECS®) is a comprehensive database of basic toxicity information for over 150,000 chemical substances, including prescription and non-prescription drugs, food additives, pesticides, fungicides, herbicides, foaming agents, diluents, chemical wastes, reaction products of chemical waste, and substances used in both industrial and household situations. Reports of the toxic effects of each compound are cited. In addition to toxic effects and general toxicology reviews, data on skin and/or eye irritation, mutation, reproductive consequences, and tumorigenicity are provided. Federal standards and regulations, NIOSH recommended exposure limits, and information on the activities of the EPA, NIOSH, NTP, and OSHA regarding the substance are also included. The toxic effects are linked to literature citations from both published and unpublished governmental reports and published articles from the scientific literature. The database corresponds to the print version of the Registry of Toxic Effects of Chemical Substances, formerly known as the Toxic Substances List, started in 1971. Originally prepared by the National Institute for Occupational Safety and Health (NIOSH), the RTECS® database is now produced and distributed by MDL Information Systems, Inc.

SAPONIFICATION VALUE
The saponification value is the number of mg of potassium hydroxide required to saponify one gram of fat.

SHIPPING NAME
The shipping name of the product as defined by the US Department of Transportation, which can be found in Hazardous Materials Table (Title CFR49).

SKIN IRRITATION
Although several mammalian species may be used, the albino rabbit is the preferred species. Exposure duration normally is four hours.

SOLUBILITY PROPERTIES
Information on the solubility of adhesion promoters in selected solvents.

SOLUBILITY IN WATER
Information on the solubility of adhesion promoters in water.

SPECIFIC GRAVITY
The ratio of the density of a substance to the density of a standard (water).

SPECIFIC HEAT
The amount of heat in Jules needed to raise the temperature of one mole of a substance by one degree of Kelvin.

STATE
State of adhesion promoters at room temperature are listed in this field.

SULFUR CONTENT
The weight percentage of sulfur is expressed as SO_2. Several methods can be used for this determination, such as combustion method, sodium peroxide fusion, or oxygen flask combustion.

SURFACE TENSION
The force acting on the surface of a liquid, tending to minimize the area of the surface; quantitatively, the force that appears to act across a line of unit length on the surface. Also known as interfacial force, interfacial tension, surface tensity.

THEORETICAL OXYGEN DEMAND
The calculated amount of oxygen that is required to oxidize a compound to its final oxidation products. However, there are some differences between standard methods that can influence the results obtained: for example, some calculations assume that nitrogen released from organics is generated as ammonia, whereas others allow for ammonia oxidation to nitrate. Therefore in expressing results, the calculation assumptions should always be stated.

THERMAL CONDUCTIVITY
The thermal conductivity, λ, is the quantity of heat transmitted, due to a unit temperature gradient, in unitary time under steady conditions in a direction normal to a surface of the unit area, when the heat transfer is dependent only on the temperature gradient.

THERMAL STABILITY BY TGA
The thermal stability of adhesion promoters is determined by TGA at constant temperature increase (usually 10°C/min), and it is recorded as a temperature at which degradation begins.

THERMOMECHANICAL ANALYSIS DATA

The thermomechanical analysis is used for determination of the beginning of decomposition temperature of an adhesion promoter, the temperature at which there is the maximum evolution of gaseous substances, and the density of the material being expanded. The method is the most frequently used in studies of microspheres which are expanded at the process temperature.

TLV-TWA 8h

The time-weighted average concentration for a conventional 8-hour workday and 40-hour workweek exposure to a substance, to which it is believed that nearly all workers may be repeatedly exposed, day after day, without adverse health effects. The data are given after the American Conference of Governmental Industrial Hygienists, ACGIH, National Institute for Occupational Safety and Health, NIOSH, and Occupational Safety & Health Administration, OSHA.

TOTAL GAS EVOLUTION

It is determined from the mass loss by the sample during thermogravimetric analysis.

UN RISK PHRASES

A list of numbers of risk phrases compiled by UN which characterize behavior of a particular compound

UN SAFETY PHRASES

A list of numbers of safety phrases compiled by UN which should be followed during transportation of a particular compound

UN/NA CLASS

A four digit number representing a particular chemical or group of chemicals. These numbers are assigned by the United Nations (UN Numbers), the U.S. Department of Transportation (NA Numbers), or Transport Canada (NA Numbers). These numbers are commonly used throughout the world to aid in the quick identification of the materials contained within bulk containers (such as rail cars, semi-trailers, and intermodal containers).

VAPOR DENSITY

The density of a gas relative to the density of air (air density=1).

VAPOR PRESSURE

Vapor pressure is a fundamental thermodynamic property of a solid or liquid. It is the pressure generated at a particular temperature by a pure component that has liquid (or solid) and vapor in equilibrium in a closed vessel. Its units are the usual units of pressure (e.g., kPa). The vapor pressure of a liquid increases with a temperature between the triple point and the critical point.

VAPOR PRESSURE TEMPERATURE

The temperature at which vapor pressure was measured.

VISCOSITY
It is a ratio of shear stress and shear strain expressed in mPa s.

3.1 Acids
Methacrylic acid

PARAMETER	UNIT	VALUE
GENERAL INFORMATION		
Name		Methacrylic acid
CAS #	-	79-41-4
EC number	-	201-204-4
Empirical formula	-	C4H6O2
Formula		
Molecular mass	daltons	86.09
RTECS number	-	OZ2975000
Purity	wt%	99.5
PHYSICAL PROPERTIES		
State	-	liquid
Odor	-	repulsive
Color	-	colorless
Boiling point	°C	162
Melting point	°C	15.4-15.5
Density at 20°C	kg/m³	1010
Henry constant	mPa m³/mol	126
pH	-	2-2.2
Refractive index at 20°C	-	1.4288
Solubility in water at 20°C	g/l	98
Surface tension at 20°C	mN/m	65.9
Vapor density	-	2.97
Vapor pressure at 20°C	hPa	0.97
Viscosity at 20°C	mPas	1.4
HEALTH & SAFETY		
NFPA classification	Flammability	2
	Health	3
	Instability	2
HMIS classification	Flammability	2
	Health	3
	Physical hazard	1
Carcinogenicity	Not listed by ACGIH, IARC, NTP, or CA Prop 65.	
Autoignition temperature	°C	400

3.1 Acids
Methacrylic acid

PARAMETER	UNIT	VALUE
Flash point	°C	67
Flash point method	-	CC
Animal testing, acute toxicity, Rat oral LD50	mg/kg	1060
Animal testing, acute toxicity, Mouse oral LD50	mg/kg	1250
Animal testing, acute toxicity, Rabbit dermal LD50	mg/kg	500
Animal testing, acute toxicity, Guinea pig dermal LD50	mg/kg	1000
Animal testing, acute toxicity, Rat inhalation, LC50	mg/l	7.1/4h
First aid, eye		In case of contact, immediately flush eyes with plenty of water for a t least 15 minutes. Get medical aid immediately.
First aid, inhalation		If swallowed, do NOT induce vomiting. Get medical aid immediately. If victim is fully conscious, give a cupful of water. Never give anything by mouth to an unconscious person.
First aid, skin		In case of contact, immediately flush skin with plenty of water for at least 15 minutes while removing contaminated clothing and shoes. Get medical aid immediately. Wash clothing before reuse.
ACGIH, TLV	ppm	20
NIOSH, REL	ppm	20/70
OSHA, PEL	ppm	20
UN/NA #	-	2531
ECOLOGICAL PROPERTIES		
Aquatic toxicity, *Daphnia magna*, 48-h LC50	mg/l	>130
Aquatic toxicity, *Rainbow trout*, 96-h LC50	mg/l	85
Biodegradation probability	68%/28d, readily biodegradable	
Partition coefficient, log P_{ow}	-	0.93
USE & PERFORMANCE		
Manufacturer	generic	
Outstanding properties	Methacrylic acid occurs in oil from the plant, *Roman chamomile*.	
Recommended for products	In cosmetics and personal care products, methacrylic acid is used as an adhesion promoter (primer) in certain artificial nail products.	
Guidelines for use	stabilized with 250 ppm MEHQ (150-76-5)	

3.2 Acrylates
Bomar BR-541S

PARAMETER	UNIT	VALUE
GENERAL INFORMATION		
Name		Bomar BR-541S
Composition		difunctional polyether urethane acrylate
Chemical class	-	acrylate
PHYSICAL PROPERTIES		
Pt-Co color	-	<50
Density at 25°C	kg/m^3	1060
Refractive index at 25°C	-	1.49
Glass transition temperature	°C	440
Viscosity at 60°C	mPas	3,000
USE & PERFORMANCE		
Manufacturer		Dymax Corporation
Outstanding properties		improved adhesion (cold rolled steel, glass, PC, stainless steel) and weatherability, optically clear, stable color, gloss finish
Recommended for polymers		PC
Recommended for products		nail coating, hard glossy coatings, scratch-resistant coating, inks
Recommended applications		ideal for nail gel application

Bomar BR-543MB

PARAMETER	UNIT	VALUE
GENERAL INFORMATION		
Name		Bomar BR-543MB
General description		difunctional oligomer
Composition		aliphatic polyether urethane methacrylate
Chemical class	-	acrylate
PHYSICAL PROPERTIES		
Pt-Co color	-	20
Density	kg/m^3	1030
Refractive index	-	1.476
Glass transition temperature	°C	-56
Viscosity at 60°C	mPas	14,000
USE & PERFORMANCE		
Manufacturer		Dymax Corporation
Outstanding properties		provides high tensile strength, excellent optical clarity, oil resistance and improved impact resistance, hydrolytic stability and offers improved adhesion
Recommended for polymers		ABS, PET
Recommended applications		nail coatings, optical coatings, adhesion-promoting oligomer

Bomar BR-741

PARAMETER	UNIT	VALUE
GENERAL INFORMATION		
Name		Bomar BR-741
General description		difunctional oligomer
Composition		aliphatic polyester urethane diacrylate
Chemical class	-	acrylate
PHYSICAL PROPERTIES		
Color, Platinum-cobalt scale	-	40
Density at 25°C	kg/m^3	1,100
Refractive index at 25°C	-	1.494
Viscosity at 50°C	mPas	150,000
USE & PERFORMANCE		
Manufacturer		Dymax Corporation
Outstanding properties		improved weatherability, enhanced hardness, abrasion resistance, gloss finish and non-yellowing effect, exceptional adhesion properties, it can be effectively formulated for adhesion to various plastics with some amount of surface energy
Recommended for polymers		PC
Recommended applications		metal and plastic coatings

Bomar BR-742S

PARAMETER	UNIT	VALUE
GENERAL INFORMATION		
Name		Bomar BR-742S
General description		difunctional oligomer
Composition		polyester urethane acrylate
Chemical class	-	acrylate
PHYSICAL PROPERTIES		
Pt-Co color	-	30
Glass transition temperature	°C	66
Density at 25°C	kg/m³	1,110
Refractive index at 25°C	-	1.48
Viscosity at 60°C	mPas	25,000
USE & PERFORMANCE		
Manufacturer		Dymax Corporation
Outstanding properties		excellent balance of hardness and flexibility, high clarity and high abrasion resistance & impact strength, non-yellowing
Recommended for polymers		PC, PMMA
Recommended for products		nail coating, 3D printing inks, wood & floor coating, electronic coatings and inks

Bomar BR-744BT

PARAMETER	UNIT	VALUE
GENERAL INFORMATION		
Name	Bomar BR-744BT	
General description	difunctional oligomer	
Composition	aliphatic polyester urethane acrylate	
Chemical class	-	acrylate
PHYSICAL PROPERTIES		
Pt-Co color	-	45
Density at 25°C	kg/m^3	1110
Refractive index at 25°C	-	1.496
Glass transition temperature	°C	8.00 (DMA)
Viscosity at 60°C	mPas	46000
USE & PERFORMANCE		
Manufacturer	Dymax Corporation	
Outstanding properties	enhanced flexibility, improved impact- and weather resistance, non-yellowing with low MEHQ levels.	
Recommended for polymers	ABS, PC, PET, PMMA, PVC	
Recommended for products	nail coatings	
Recommended applications	adhesion promoter	

Bomar BR-941

PARAMETER	UNIT	VALUE
GENERAL INFORMATION		
Name		Bomar BR-941
General description		dendritic oligomer
Composition		hexafunctional aliphatic urethane acrylate
Molecular mass	daltons	2000
Chemical class	-	acrylate
PHYSICAL PROPERTIES		
Pt-Co color	-	30
Glass transition temperature	°C	200 (DMA)
Density at 25°C	kg/m^3	1200
Refractive index at 25°C	-	1.496
Viscosity at 25°C	mPas	4000
USE & PERFORMANCE		
Manufacturer		Dymax Corporation
Outstanding properties		excellent chemical and thermal resistance, weatherability, hydrolytic stability, chemical and scratch resistance, exceptional adhesion properties to a variety of substrates
Recommended for polymers		polycarbonate
Recommended applications		floor, plastics, and wood coatings

Bomar BR-970BT

PARAMETER	UNIT	VALUE
GENERAL INFORMATION		
Name	Bomar BR-970BT	
General description	difunctional oligomer	
Composition	aliphatic urethane acrylate	
Molecular mass	daltons	2000
Chemical class	-	acrylate
PHYSICAL PROPERTIES		
Color, Platinum-cobalt scale	-	30
Density at 25°C	kg/m^3	1120
Refractive index at 25°C	-	1.489
Viscosity at 25°C	mPas	10,000
USE & PERFORMANCE		
Manufacturer	Dymax Corporation	
Outstanding properties	provides hydrolytic stability, gloss finish, abrasion-, chemical,- and stain resistance, can be blended with other oligomers to achieve a balance of toughness and flexibility	
Recommended for products	inkjet printing, graphic arts, rapid prototyping 3D printing resins	
Recommended for polymers	ABS, PC, PET, PMMA, PVC	

Bomar BR-3641AJ

PARAMETER	UNIT	VALUE
GENERAL INFORMATION		
Name		Bomar BR-3641AJ
General description		difunctional oligomer
Composition		1.3 functional aliphatic polyether urethane acrylate
Chemical class	-	acrylate
PHYSICAL PROPERTIES		
Color, Platinum-cobalt scale	-	15
Glass transition temperature	°C	-36 (DMA)
Density at 25°C	kg/m^3	1,010
Refractive index at 25°C	-	1.46
Viscosity at 50°C	mPas	10,000
USE & PERFORMANCE		
Manufacturer		Dymax Corporation
Outstanding properties		excellent optical clarity, resilience, tenacious adhesion, and non-yellowing effect, offers hydrolytic stability and enhances weatherability. Films formulated with BR-3641AJ are highly elastomeric in nature, adhesion promoter, and also as a reactive tackifier
Recommended applications		adhesion promoter, pressure-sensitive adhesive, reactive tackifier
Recommended for polymers		PET, PMMA, PVC

Bomar BR-7432GB

PARAMETER	UNIT	VALUE
GENERAL INFORMATION		
Name		Bomar BR-7432GB
General description		difunctional oligomer
Composition		aliphatic polyester urethane diacrylate
Chemical class	-	acrylate
PHYSICAL PROPERTIES		
Color, Platinum-cobalt scale	-	15
Glass transition temperature	°C	-4 (DMA)
Density at 25°C	kg/m^3	1,160
Refractive index at 25°C	-	1.549
Viscosity at 50°C	mPas	88,000
USE & PERFORMANCE		
Manufacturer		Dymax Corporation
Outstanding properties		imparts toughness, high tensile strength, impact resistance, and abrasion resistance. It offers excellent flexibility and provides an excellent balance of weathering, solvent resistance, tensile strength, and adhesion properties.
Recommended for polymers		ABS, PC, PET, PMMA, PVC
Recommended applications		impact resistant coatings, thermoforming coatings and inks, nail gel coatings, high-impact 3D printing resins

Bomar BR-7432GI30

PARAMETER	UNIT	VALUE
GENERAL INFORMATION		
Name		Bomar BR-7432GI30
General description		difunctional oligomer
Composition		aliphatic polyester urethane diacrylate
Chemical class	-	acrylate
PHYSICAL PROPERTIES		
Color, Platinum-cobalt scale	-	35
Density at 25°C	kg/m^3	1110
Glass transition temperature	°C	40 (DMA)
Refractive index at 25°C	-	1.483
Viscosity at 50°C	mPas	69,000
USE & PERFORMANCE		
Manufacturer		Dymax Corporation
Outstanding properties		provides an excellent balance of weathering, solvent resistance, tensile strength, and adhesion properties. Imparted toughness, high tensile strength, impact and abrasion resistance.
Recommended applications		coatings for metal and plastics, thermoforming coatings and inks, impact modifier for coatings, 3D printing resins, and peroxide-cure composites
Recommended for polymers		ABS, PC, PET, PMMA, PVC

Bomar BR-14320S

PARAMETER	UNIT	VALUE
GENERAL INFORMATION		
Name		Bomar BR-14320S
General description		difunctional oligomer
Composition		silicone urethane acrylate
Acronym	-	PBDUA
Chemical class	-	acrylate
PHYSICAL PROPERTIES		
Pt-Co color	-	40
Density at 25°C	kg/m³	1120
Refractive index at 25°C	-	1.425
Glass transition temperature	°C	-112
Viscosity at 60°C	mPas	18,000
USE & PERFORMANCE		
Manufacturer		Dymax Corporation
Outstanding properties		provides low shrinkage, hydrophobicity and chemical, and high-temperature resistance. Possesses hydrolytic stability and offers enhanced, improved adhesion, adhesion promoter, ideally suited for producing films that are compounded for flexible systems
Recommended applications		low polarity coatings, high-temperature applications, thermoforming coatings & inks
Recommended for polymers		ABS, HDPE, PC, PMMA

Bomar XR-741MS

PARAMETER	UNIT	VALUE
GENERAL INFORMATION		
Name		Bomar XR-741MS
General description		difunctional oligomer
Composition		aliphatic polyester urethane methacrylate
Chemical class	-	acrylate
PHYSICAL PROPERTIES		
Color, Platinum-cobalt scale	-	30
Density at 25°C	kg/m^3	1240
Glass transition temperature	°C	107
Refractive index at 25°C	-	1.500
Viscosity at 60°C	cP	52,000
USE & PERFORMANCE		
Manufacturer		Dymax Corporation
Outstanding properties		Improved hardness, low skin irritation and non-yellowing effect
Recommended applications		hard gel nail coatings, high-hardness coatings, dental resins
Recommended for polymers		ABS, PC, PET, PMMA, PVC

Elvaloy AC 2103

PARAMETER	UNIT	VALUE
GENERAL INFORMATION		
Name		Elvaloy AC 2103
CAS #	-	9010-86-0
Composition		>99% acrylic acid ethyl ester, polymer with ethylene
Acronym	-	EEA
Formula		
Chemical class	-	acrylate
Mixture	-	no
Composition		>99.0% acrylic acid ethyl ester, polymer with ethylene (19.5 wt% ethyl acrylate content)
PHYSICAL PROPERTIES		
State	-	solid/pellets
Odor	-	pungent
Color	-	white
Freezing point	°C	78
Melting point	°C	95
Density at 25°C	kg/m^3	930
Melt flow rate at 190°C/2.16 kg	g/10 min	21
HEALTH & SAFETY		
Carcinogenicity	no data available	
Mutagenicity	no data available	
Teratogenicity	no data available	
DOT class	-	not regulated
TDG class	-	not regulated
ICAO/IATA class	-	not regulated
IMDG class	-	not regulated
Agency rating, listed		AICS Australia, DSL Canada, IECSC China, KECI Korea, RECH Europa, NZIoC New Zealand, PICCS Philippines, TSCA USA, TCSCA Taiwan
Animal testing, acute toxicity, Rat oral LD50	mg/kg	>5000
Animal testing, acute toxicity, Rabbit dermal LD50	mg/kg	>2000

Elvaloy AC 2103

PARAMETER	UNIT	VALUE
Effect of exposure, eye (human)		Solid or dust may cause irritation or corneal injury due to mechanical action.
Effect of exposure, inhalation (human)		Solid or dust may cause irritation.
Effect of exposure, skin (human)		Nonirritating to skin. Contact with molten material may cause thermal burns.
Effect of exposure, swallowing (human)		Very low toxicity if swallowed.
Effect of repeated or overexposure (human)		Prolonged contact is essentially nonirritating to skin.
Exposure, personal protection		Safety glasses, protective clothing based on chemical resistance data, chemical-resistant gloves, general and local exhaust ventilation.
First aid, eye		Flush eyes thoroughly with water for several minutes. Remove contact lenses after the initial 1-2 minutes and continue flushing for several additional minutes. If effects occur, consult a physician, preferably an ophthalmologist.
First aid, inhalation		Move person to fresh air, if effects occur, consult a physician.
First aid, skin		Wash off with plenty of water. Seek first aid or medical attention as needed. If molten material comes in contact with the skin, do not apply ice but cool under ice water or running stream of water. DO NOT attempt to remove the material from skin. Seek medical attention.
ECOLOGICAL PROPERTIES		
Bioaccumulation potential		no bioconcentration is expected because of the relatively high molecular weight (MW greater than 1000).
USE & PERFORMANCE		
Manufacturer		Dow Chemicals
Outstanding properties		excellent thermal stability, exhibits high flexibility, imparts low-temperature toughness to a wide range of resins, excellent blend compatibility with other polyolefins
Recommended for polymers		PVDC, PO, cellulose, PS, PC, glass, foil, PVC, PET
Recommended for products		high performance packaging

Elvaloy AC 2103

PARAMETER	UNIT	VALUE
Recommended applications		utilized as a tie layer between polyolefins and a variety of polar substrates such as metals, polyvinylidene chloride, polyolefins, cellulose, polyester, polycarbonate, glass, foil, PVC, PET, polystyrene.
Guideline for use		310°C (maximum processing emperature)
Food approval (FDA)		Elvaloy™ AC 2103 acrylate copolymer resin complies with Food and Drug Administration Regulation 21 CFR 177.1320(a) - - Ethylene-ethyl acrylate copolymer resins, subject to the limitations and requirements therein. This Regulation describes polymers that may be used in contact with food, except for holding food during cooking, subject to the blend and finished food-contact article meeting the extractive limitations, as shown in paragraph (c)(2) of the Regulation. This resin must be blended with polyethylene or with one or more olefin polymers complying with Food and Drug Administration Regulation 21 CFR 177.1520, or used in a coating complying with 21 CFR 175.300 or 21 CFR 176.170 in such proportions that the ethyl acrylate content of the blend or finished coating does not exceed 8 percent by weight, in accordance with paragraph (b) of 21 CFR 177.1320.

Sarbox SB401

PARAMETER	UNIT	VALUE
GENERAL INFORMATION		
Name		Sarbox SB401
Composition		carboxylic acid and anhydride containing methacrylate oligomer blended with ethyl-3-ethoxy propionate solvent
Chemical class	-	acrylate
Functionality, average	-	8
Moisture content	wt%	0.5
PHYSICAL PROPERTIES		
State	-	liquid
Color	-	clear
Color, Platinum-cobalt scale	-	195
Acid number	mg KOH/g	138
Density at 25°C	kg/m³	1114.4
Refractive index at 25°C	-	1.491
Solubility in water at 25°C	g/l	slightly soluble
Specific gravity at 40°C	-	1.14
Viscosity at 25°C	mPas	1,000
USE & PERFORMANCE		
Manufacturer		Sartomer/Arkema Group
Outstanding properties		provides good solvent and acid resistance and excellent high gloss aqueous development
Recommended for products		metal, plastic, and wood coatings, photoresists, solder masks, and inks
Recommended applications		excellent adhesion to copper, offers fast cure

Sarbox SB510M35

PARAMETER	UNIT	VALUE
GENERAL INFORMATION		
Name		Sarbox SB510M35
Composition		aromatic acid methacrylate half ester blended with 2-phenoxyethyl acrylate
Chemical class	-	acrylate
PHYSICAL PROPERTIES		
State	-	liquid
Acid number	mg KOH/g	115
Density at 25°C	kg/m^3	1153.6
pH	-	6
Solubility in water at 25°C	g/l	insoluble
Specific gravity at 25°C	-	1.16
Viscosity at 25°C	mPas	800
HEALTH & SAFETY		
Agency rating, listed		IECSC China, EINECS Europe, ENCS Japan.
USE & PERFORMANCE		
Manufacturer		Sartomer/Arkema Group
Outstanding properties		provides negligible volatiles, fast cure response, good flexibility, good hardness, and high impact strength. Possesses acid- and oil resistance.
Recommended applications		excellent adhesion to metals and good pigment-wetting

Sarbox SB520E35

PARAMETER	UNIT	VALUE
GENERAL INFORMATION		
Name		Sarbox SB520E35
Composition		moderately functional, carboxylic acid containing acrylate oligomer blended in ethoxylated trimethylolpropane triacrylate monomer
Chemical class	-	acrylate
Reactive solids	wt%	100
Functionality	-	8
Monomer	wt%	65
Oligomer	wt%	35
PHYSICAL PROPERTIES		
State	-	liquid
Color, Platinum-cobalt scale	-	215
Acid number	mg KOH/g	179
Density at 25°C	kg/m^3	1153.8
Refractive index at 25°C	-	1.5021
Solubility in water at 25°C	g/l	insoluble
Specific gravity at 25°C	-	1.16-1.20
Viscosity at 60°C	mPas	1000
USE & PERFORMANCE		
Manufacturer		Sartomer/Arkema Group
Outstanding properties		provides good hardness, high impact strength, and high flexibility. Possesses chemical, heat, water, and abrasion resistance. Offers good pigment wetting & flow characteristics.
Recommended for products		adhesives, coatings on metal, paper, plastics, and wood, and inks
Recommended applications		adhesion promoter for metals and plastics

Sarbox SB520M35

PARAMETER	UNIT	VALUE
GENERAL INFORMATION		
Name		Sarbox SB520M35
Composition		carboxylic acid containing acrylate oligomer blended in phenoxy ethyl acrylate monomer
Chemical class	-	acrylate
Active matter	wt%	100
Functionality, average	-	8
Monomer	wt%	65
Oligomer	wt%	35
PHYSICAL PROPERTIES		
State	-	liquid
Color	-	clear
Color, Platinum-cobalt scale	-	210
Acid number	mg KOH/g	181
Density at 25°C	kg/m^3	1160
Refractive index at 25°C	-	1.5338
Solubility in water at 25°C	g/l	insoluble
Viscosity at 60°C	mPas	500
HEALTH & SAFETY		
Agency rating, listed		IECSC China, EINECS Europe, ENCS Japan, NZIoC New Zealand.
USE & PERFORMANCE		
Manufacturer		Sartomer/Arkema Group
Outstanding properties		provides good chemical, heat, water, and abrasion resistance, acrylate functionality.
Recommended for products		adhesives, coatings for metal, paper, plastics, and woods, and inks
Recommended applications		excellent adhesion to metals and plastics, offers one of the fastest cure rates in the Sarbox product line

Sarbox SR9050

PARAMETER	UNIT	VALUE
GENERAL INFORMATION		
Name		Sarbox SR9050
General description		monofunctional
Composition		mixture of methacrylate acid ester and 2-(2-ethoxyethoxy)ethyl acrylate
Chemical class	-	acrylate
PHYSICAL PROPERTIES		
State	-	liquid
Acid number	mg KOH/g	130-195
Refractive index at 25°C	-	1.4513
Solubility in water at 25°C	g/l	insoluble
Specific gravity at 25°C	-	1.132
Viscosity at 20°C	mPas	20
HEALTH & SAFETY		
Flash point	°C	93
Agency rating, listed		AICS Australia, DSL Canada, IECSC China, KECI Korea, EINECS Europa, Philippines PICCS
USE & PERFORMANCE		
Manufacturer		Sartomer/Arkema Group
Outstanding properties		provides good adhesion, high flexibility, high impact strength, and low shrinkage.
Recommended applications		monofunctional adhesion promoting monomer that provides exceptional adhesion to metal substrates, including aluminum, brass, tin-free steel and steel. Recommended for use in formulations containing tertiary amines.
Concentrations used	wt%	3-7

Sarbox SR9051

PARAMETER	UNIT	VALUE
GENERAL INFORMATION		
Name	Sarbox SR9051	
CAS #	-	308360-93-2
General description	trifunctional acid ester	
Chemical class	-	acrylate
Functionality	-	3
PHYSICAL PROPERTIES		
State	-	liquid
Color	-	clear, yellowish
Color, Gardner	-	5
Acid value	mg KOH/g	120-180
Refractive index at 25°C	-	1.4696
Specific gravity at 25°C	-	1.187
Viscosity at 25°C	mPas	250
USE & PERFORMANCE		
Manufacturer	Sartomer/Arkema Group	
Outstanding properties	provides faster cure response and greater hardness than SR9050. Provides good adhesion, high flexibility, high impact strength and low shrinkage.	
Recommended for products	adhesives, lamination, coatings on glass, and metals	
Recommended applications	trifunctional adhesion promoter. Not recommended for use in formulations containing tertiary amines.	
Concentrations used	wt%	3-7

Sarbox SR9053

PARAMETER	UNIT	VALUE
GENERAL INFORMATION		
Name	Sarbox SR9053	
General description	trifunctional	
Composition	trifunctional acid ester	
Chemical class	-	acrylate
PHYSICAL PROPERTIES		
State	-	liquid
Color	-	clear
Color, Gardner scale	-	5
Acid number	mg KOH/g	140
Density at 25°C	kg/m³	1200
Solubility in water at 25°C	g/l	slightly soluble
Specific gravity at 25°C	-	1.2
Viscosity at 25°C	mPas	950
USE & PERFORMANCE		
Manufacturer	Sartomer/Arkema Group	
Recommended for products	coatings	
Recommended applications	enhances adhesion of adhesive and coating formulations to metal, plastic, rubber, and glass. SR9053 is not recommended for use with amines. Suggested applications include coatings for automotive, electronics, paper, plastic, and wood. These products are UV/EB curable, peroxide curable, amine curable, and can enhance the performance of two-part epoxy/amine coating systems.	

Sarbox SR9054

PARAMETER	UNIT	VALUE
GENERAL INFORMATION		
Name	Sarbox SR9054	
General description	difunctional	
Chemical class	-	acrylate
PHYSICAL PROPERTIES		
State	-	liquid
Color	-	clear
Acid value	mg KOH/g	280-310
Viscosity at 25°C	mPas	1600
USE & PERFORMANCE		
Manufacturer	Arkema-Sartomer Americas	
Outstanding properties	enhanced performance of advanced materials by reducing VOC, increasing gel time, improving surface hardness and impact resistance.	
Recommended for products	adhesives, coatings	
Recommended applications	unique acid acrylate adhesion promoter, recommended to maximize adhesion of coatings and adhesives on glass and metals. Suitable for use in ultraviolet and electron beam curing compositions. Possible applications include plastic coatings.	

3.3 Amides, amines, amidoamines
Ancamide 261A

PARAMETER	UNIT	VALUE
GENERAL INFORMATION		
Name		Ancamide 261A
General description		modified polyamide
Chemical class	-	amide
Amine value	mg KOH/g	320-380
Equivalent weight	-	120
PHYSICAL PROPERTIES		
State	-	liquid
Color	-	amber
Color, Gardner scale	-	7
Glass transition temperature	°C	131/2H at 71°C
Specific gravity at 25°C	-	0.96
Viscosity at 25°C	mPas	35,000-45,000
HEALTH & SAFETY		
Flash point	°C	>93
USE & PERFORMANCE		
Manufacturer		Evonik
Outstanding properties		good corrosion resistance, excellent adhesion to a variety of substrates, good color and light stability, compatible with wide range of solvents
Recommended for polymers		epoxy
Recommended applications		adhesives, sealants, solvent based marine and protective coatings, primers, sealers and coatings for concrete, casting, concrete repair compounds
Concentrations used		66 phr (EEW=190)

Ancamide 2353

PARAMETER	UNIT	VALUE
GENERAL INFORMATION		
Name		Ancamide 2353
General description		modified polyamide
Chemical class	-	amide
Amine value	mg KOH/g	330
PHYSICAL PROPERTIES		
State	-	liquid
Color	-	amber
Color, Gardner scale	-	9
Glass transition temperature	°C	131/2H at 71°C
Specific gravity at 25°C	-	1.01
Viscosity at 25°C	mPas	3,000
HEALTH & SAFETY		
Flash point	°C	>93
USE & PERFORMANCE		
Manufacturer		Evonik
Outstanding properties		provides hard films with very good solvent and corrosion resistance. Good for cathodic disbondment resistance. Best chemical resistance of polyamides.
Recommended for polymers		epoxy
Recommended applications		used for high-solid marine coating and maintenance coatings, concrete primers and coatings, pipeline coatings. Excellent high-gloss film formation and fast dry with no induction time.
Concentrations used		60 phr (EEW=190); 23 phr (EEW=500)

Ancamide 2396

PARAMETER	UNIT	VALUE
GENERAL INFORMATION		
Name	Ancamide 2396	
General description	modified amide/imidazoline (amidoamine)	
Chemical class	-	amide
Amine value	mg KOH/g	350
Equivalent weight	-	93
PHYSICAL PROPERTIES		
State	-	liquid
Color	-	amber
Color, Gardner scale	-	8
Glass transition temperature	°C	137
Density at 25°C	kg/m^3	993
Specific gravity at 25°C	-	0.99
Viscosity at 25°C	mPas	680
HEALTH & SAFETY		
Flash point	°C	130
Flash point method	-	CC
USE & PERFORMANCE		
Manufacturer	Evonik	
Outstanding properties	provides long pot life, good strength and modulus, and improved film appearance. Provides very good chemical resistance, and very good adhesion to cold, damp concrete. Possesses low temperature cure and resistance to blush.	
Recommended for polymers	epoxy	
Recommended applications	curing and coupling agent used for concrete primers, self-leveling and trowelable flooring, tile grouts.	
Concentrations used	phr	49 (EEW=190)

Ancamide 2424

PARAMETER	UNIT	VALUE
GENERAL INFORMATION		
Name		Ancamide 2424
General description		modified polyamide
Chemical class	-	amide
Amine value	mg KOH/g	327
Equivalent weight	-	114
PHYSICAL PROPERTIES		
State	-	liquid
Color	-	amber
Color, Gardner scale	-	9
Glass transition temperature	°C	129/2H at 71.0°C
Specific gravity at 25°C	-	1
Viscosity at 25°C	mPas	14,000
HEALTH & SAFETY		
Flash point	°C	135
Flash point method	-	CC
USE & PERFORMANCE		
Manufacturer		Evonik
Outstanding properties		excellent adhesion to metal and plastic substrates. Low-temperature cure and excellent environmental resistance. Can reduce or eliminate the need for accelerators.
Recommended for polymers		epoxy
Recommended applications		excellent adhesives for metal or plastic where rapid development of handling strength is required.
Concentrations used	phr	60 (EEW=190)

Ancamide 2482

PARAMETER	UNIT	VALUE
GENERAL INFORMATION		
Name	Ancamide 2482	
General description	modified polyamide	
Chemical class	-	amide
Amine value	mg KOH/g	370
Hydroxyl equivalent weight	-	125
PHYSICAL PROPERTIES		
State	-	liquid
Color	-	amber
Color, Gardner scale	-	7
Glass transition temperature	°C	172/2H at 71.0°C
Density at 25°C	kg/m³	967
Specific gravity at 25°C	-	0.967
Viscosity at 25°C	mPas	5,500
HEALTH & SAFETY		
Flash point	°C	>392
Flash point method	-	SFCC
USE & PERFORMANCE		
Manufacturer	Evonik	
Outstanding properties	excellent adhesion to metal and plastic substrates. Provides good flexibility and very good environmental resistance. Low cost	
Recommended for polymers	epoxy	
Recommended applications	curing agent, provides very good adhesion to metal and plastic substrates. Adhesives applications where high filler loading with good handling is needed.	
Concentrations used	phr	60-65 (EEW=190)

Ancamine K54

PARAMETER	UNIT	VALUE
GENERAL INFORMATION		
Name	Ancamine K54	
CAS #	-	90-72-2
EC number	-	202-013-9
Composition	>95% tris-2,4,6-(dimethylaminomethyl) phenol, <5% bis(dimethylaminomethyl) phenol	
Formula		
Chemical class	-	amine
PHYSICAL PROPERTIES		
State	-	liquid
Odor	-	amine-like
Color	-	amber
Color, Gardner scale	-	6
Density at 25°C	kg/m^3	980
Viscosity at 25°C	mPas	200
HEALTH & SAFETY		
HMIS classification	Flammability	1
	Health	3
	Reactivity	0
Carcinogenicity	no data available	
Flash point	°C	140
Flash point method	-	CC
Animal testing, acute toxicity, Rabbit dermal LD50	mg/kg	1400
Effect of exposure, eye (human)	Causes serious eye damage	
Effect of exposure, inhalation (human)	May cause irritation of respiratory tract. May cause irritation of the mucous membranes.	
Effect of exposure, skin (human)	Causes skin irritation.	
Effect of exposure, swallowing (human)	Harmful if swallowed.	
Exposure, personal protection	Safety glasses, protective clothing based on chemical resistance data, chemical-resistant gloves, general and local exhaust ventilation.	

Ancamine K54

PARAMETER	UNIT	VALUE
First aid, eye		Immediately flush eyes with plenty of water for at least 15 minutes. If eye irritation persists, consult a specialist.
First aid, inhalation		Move person to fresh air. If symptoms persist, call a physician.
First aid, skin		Immediately flush skin with plenty of water for at least 15 minutes while removing contaminated clothing. If symptoms persist, call a physician.
ECOLOGICAL PROPERTIES		
Aquatic toxicity, *Rainbow trout*, 96-h LC50	mg/l	240
USE & PERFORMANCE		
Manufacturer	Evonik	
Outstanding features	Ancamine K54 curing agent is a technical grade of tris-(dimethylaminomethyl) phenol—a versatile Lewis base catalyst for curing epoxy resins.	
Recommended applications	used in coatings, flooring, adhesives, castings, potting, and encapsulation.	
Concentrations used	1.0-15 phr, 1-5% as an accelerator	

Cohedur H 30

PARAMETER	UNIT	VALUE
GENERAL INFORMATION		
Name		Cohedur H 30
Composition		>95% hexamethylenetetramine, 3% amorphous silica
Acronym	-	HMTA
Empirical formula	-	C6H12N4
Formula		
Chemical class	-	melamine
Mixture	-	yes
Active matter	wt%	>95
PHYSICAL PROPERTIES		
State	-	solid/powder
Color	-	white
Ash content	wt%	2.5
Sieve residue (0.063 mm)	%	<=0.8
Density at 25°C	kg/m³	1,300
Solubility (diluents)		ethanol, methylene chloride, slightly soluble in ethyl acetate, and acetone, insoluble in aliphatic hydrocarbons
Solubility in water at 25°C	g/l	soluble
USE & PERFORMANCE		
Manufacturer		RheinChemie Additives/Lanxess
Outstanding properties		Cohedur H 30 is a component of the direct bonding system, also known as RFS system. RFS bonding systems are multi-component systems. They are created by providing the rubber component with a resorcinol component, a methylene component, and reinforcing silica, e.g., Vulkasil® S.
Recommended for polymers		rubber, latex
Recommended for products		tires, conveyor belting, V-belts, reinforcing hose, air springs, flexible containers and fabric proofing's

Cohedur H 30

PARAMETER	UNIT	VALUE
Recommended applications		used as direct bonding agent for rubber to reinforcing materials, e.g. fabrics, steel cord and glass fibers. It is employed as an additive to the rubber compound. Used with most types of rubber. Best adhesion is obtained either with polar and nonpolar elastomers, such as NBR, HNBR, CR, NR, IR, SBR and BR.
Concentrations used	phr	1.5

Jaylink JL-103M

PARAMETER	UNIT	VALUE
GENERAL INFORMATION		
Name		Jaylink JL-103M
Composition		acrylamidomethyl substituted cellulose ester polymer
Chemical class	-	amide
PHYSICAL PROPERTIES		
State	-	solid/powder free flowing
Color	-	white
Density at 25°C	kg/m^3	1310
Nitrogen content	wt%	0.65
Viscosity at 50°C in 50% N,N'-DMA	mPas	29,000
USE & PERFORMANCE		
Manufacturer		Dymax Corporation
Outstanding properties		enhanced surface hardness without impacting clarity. Provides low haze to the coating system. Offers excellent abrasion, and chemical and impact resistance. Coatings have less turbidity and improved clarity due to the material solubility.
Recommended applications		adhesion promoter, which once cured, will chemically crosslink into the matrix keeps the material from leaching out. The hydrophobic and hydrophilic segments make it an excellent compatibilizer for materials of differing polarity. Recommended for UV or radiation cured coating system. Suitable for flexoprinting, overprint varnish, and UV printing ink.
Concentrations used	wt%	2-10

Jaylink JL-106E

PARAMETER	UNIT	VALUE
GENERAL INFORMATION		
Name		Jaylink JL-106E
Composition		acrylamidomethyl substituted cellulose ester
Chemical class	-	amide
PHYSICAL PROPERTIES		
State	-	solid/powder free flowing
Color	-	white
Moisture content	%	0.5
Specific gravity	-	1.28
Glass transition temperature	°C	118
Viscosity at 50°C	mPas	59,000
USE & PERFORMANCE		
Manufacturer		Dymax Corporation
Outstanding properties		accelerates cure, improves impact resistance, enhances surface hardness without impacting clarity (offers low haze). Exhibits chemical-, abrasion- and impact resistance. Chemically crosslinks within the matrix, which prevents it from leaching out. Shows mild thixotropic behavior, less turbidity, rapid UV cure response and higher rates of photocure.
Recommended for polymers		PC
Recommended applications		adhesion promoter suitable for fiber glass composition
Concentrations used	wt%	2-10

Markoba HMT

PARAMETER	UNIT	VALUE
GENERAL INFORMATION		
Name		Markoba HMT
CAS #	-	100-97-0
EC number	-	202-905-8
Composition	hexamethylenetetramine	
Acronym	-	HMT
Empirical formula	C6H12N4	
Formula		

PARAMETER	UNIT	VALUE
Molecular mass	daltons	140.19
RTECS number	-	MN4725000
Chemical class	-	amine
PHYSICAL PROPERTIES		
State	-	solid/crystals
Color	-	white
Melting point	°C	280
HEALTH & SAFETY		
UN risk phrases, R	-	R11,R42/43,R43
US safety phrases, S	-	S16,S22,S24,S37
USE & PERFORMANCE		
Manufacturer	Wholemark Fine Chemical	
Recommended for polymers	rubber	
Recommended applications	adhesion promoter	

Nourybond 276

PARAMETER	UNIT	VALUE
GENERAL INFORMATION		
Name		Nourybond 276
Composition		modified polyamidoamine
Chemical class	-	amidoamine
Amine value	mg KOH/g	110 -130
PHYSICAL PROPERTIES		
State	-	liquid
Color	-	clear, amber
Color, Gardner scale	-	10
Viscosity at 25°C	mPas	8000-28,000
USE & PERFORMANCE		
Manufacturer	Evonik	
Outstanding properties	provides unique rheological properties and excellent color stability. Excellent color stability. Low-temperature cure (120°C)	
Recommended for polymers	PVC plastisols	
Recommended for products	automobile industry, trucks, buses	
Recommended applications	a modified polyamidoamine adhesion promoter is for PVC plastisol intended for use in automotive sealants, under-body coatings, and anti-chip primers. It is designed to provide adhesion to electrodeposition primers used in the manufacture of automobiles, trucks, and buses.	

Priamine 1073-LQ-(GD)

PARAMETER	UNIT	VALUE
GENERAL INFORMATION		
Name	Priamine 1073-LQ-(GD)	
CAS #	-	68955-56-6
Composition	<100% dimer diamine	
Chemical class	-	amine
Renewable carbon content	%	100
PHYSICAL PROPERTIES		
State	-	liquid
Color	-	yellow
Color Gardner	-	7
Amine value	mg KOH/g	205
Boiling point	°C	>350
Melting point	°C	<-30
Density at 25°C	kg/m³	900
Vapor pressure at 20°C	kPa	< 2E-09
Viscosity at 25°C	mPas	250
HEALTH & SAFETY		
NFPA classification	Flammability	1
	Health	3
	Reactivity	0
HMIS classification	Flammability	1
	Health	3
	Reactivity	0
Carcinogenicity	no data available	
Mutagenicity	no data available	
Teratogenicity	no data available	
DOT class	domestic regulation 49CFR: not regulated as a dangerous goods	
ICAO/IATA class	Environmentally hazardous substance, liquid, n.o.s. (alkyldiamine) 9, III	
IMDG class	Environmentally hazardous substance, liquid, n.o.s. (alkyldiamine) 9, III	
Flash point	°C	>200
Agency rating, listed	AICS Australia, DSL Canada, IECSC China, ENCS Japan, ISHL Japan, KECI Korea, NZIoC New Zealand, PICCS Philippines, TSCA USA	
Animal testing, acute toxicity, Rat oral LD50	mg/kg	>5000

Priamine 1073-LQ-(GD)

PARAMETER	UNIT	VALUE
Animal testing, acute toxicity, Rabbit dermal LD50	mg/kg	>2000
Effect of exposure, eye (human)	Causes serious eye damage	
Effect of exposure, inhalation (human)	May cause irritation of respiratory tract. May cause irritation of the mucous membranes.	
Effect of exposure, skin (human)	Causes skin irritation.	
Effect of exposure, swallowing (human)	Harmful if swallowed.	
Exposure, personal protection	Safety glasses, protective clothing based on chemical resistance data, chemical-resistant gloves, general and local exhaust ventilation.	
First aid, eye	Immediately flush eyes with plenty of water for at least 15 minutes. If eye irritation persists, consult a specialist.	
First aid, inhalation	Move person to fresh air. If symptoms persist, call a physician.	
First aid, skin	Immediately flush skin with plenty of water for at least 15 minutes while removing contaminated clothing. If symptoms persist, call a physician.	
UN/NA class	-	3082
ECOLOGICAL PROPERTIES		
Aquatic toxicity, *Green algae*, 96-h EC50	mg/l	0.0443/72H
Partition coefficient, log K_{ow}	-	14.11
USE & PERFORMANCE		
Manufacturer	Croda	
Outstanding properties	provides high flexibility, moisture repellency, and improved adhesion to plastics	
Recommended for polymers	epoxy, PA, PU	
Recommended for products	adhesives & sealants	
Recommended applications	modification with Priamine 1074 gives formulation freedom in the choice of diacids used.	
Food contact approval	EU 10/2011, FDA 175.105 and 175.300	

Priamine 1074

PARAMETER	UNIT	VALUE
GENERAL INFORMATION		
Name	Priamine 1074	
Composition	dimer diamine	
Chemical class	-	amine
Renewable carbon content	%	100
PHYSICAL PROPERTIES		
State	-	liquid
Color	-	yellow
Color Gardner	-	4
Amine value	mg KOH/g	202-212
Viscosity at 25°C	mPas	200
HEALTH & SAFETY		
HMIS classification	Flammability	1
	Health	3
	Reactivity	0
Carcinogenicity	no data available	
ICAO/IATA class	Environmentally hazardous substance, liquid, n.o.s. (alkyldiamine) 9, III	
IMDG class	Environmentally hazardous substance, liquid, n.o.s. (alkyldiamine) 9, III	
Effect of exposure, eye (human)	Causes serious eye damage	
Effect of exposure, inhalation (human)	May cause irritation of respiratory tract. May cause irritation of the mucous membranes.	
Effect of exposure, skin (human)	Causes skin irritation.	
Effect of exposure, swallowing (human)	Harmful if swallowed.	
Exposure, personal protection	Safety glasses, protective clothing based on chemical resistance data, chemical-resistant gloves, general and local exhaust ventilation.	
First aid, eye	Immediately flush eyes with plenty of water for at least 15 minutes. If eye irritation persists, consult a specialist.	
First aid, inhalation	Move person to fresh air. If symptoms persist, call a physician.	
First aid, skin	Immediately flush skin with plenty of water for at least 15 minutes while removing contaminated clothing. If symptoms persist, call a physician.	

Priamine 1074

PARAMETER	UNIT	VALUE
USE & PERFORMANCE		
Manufacturer		Croda
Outstanding properties		provides high flexibility, moisture repellency, and improved adhesion to plastics.
Recommended for polymers		PA
Recommended for products		3D printing, adhesives & sealants, bio-polymers, engineering plastics
Recommended applications		used in polyamides. Modification with Priamine 1074 gives formulation freedom in the choice of diacids used. This gives formulators the flexibility to adjust the melting point allowing for higher temperature exposure of the end product without compromise on performance.
Food contact approval		FDA 175.105 and 175.300

3.4 Benzene derivatives
Ekaland BQD 30 DS

PARAMETER	UNIT	VALUE
GENERAL INFORMATION		
Name	Ekaland BQD 30 DS	
CAS #	-	105-11-3
EC number	-	203-271-5
Composition	50% 1,4-benzoquinone dioxime in ethyl benzene or 75% 1,4-benzoquinone dioxime in water	
Empirical formula	$C_6H_6N_2O_2$	
Formula		
RTECS number	-	DK4900000
Chemical class	benzene derivatives	
Purity	wt%	>=90
PHYSICAL PROPERTIES		
State	-	liquid
Color	-	gray to ochre
Melting point	°C	245-255
USE & PERFORMANCE		
Manufacturer	Arkema/MLPC	
Recommended for polymers	rubber	
Recommended applications	adhesion promoter and curative for rubber to metal bonding	

Ekaland PPDN 30 DT

PARAMETER	UNIT	VALUE
GENERAL INFORMATION		
Name	Ekaland PPDN 30 DT	
CAS #	-	9003-34-3
EC number	-	203-272-0
Composition	1,4-dinitrosobenzene homopolymer	
Common synonym	poly-(p-dinitrosobenzol)	
Empirical formula	$(C6H4N2O2)x$	
Chemical class	benzene derivatives	
PHYSICAL PROPERTIES		
State	-	liquid
USE & PERFORMANCE		
Manufacturer	Arkema/MLPC	
Recommended for polymers	rubber	
Recommended applications	adhesion promoter and curative for rubber to metal bonding	

Ekaland PPDN 30 DX

PARAMETER	UNIT	VALUE
GENERAL INFORMATION		
Name		Ekaland PPDN 30 DX
CAS #	-	9003-34-3
EC number	-	203-272-0
Composition		1,4-dinitrosobenzene homopolymer
Common synonym		poly(p-dinitrosobenzol)
Empirical formula		(C6H4N2O2)x
Chemical class		benzene derivatives
Mixture	-	no
PHYSICAL PROPERTIES		
State	-	liquid
Color	-	brown
USE & PERFORMANCE		
Manufacturer		Arkema/MLPC
Recommended for polymers		rubber
Recommended applications		adhesion promoter and curative for rubber to metal bonding

Ekaland PPDN 50 HU

PARAMETER	UNIT	VALUE
GENERAL INFORMATION		
Name	Ekaland PPDN 50 HU	
CAS #	-	9003-34-3
EC number	-	203-272-0
Composition	1,4-dinitrosobenzene homopolymer	
Common synonym	poly-(p-dinitrosobenzol)	
Formula		
Empirical formula	(C6H4N2O2)x	
Chemical class	benzene derivatives	
PHYSICAL PROPERTIES		
State	-	liquid
Color	-	brown
HEALTH & SAFETY		
Animal testing, acute toxicity, Rat oral LD50	mg/kg	1500
USE & PERFORMANCE		
Manufacturer	Arkema/MLPC	
Recommended for polymers	rubber	
Recommended applications	adhesion promoter and curative for rubber to metal bonding	

3.5 Carbamic resin
Alnovol UF 410 RPC

PARAMETER	UNIT	VALUE	
GENERAL INFORMATION			
Name		Alnovol UF 410 RPC	
Composition		carbamic resin on silica (65 wt% silica)	
Chemical class		carbamic resin	
Moisture content	%	<=4	
PHYSICAL PROPERTIES			
State	-		free flowing powder
Color	-		white
Particle size, passing through 80 mesh	%		=>99.7
Ash content (1 h at 800°C)	%		31-35
Viscosity at 25°C	mPas		3050-13500
USE & PERFORMANCE			
Manufacturer		Allnex	
Outstanding properties		improved hardness, low viscosity, phenolic crosslinker. Plasticizing component and compatibility promoter for rubber compounds.	
Recommended for polymers		natural and synthetic rubbers	
Recommended applications		adhesion promoter for textile cord in tires	
Processing		Alnovol UF 410 RPC can be added to the internal mixer together with carbon black directly. Partial or full dosage at the final mixing step is possible due to its easy incorporation. It efficiently reduces Mooney viscosity. Because of its viscosity effect, extrusion works well, and shape and surface are improved.	

3.6 Chlorinated polyolefins
Eastman Advantis 510W

PARAMETER	UNIT	VALUE
GENERAL INFORMATION		
Name		Eastman Advantis 510W
CAS #	-	7732-18-5
Composition		70-75% water, 22-24% modified polyolefin, 2-3% substituted aliphatic diol, 1-2% C12-C14 secondary alcohols, 0.1-0.5% 2-dimethylaminoethanol
Chemical class		chlorinated polyolefin
Mixture	-	yes
PHYSICAL PROPERTIES		
State	-	liquid (dispersion)
Odor	-	mild
Color	-	off-white milky
Density	g/l	970
Boiling point	°C	100/water
Decomposition temperature	°C	347
Neutralizing agent	-	amine
pH	-	8
Particle size	μm	0.03
Glass transition temperature	°C	<0
Viscosity at 25°C	mPas	<250
HEALTH & SAFETY		
HMIS classification	Flammability	1
	Health	2
	Physical hazard	1
NFPA classification	Flammability	1
	Health	2
	Instability	1
Carcinogenicity		No ingredient of this product present at levels greater than or equal to 0.1% is identified as probable, possible or confirmed human carcinogen by IARC.
Mutagenicity		no data available
DOT class		not regulated
TDG class		not regulated
ICAO/IATA class		not regulated
IMDG class		not regulated

3.6 Chlorinated polyolefins
Eastman Advantis 510W

PARAMETER	UNIT	VALUE
Flash point	°C	98/no flash observed up to boiling point
Hazardous combustion products	Carbon monoxide, carbon dioxide	
Agency rating, listed	TSCA USA, DSL Canada, AICS Australia, MITI Japan, ECL Korea, IECSC China	
Animal testing, acute toxicity, Rat oral LD50	mg/kg	>2000/modified polyolefin, >3250/surfactant, 1182.7/2-dimethylaminoethanol
Animal testing, acute toxicity, Rabbit dermal LD50	mg/kg	3180/surfactant
Animal testing, acute toxicity, Guinea pig dermal LD50	mg/kg	no skin sensitizer/ 2-dimethylaminoethanol
Animal testing, acute toxicity, Rat inhalation, LC50	mg/m³	1641ppm/4H/ 2-dimethylaminoethanol
Effect of exposure, eye (human)	Causes serious eye irritation. May irritate and cause redness and pain.	
Exposure, personal protection	Safety glasses, protective clothing based on chemical resistance data, chemical-resistant gloves, general and local exhaust ventilation.	
First aid, eye	In case of contact, immediately flush eyes with plenty of water for at least 15 minutes. Call a physician or poison control center immediately. Remove person to fresh air. If signs/symptoms continue, get medical attention.	
First aid, inhalation	Remove to fresh air. Treat symptomatically. If symptoms persist, call a physician.	
First aid, skin	Wash off with soap and water. Get medical attention if symptoms occur.	
ECOLOGICAL PROPERTIES		
Aquatic toxicity, *Golden orfe*, 96-h EC50	mg/l	146.63
Aquatic toxicity, *Green algae*, 72-h LC50	mg/l	34.47
Aquatic toxicity, *Daphnia magna*, 48-h LC50	mg/l	98.37
Biodegradability	readily biodegradable	
Bioaccumulative potential	does not bioaccumulate	
Partition coefficient, log K_{oc}	-	0.848

3.6 Chlorinated polyolefins
Eastman Advantis 510W

PARAMETER	UNIT	VALUE
Biodegradation probability		
USE & PERFORMANCE		
Manufacturer		Eastman Chemical Company
Outstanding properties		excellent adhesion to non-flame treated EPDM rubber modified polypropylene substrates; chlorine-free and APEO-free
Recommended for polymers		TPO, PP, EPDM rubber modified polypropylene
Recommended for products		automotive OEM, waterborne primers and waterborne basecoats for automotive bumper applications, coatings for automotive plastics, auto refinish, graphic arts, adhesion promoter for applications beyond automotive
Recommended applications		adhesion promoter for paints; primary application as a blend-in resin in waterborne primers and waterborne base coats for automotive bumper applications; can also be used as a wash primer and is suitable as an adhesion promoter for applications beyond automotive

Eastman AP 550-1 (25% in Aromatic 100)

PARAMETER	UNIT	VALUE
GENERAL INFORMATION		
Name		Eastman AP 550-1 (25% in Aromatic 100)
CAS #		64742-95-6, 95-63-6, 98-82-8
Composition		>25% modified polyolefin, 30-50% light aromatic solvent naphtha, petroleum, 20-30% 1,2,4-trimethylbenzene, 5% cumene
Acronym		CP
Chemical class		chlorinated polyolefin
Mixture	-	yes
Active matter	wt%	23-26
PHYSICAL PROPERTIES		
State	-	viscous liquid
Odor	-	aromatic
Color	-	yellow
Color, Gardner scale	-	4
Solubility (diluents)		soluble in: xylene, toluene, aromatic 100, not soluble in aliphatic hydrocarbons, ketones, esters, alcohols but can be diluted with long chain ketones n-butylpropionate, and solutions may become hazy. Provide excellent redissolve resistance.
Solubility in water at 25°C	g/l	negligible
Specific gravity at 25°C	-	1
Viscosity at 25°C	mPas	<4000
HEALTH & SAFETY		
HMIS classification	Flammability	2
	Health	3
	Reactivity	0
Carcinogenicity		cumene/IARC 2B: possibly carcinogenic to humans, ethylbenzene/IARC 2B: possibly carcinogenic to humans. NTP Not Listed. OSHA Not Listed. Expert judgment and weight of evidence determination: Not classified
Mutagenicity		no data available
Teratogenicity		no data available

Eastman AP 550-1 (25% in Aromatic 100)

PARAMETER	UNIT	VALUE
DOT class	Class combustible liquid, III. Not regulated for quantities of 450 liters (119 gallons).	
ICAO/IATA class	Coating solution 3, III	
IMDG class	Coating solution 3, III	
Flash point	°C	46.8
Flash point method	-	PMCC
Animal testing, acute toxicity, Rat oral LD50	mg/kg	>5000/ solvent naphtha, 6000/1,2,4-trimethylbenzene, 1400/ 2-ethylhexane-1,3-diol, 2910/cumene, 3500/ethylbenzene
Animal testing, acute toxicity, Rabbit dermal LD50	mg/kg	>2000/ solvent naphtha, 2000/2-ethylhexane-1,3-diol, 4200/ xylene,15400/ ethylbenzene ; >10000 cumene
Animal testing, acute toxicity, Rat inhalation, LC50	mg/l	76.3/4h/solvent naphtha (petroleum), light arom, 18/4H/1,2,4-trimethylbenzene, 6.7 ppm/4H/ xylene, 4000ppm/4H/ethylbenzene, 41.6/4H/ cumene
Effect of exposure, eye (human)	May irritate and cause redness and pain.	
Effect of exposure, inhalation (human)	May cause respiratory irritation.	
Effect of exposure, skin (human)	Causes skin irritation.	
Effect of exposure, swallowing (human)	May be fatal if swallowed and enters airways.	
Exposure, personal protection	Safety glasses, protective clothing based on chemical resistance data, chemical-resistant gloves, general and local exhaust ventilation.	

Eastman AP 550-1 (25% in Aromatic 100)

PARAMETER	UNIT	VALUE
First aid, eye		Rinse cautiously with water for several minutes. Remove contact lenses, if present and easy to do. Continue rinsing. If molten material contacts the eye, immediately flush with plenty of water for at least 15 minutes. Get medical attention immediately.
First aid, inhalation		Remove to fresh air. If not breathing, give artificial respiration. If breathing is difficult, give oxygen. If irritation persists, obtain medical advice.
First aid, skin		Immediately flush with plenty of water for at least 15 minutes while removing contaminated clothing and shoes. Get medical attention. Wash contaminated clothing before reuse. Destroy or thoroughly clean contaminated shoes.
NIOSH, REL	mg/m^3	245/cumene
OSHA, PEL	mg/m^3	435/xylene, 245/cumene, 435/ethylbenzene
ACGIH, TLV	ppm	25/1,2,4-trimethylbenzene, 100/xylene, 50/cumene
OSHA, PEL	ppm	100/xylene
UN/NA class	-	1139
ECOLOGICAL PROPERTIES		
Aquatic toxicity, *Green algae*, 96-h EC50	mg/l	2.356/72d/1,2,4-trimethylbenzen, 3.1/72H/solvent naphtha (petroleum), light arom, 2.1/72H/cumene
Aquatic toxicity, *Daphnia magna*, 48-h LC50	mg/l	3.6/1,2,4-trimethylbenzene, 4.3/chlorobenzene
Aquatic toxicity *Fathead minnow*, 96-h LC50	mg/l	8.2/solvent naphtha (petroleum) light arom,7.72/1,2,4-trimethylbenzene, 42.3-48.5/ethylbenzene
Aquatic toxicity, *Rainbow trout*, 96-h LC50	mg/l	2.6/xylene

Eastman AP 550-1 (25% in Aromatic 100)

PARAMETER	UNIT	VALUE
Bioconcentration factor	BCF	33-275/1,2,4-tri-methylbenzene, 94.69/cumene
Biodegradation probability	74.0%/28d/solvent naphtha (petroleum) light arom, 8.0-14%/28d/1,2,4-trimethyl-benzene, 70.0%/20d/cumene	
Chemical oxygen demand	g/g	1.13
USE & PERFORMANCE		
Manufacturer	Eastman Chemical Company	
Outstanding properties	excellent gasoline resistance, good humidity resistance	
Recommended for polymers	PP, TPO, 2-part urethane	
Recommended for products	aerosol coatings, graphic arts, automo-tive, automotive refinish, coatings for automotive plastics	
Recommended applications	adhesion Promoter 550-1 is Eastman's second-generation non-chlorinated product for adhesion to TPO and PP it demonstrates excellent performance in 2-part urethane (2K) coatings but often their application is limited to 2Ks. AP 550-1 expands non-chlorinated per-formance to applications under many melamine-cured systems. Used for automotive and automotive refinish, coatings for automotive plastics, and graphic arts	

Eastman AP 550-1 (25% in xylene)

PARAMETER	UNIT	VALUE
GENERAL INFORMATION		
Name		Eastman AP 550-1 (25% in xylene)
CAS #		1330-20-7, 100-41-1, 94-96-2
Composition		23-27% modified polyolefin, 60-65% xylene, 12-18% ethylbenzene
Acronym		CP
Chemical class		chlorinated polyolefin
Mixture	-	yes
PHYSICAL PROPERTIES		
State	-	viscous liquid
Odor	-	slight aromatic
Color	-	yellow
Color, Gardner scale	-	4
Boiling point	°C	137-144
Solubility (diluents)		soluble in: xylene, toluene, aromatic 100, not soluble in aliphatic hydrocarbons, ketones, esters, alcohols but can be diluted with long chain ketones n-butylpropionate, and solutions may become hazy. Provide excellent redissolve resistance.
Solubility in water at 25°C	g/l	negligible
Specific gravity at 25°C	-	<1
HEALTH & SAFETY		
HMIS classification	Flammability	3
	Health	2
	Reactivity	0
Carcinogenicity		IARC, OSHA, NTP: no ingredient of this product present at levels greater than or equal to 0.1% is identified as probable, possible or confirmed human carcinogen
Mutagenicity		no data available
Teratogenicity		no data available
ICAO/IATA class		Coating solution 3, III
IMDG class		Coating solution 3, III
Autoignition temperature	°C	450
Flash point	°C	27
Flash point method	-	PMCC
Animal testing, acute toxicity, Rat oral LD50	mg/kg	3523-4000/xylene, 3500/ethylbenzene

Eastman AP 550-1 (25% in xylene)

PARAMETER	UNIT	VALUE
Animal testing, acute toxicity, Rabbit dermal LD50	mg/kg	4200/xylene, 15400/ethylbenzene
Animal testing, acute toxicity, Rat inhalation, LC50	mg/m³	6700 ppm/4H/xylene, 4000ppm/4H/ethylbenzene
Effect of exposure, eye (human)		Causes serious eye irritation. May cause redness and pain.
Effect of exposure, inhalation (human)		May cause respiratory irritation. May cause drowsiness or dizziness. Narcotic effect. Symptoms may be delayed.
Effect of exposure, skin (human)		Causes skin irritation. May cause redness and pain.
Effect of exposure, swallowing (human)		May be fatal if swallowed and enters airways.
Effect of repeated or overexposure (human)		Contains ethylbenzene. May cause damage to organs (auditory organ) through prolonged or repeated exposure.
Exposure, personal protection		Safety glasses, protective clothing based on chemical resistance data, chemical-resistant gloves, general and local exhaust ventilation.
First aid, eye		Rinse cautiously with water for several minutes. Remove contact lenses, if present and easy to do. Continue rinsing. If molten material contacts the eye, immediately flush with plenty of water for at least 15 minutes. Get medical attention immediately.
First aid, inhalation		Remove to fresh air. If not breathing, give artificial respiration. If breathing is difficult, give oxygen. If irritation persists, obtain medical advice.
First aid, skin		Wash with soap and water. Get medical attention if symptoms occur. If burned by contact with hot material, cool molten material adhering to skin as quickly as possible with water, and see a physician for removal of adhering material and treatment of burn.
OSHA, PEL	mg/m³	435/xylene, 435/ethylbenzene, 350/chlorobenzene

Eastman AP 550-1 (25% in xylene)

PARAMETER	UNIT	VALUE
ACGIH, TLV	ppm	100/xylene, 20/ethylbenzene, 10/chlorobenzene, STEL150/xylene
NIOSH, REL	ppm	100/ethylbenzene
OSHA, PEL	ppm	100/xylene, 100/ethylbenzene
UN/NA class	-	1139
ECOLOGICAL PROPERTIES		
Aquatic toxicity, *Green algae*, 96-h EC50	mg/l	2.2/72H/xylene
Aquatic toxicity *Fathead minnow*, 96-h LC50	mg/l	42.3-48.5/ethylbenzene
Aquatic toxicity, *Rainbow trout*, 96-h LC50	mg/l	2.6/xylene
Biodegradation probability	readily biodegradable/xylene	
Partition coefficient, log K_{ow}	-	3.12-3.30/xylene, 3.15/ethylbenzene
USE & PERFORMANCE		
Manufacturer	Eastman Chemical Company	
Outstanding properties	excellent gasoline resistance, good humidity resistance, excellent redissolve resistance.	
Recommended for polymers	PP, TPO	
Recommended for products	consumer electronics, automotive, automotive refinish, coatings for automotive plastics, graphic arts	
Recommended applications	adhesion Promoter 550-1 is a non-chlorinated product for adhesion to TPO and polypropylene-based substrates. Non-chlorinated systems are known to demonstrate excellent performance in 2-part urethane (2K) coatings, but often their application is limited to 2Ks. AP 550-1 expands non-chlorinated performance to applications under many melamine-cured systems. Used for automotive, automotive refinish, coatings for automotive plastics, graphic arts.	

Eastman CP 153-2 (25% solids in xylene)

PARAMETER	UNIT	VALUE
GENERAL INFORMATION		
Name		Eastman CP 153-2 (25% solids in xylene)
CAS #		1330-20-7, 100-41-4, 108-90-7, 61789-01-3
Composition		>10% chlorinated polyolefin, 49.2-75% xylene, 0-18.8% ethylbenzene, <5% chlorobenzene, <3% epoxidized oil
Acronym		CP
Chemical class		chlorinated polyolefin
Mixture	-	yes
PHYSICAL PROPERTIES		
State	-	viscous liquid
Odor	-	aromatic
Color	-	amber
Color, Gardner scale	-	12-15
Boiling point	°C	138-140
Chloride content	wt%	21-25
Kinematic viscosity at 25°C	cSt	103-206
Solubility (diluents)		xylene, toluene, selected ketones, and ester solvents
Solubility in water at 25°C	g/l	negligible
Specific gravity at 25°C	-	0.97
Vapor pressure at 20°C	mbar	8.6
Viscosity at 25°C	mPas	100-200
HEALTH & SAFETY		
HMIS classification	Flammability	3
	Health	2
	Reactivity	0
Carcinogenicity		ethylbenzene IARC 2B: possible carcinogenic to humans. NTP Not Listed, OSHA Not Listed. Expert judgment and weight of evidence determination: Not classified
Mutagenicity		no data available
Teratogenicity		no data available
DOT class		Coating solution 3, III Shipping descriptions may vary based on mode of transport, quantities, package size, and/or origin and destination.

Eastman CP 153-2 (25% solids in xylene)

PARAMETER	UNIT	VALUE
ICAO/IATA class		Coating solution 3, III. Shipping descriptions may vary based on mode of transport, quantities, package size, and/or origin and destination.
IMDG class		Coating solution 3, III. Shipping descriptions may vary based on mode of transport, quantities, package size, and/or origin and destination.
Flash point	°C	27
Flash point method	-	TCC
Hazardous ingredients, labelling	COMBUSTIBLE LIQUID, N.O.S.	
Animal testing, acute toxicity, Rat oral LD50	mg/kg	3523-4000/xylene, 3500/ethylbenzene, 2262/chlorobenzene, >3200/ epoxidized oil
Animal testing, acute toxicity, Mouse oral LD50	mg/kg	
Animal testing, acute toxicity, Rabbit dermal LD50	mg/kg	4200/xylene, 15400/ ethylbenzene
Animal testing, acute toxicity, Guinea pig dermal LD50	mg/kg	>20000/ chlorobenzene
Animal testing, acute toxicity, Rat dermal LD50	mg/kg	>3200
Animal testing, acute toxicity, Rat inhalation, LC50	mg/m^3	6700 ppm/4H/xylene, 4000ppm/4H/ ethylbenzene, 29700/4H/ chlorobenzene
Effect of exposure, eye (human)		Causes serious eye irritation.
Effect of exposure, inhalation (human)		May cause respiratory irritation. Avoid inhalation of vapor or mist.
Effect of exposure, skin (human)		Causes skin irritation.
Effect of exposure, swallowing (human)		May be fatal if swallowed and enters airways.
Exposure, personal protection		Safety glasses, protective clothing based on chemical resistance data, chemical-resistant gloves, general and local exhaust ventilation.

Eastman CP 153-2 (25% solids in xylene)

PARAMETER	UNIT	VALUE
First aid, eye		Rinse cautiously with water for several minutes. Remove contact lenses, if present and easy to do. Continue rinsing. If molten material contacts the eye, immediately flush with plenty of water for at least 15 minutes. Get medical attention immediately.
First aid, inhalation		Remove to fresh air. If not breathing, give artificial respiration. If breathing is difficult, give oxygen. If irritation persists, obtain medical advice.
First aid, skin		Immediately flush with plenty of water for at least 15 minutes while removing contaminated clothing and shoes. Get medical attention. Wash contaminated clothing before reuse. Destroy or thoroughly clean contaminated shoes.
UN/NA class	-	1139
ECOLOGICAL PROPERTIES		
Aquatic toxicity, *Green algae*, 96-h EC50	mg/l	2.2/72H/xylene
Aquatic toxicity, *Daphnia magna*, 48-h LC50	mg/l	4.3/chlorobenzene
Aquatic toxicity *Fathead minnow*, 96-h LC50	mg/l	42.3-48.5/ ethylbenzene
Aquatic toxicity, *Rainbow trout*, 96-h LC50	mg/l	2.6/xylene
Bioconcentration factor	BCF	7.4-18.5/xylene
Biodegradation probability	readily biodegradable/xylene	
BOC/COD ratio	%	7.32/ chlorobenzene
Biological oxygen demand, 5 days	g/g	0.030/ chlorobenzene
Theoretical oxygen demand	g/g	0.410/ chlorobenzene
USE & PERFORMANCE		
Manufacturer	Eastman Chemical Company	
Outstanding properties	excellent heat stability, average humidity resistance and compatibility, excellent redissolve resistance	
Recommended for polymers	PE	
Recommended for products	automotive, graphic arts, protective coatings	
Recommended applications	used as a primer or additive for paints, coatings, coatings for plastic, laminating adhesives, and printing inks	

Eastman CP 310W

PARAMETER	UNIT	VALUE
GENERAL INFORMATION		
Name		Eastman CP 310W
CAS #		7732-18-5, 68609-36-9, 84133-50-6
Composition		65-75% water, 22-25% 2,5-furandione, reaction products with polypropylene, chlorinated, 5-7% C12-C14 secondary alcohols ethoxylated, 1-5% chlorobenzene, 1-3% ammonium hydroxide
Acronym		CP
Mixture	-	yes
PHYSICAL PROPERTIES		
State	-	liquid (dispersion)
Odor	-	mild, ammonia
Color	-	off-white milky
Boiling point	°C	90
Melting point	°C	0
Chloride content	wt%	18-23
Solids	wt%	30
pH	-	9-10
Solubility in water at 25°C	g/l	completely soluble
Specific gravity at 25°C	-	1.008
Viscosity at 25°C	mPas	8
HEALTH & SAFETY		
HMIS classification	Flammability	1
	Health	1
	Reactivity	0
Carcinogenicity		chlorobenzene, IARC, OSHA, NTP: no ingredient of this product present at levels greater than or equal to 0.1% is identified as probable, possible or confirmed human carcinogen
Mutagenicity		no data available
Teratogenicity		no data available
DOT class		9, III when material is shipped in quantities in one the Reportable Quantity and when no other hazard class applies; otherwise, not regulated.
TDG class		not regulated
ICAO/IATA class		not regulated
IMDG class		not regulated

Eastman CP 310W

PARAMETER	UNIT	VALUE
Hazardous combustion products	Carbon monoxide (CO), Carbon dioxide (CO2)	
Agency rating, listed	TSCA USA, DSL Canada, AICS Australia, MITI Japan, ECL Korea, IECSC China	
Animal testing, acute toxicity, Rat oral LD50	mg/kg	>2000/product, 20000/chlorobenzene, 3180/C12-C14 ethoxylated secondary alcohols
Animal testing, acute toxicity, Rabbit dermal LD50	mg/kg	non irritant
Animal testing, acute toxicity, Guinea pig dermal LD50	mg/kg	>20000/ chlorobenzene
Animal testing, acute toxicity, Rat inhalation, LC50	mg/m^3	29700/4H/ chlorobenzene
Effect of exposure, eye (human)	no data available	
Effect of exposure, inhalation (human)	no data available	
Effect of exposure, skin (human)	no data available	
Exposure, personal protection	Safety glasses, protective clothing based on chemical resistance data, chemical-resistant gloves, general and local exhaust ventilation.	
First aid, eye	Immediately flush with plenty of water for at least 15 minutes. If easy to do, remove contact lenses. Call a physician or poison control center immediately. In case of irritation from airborne exposure, move to fresh air. Get medical attention if symptoms persist.	
First aid, inhalation	Remove to fresh air. If not breathing, give artificial respiration. If breathing is difficult, give oxygen. If irritation persists, obtain medical advice.	
First aid, skin	Wash with soap and water. Get medical attention if symptoms occur.	
OSHA, PEL	mg/m^3	350/chlorobenzene, 35/ammonium hydroxide
ACGIH, TLV	ppm	10/chlorobenzene, 25/ammonium hydroxide
OSHA, PEL	ppm	75/chlorobenzene, 50/ammonium hydroxide

Eastman CP 310W

PARAMETER	UNIT	VALUE
ECOLOGICAL PROPERTIES		
Aquatic toxicity *Fathead minnow*, 96-h LC50	mg/l	>100/product
BOC/COD ratio	%	7.32/ chlorobenzene
USE & PERFORMANCE		
Manufacturer		Eastman Chemical Company
Outstanding properties		APEO free
Recommended for products		automotive OEM, auto refinish, general industrial coatings, graphic arts, protective coatings, trucks/buses/RVs
Recommended applications		used as an adhesion promoter for paints, automotive OEM, coatings for automotive plastics, and coatings for plastics, it is more useful as adhesion-promoting primer that dried prior to the application of topcoat.

Eastman CP 343-1 (25% solids in xylene)

PARAMETER	UNIT	VALUE
GENERAL INFORMATION		
Name		Eastman CP 343-1 (25% solids in xylene)
CAS #		68609-36-9, 1330-20-7, 100-41-4, 108-90-7, 61789-01-3
Composition		>23% modified chlorinated polyolefin, 54.2-75% xylene, 0-18.80% ethyl-benzene, <2% chlorobenzene, <2% epoxidized oil,
Acronym		CP
Chemical class		chlorinated polyolefin
Mixture	-	yes
Active matter	wt%	24.0-26.5
PHYSICAL PROPERTIES		
State	-	viscous liquid
Odor	-	aromatic
Color	-	amber
Color, Gardner scale	-	7
Boiling point	°C	138-140
Chloride content	wt%	18-23
Solubility (diluents)		limited solubility in toluene and xylene, solutions of CP 343-1 may become hazy partially precipitate from solution or gel with exposure to low temperatures.
Solubility in water at 25°C	g/l	negligible
Specific gravity at 25°C	-	0.898
Vapor density	-	3.7
Viscosity at 25°C	mPas	100-200
HEALTH & SAFETY		
HMIS classification	Flammability	3
	Health	2
	Reactivity	0
Carcinogenicity		ethylbenzene IARC 2B: possibly carcinogenic to humans. NTP Not Listed, OSHA Not Listed. Expert judgment and weight of evidence determination: Not classified
Mutagenicity		no data available
Teratogenicity		no data available

Eastman CP 343-1 (25% solids in xylene)

PARAMETER	UNIT	VALUE
DOT class		Coating solution 3, III. Shipping descriptions may vary based on mode of transport, quantities, package size, and/or origin and destination.
ICAO/IATA class		Coating solution 3, III
IMDG class		Coating solution 3, III
Autoignition temperature	°C	485
Flash point	°C	27
Flash point method	-	TCC
Animal testing, acute toxicity, Rat oral LD50	mg/kg	3523-4000/xylene, 3500/ethylbenzene, 2262/chlorobenzene, >3200/ epoxidized oil
Animal testing, acute toxicity, Rabbit dermal LD50	mg/kg	4200/xylene, 15400/ ethylbenzene
Animal testing, acute toxicity, Guinea pig dermal LD50	mg/kg	>20000/ chlorobenzene
Animal testing, acute toxicity, Rat dermal LD50	mg/kg	>3200
Animal testing, acute toxicity, Rat inhalation, LC50	mg/m^3	6700 ppm/4H/xylene, 4000ppm/4H/ ethylbenzene, 29700/4H/ chlorobenzene
Effect of exposure, eye (human)		Causes serious eye irritation. May cause redness and pain.
Effect of exposure, inhalation (human)		May cause respiratory irritation. May cause drowsiness or dizziness. Narcotic effect.
Effect of exposure, skin (human)		Causes skin irritation.
Effect of exposure, swallowing (human)		May be fatal if swallowed and enters airways.
Exposure, personal protection		Safety glasses, protective clothing based on chemical resistance data, chemical-resistant gloves, general and local exhaust ventilation.
First aid, eye		Rinse cautiously with water for several minutes. Remove contact lenses, if present and easy to do. Continue rinsing. If molten material contacts the eye, immediately flush with plenty of water for at least 15 minutes. Get medical attention immediately.

Eastman CP 343-1 (25% solids in xylene)

PARAMETER	UNIT	VALUE
First aid, inhalation		Remove to fresh air. If not breathing, give artificial respiration. If breathing is difficult, give oxygen. If irritation persists, obtain medical advice.
First aid, skin		Immediately flush with plenty of water for at least 15 minutes while removing contaminated clothing and shoes. Get medical attention. Wash contaminated clothing before reuse. Destroy or thoroughly clean contaminated shoes.
OSHA, PEL	mg/m^3	435/xylene, 435/ ethylbenzene, 350/ chlorobenzene
ACGIH, TLV	ppm	100/xylene, 20/ ethylbenzene,10/ chlorobenzene, STEL150/xylene
NIOSH, REL	ppm	100/ethylbenzene
OSHA, PEL	ppm	100/xylene, 100/ ethylbenzene, 75/ chlorobenzene
UN/NA class	-	1139
ECOLOGICAL PROPERTIES		
Aquatic toxicity, *Green algae*, 96-h EC50	mg/l	2.2/72H/xylene
Aquatic toxicity *Fathead minnow*, 96-h LC50	mg/l	42.3-48.5/ ethylbenzene
Aquatic toxicity, *Rainbow trout*, 96-h LC50	mg/l	2.6/xylene
Bioconcentration factor	BCF	7.4-18.5/xylene
Biodegradation probability	readily biodegradable/xylene	
BOC/COD ratio	%	7.32/ chlorobenzene
Partition coefficient, log K_{ow}	-	3.12-3.30/xylene, 3.15/ethylbenzene
USE & PERFORMANCE		
Manufacturer	Eastman Chemical Company	
Outstanding properties		excellent humidity resistance, fair fuel resistance, average compatibility, excellent redissolve resistance.
Recommended for polymers	PP, TPO, non-olefin plastics, aluminum, and galvanized steel	
Recommended for products		architectural coatings, automotive, commercial printing inks, graphic arts, protective coatings, trucks/bus/RVs

Eastman CP 343-1 (25% solids in xylene)

PARAMETER	UNIT	VALUE
Recommended applications		adhesion promoter for automotive OEM, coatings for automotive plastics

Eastman CP 343-1 (50% solids in xylene)

PARAMETER	UNIT	VALUE
GENERAL INFORMATION		
Name		Eastman CP 343-1 (50% solids in xylene)
CAS #		68609-36-9, 1330-20-7, 100-41-4, 108-90-7, 61789-01-3
Composition		>46% modified chlorinated polyolefin, 32.5-50% xylene, 0-12.5% ethylbenzene, <2.5% chlorobenzene, <4% epoxidized oil,
Acronym		CP
Chemical class		chlorinated polyolefin
Mixture	-	yes
Active matter	wt%	>46.00
PHYSICAL PROPERTIES		
State	-	viscous liquid at room temperature, gel at lower temperature
Odor	-	aromatic
Color	-	amber
Color, Gardner scale	-	8
Boiling point	°C	138-140
Chloride content	wt%	18-23
Solubility (diluents)		limited solubility in toluene and xylene, solutions of CP 343-1 may become hazy partially precipitate from solution or gel with exposure to low temperatures.
Solubility in water at 25°C	g/l	negligible
Specific gravity at 25°C	-	0.935
HEALTH & SAFETY		
HMIS classification	Flammability	3
	Health	2
	Reactivity	0
Carcinogenicity		ethylbenzene IARC 2B: possibly carcinogenic to humans. NTP Not Listed, OSHA Not Listed. Expert judgment and weight of evidence determination: Not classified
Mutagenicity		no data available
Teratogenicity		no data available

Eastman CP 343-1 (50% solids in xylene)

PARAMETER	UNIT	VALUE
DOT class	Coating solution 3, III. Shipping descriptions may vary based on mode of transport, quantities, package size, and/or origin and destination.	
ICAO/IATA class	Coating solution 3, III	
IMDG class	Coating solution 3, III	
Autoignition temperature	°C	460
Flash point	°C	28
Flash point method	-	TCC
Animal testing, acute toxicity, Rat oral LD50	mg/kg	3523-4000/xylene, 3500/ethylbenzene, 2262/chlorobenzene, >3200/epoxidized oil
Animal testing, acute toxicity, Rabbit dermal LD50	mg/kg	4200/xylene, 15400/ethylbenzene
Animal testing, acute toxicity, Guinea pig dermal LD50	mg/kg	>20000/ chlorobenzene
Animal testing, acute toxicity, Rat dermal LD50	mg/kg	>3200
Animal testing, acute toxicity, Rat inhalation, LC50	mg/m^3	6700 ppm/4H/xylene, 4000ppm/4H/ ethylbenzene, 29700/4H/ chlorobenzene
Effect of exposure, eye (human)	Causes serious eye irritation. May cause redness and pain.	
Effect of exposure, inhalation (human)	May cause respiratory irritation. May cause drowsiness or dizziness. Narcotic effect.	
Effect of exposure, skin (human)	Causes skin irritation.	
Effect of exposure, swallowing (human)	May be fatal if swallowed and enters airways.	
Effect of repeated or overexposure (human)	May cause damage to organs (auditory organ) through prolonged or repeated exposure.	
Exposure, personal protection	Safety glasses, protective clothing based on chemical resistance data, chemical-resistant gloves, general and local exhaust ventilation.	

Eastman CP 343-1 (50% solids in xylene)

PARAMETER	UNIT	VALUE
First aid, eye		Rinse cautiously with water for several minutes. Remove contact lenses, if present and easy to do. Continue rinsing. If molten material contacts the eye, immediately flush with plenty of water for at least 15 minutes. Get medical attention immediately.
First aid, inhalation		Remove to fresh air. If not breathing, give artificial respiration. If breathing is difficult, give oxygen. If irritation persists, obtain medical advice.
First aid, skin		Immediately flush with plenty of water for at least 15 minutes while removing contaminated clothing and shoes. Get medical attention. Wash contaminated clothing before reuse. Destroy or thoroughly clean contaminated shoes.
OSHA, PEL	mg/m^3	435/xylene, 435/ethylbenzene, 350/chlorobenzene
ACGIH, TLV	ppm	100/xylene, 20/ethylbenzene,10/chlorobenzene, STEL150/xylene
NIOSH, REL	ppm	100/ethylbenzene
OSHA, PEL	ppm	100/xylene, 100/ethylbenzene, 75/chlorobenzene
UN/NA class	-	1139
ECOLOGICAL PROPERTIES		
Aquatic toxicity, *Green algae*, 96-h EC50	mg/l	2.2/72H/xylene
Aquatic toxicity *Fathead minnow*, 96-h LC50	mg/l	42.3-48.5/ethylbenzene
Aquatic toxicity, *Rainbow trout*, 96-h LC50	mg/l	2.6/xylene
Bioconcentration factor	BCF	7.4-18.5/xylene
Biodegradation probability		readily biodegradable/xylene
BOC/COD ratio	%	7.32/chlorobenzene
Partition coefficient, log K_{ow}	-	3.12-3.30/xylene, 3.15/ethylbenzene
USE & PERFORMANCE		
Manufacturer		Eastman Chemical Company

Eastman CP 343-1 (50% solids in xylene)

PARAMETER	UNIT	VALUE
Outstanding properties		excellent humidity resistance, fair fuel resistance, average compatibility, excellent redissolve resistance.
Recommended for polymers		PP, TPO, non-olefin plastics, aluminum, and galvanized steel
Recommended for products		automotive, auto plastics, general industrial coatings, trucks/buses/RTs
Recommended applications		adhesion promoter for inks and graphic arts, automotive OEM, coatings for automotive plastics, coatings for plastics, and printing inks

Eastman CP 343-1 (100% solids)

PARAMETER	UNIT	VALUE
GENERAL INFORMATION		
Name		Eastman CP 343-1 (100% solids)
CAS #		68609-36-9, 61789-01-3, 108-90-7
Composition		>97% modified chlorinated polyolefin, <3% epoxidized oil, <0.4% chlorobenzene
Acronym		CP
Chemical class		chlorinated polyolefin
Mixture	-	yes
Active matter	wt%	>97.00
PHYSICAL PROPERTIES		
State	-	solid/powder
Odor	-	odorless
Color	-	tan
Softening point	°C	97
Chloride content	wt%	18-23
Solubility (diluents)		limited solubility in toluene and xylene, solutions of CP 343-1 may become hazy partially precipitate from solution or gel with exposure to low temperatures.
Solubility in water at 25°C	g/l	negligible
Specific gravity at 25°C	-	1.03
HEALTH & SAFETY		
HMIS classification	Flammability	1
	Health	1
	Reactivity	0
Carcinogenicity		IARC, OSHA, NTP: no ingredient of this product present at levels greater than or equal to 0.1% is identified as probable, possible or confirmed human carcinogen
Teratogenicity		no data available
Flash point	°C	234
Flash point method	-	TCC
Animal testing, acute toxicity, Rat oral LD50	mg/kg	2262/chlorobenzene, >3200/epoxidized oil
Animal testing, acute toxicity, Guinea pig dermal LD50	mg/kg	>20000/chlorobenzene

Eastman CP 343-1 (100% solids)

PARAMETER	UNIT	VALUE
Animal testing, acute toxicity, Rat dermal LD50	mg/kg	>3200
Animal testing, acute toxicity, Rat inhalation, LC50	mg/m^3	>140/6H
Effect of exposure, inhalation (human)		Avoid breathing vapor from heated material. At elevated temperatures, vapor may cause allergic respiratory reaction.
Effect of exposure, skin (human)		Contact with molten substance/product may cause severe burns to skin.
Exposure, personal protection		Safety glasses, protective clothing based on chemical resistance data, chemical-resistant gloves, general and local exhaust ventilation.
First aid, eye		Rinse cautiously with water for several minutes. Remove contact lenses, if present and easy to do. Continue rinsing. If molten material contacts the eye, immediately flush with plenty of water for at least 15 minutes. Get medical attention immediately.
First aid, inhalation		Remove to fresh air. If not breathing, give artificial respiration. If breathing is difficult, give oxygen. If irritation persists, obtain medical advice.
First aid, skin		Wash with soap and water. Get medical attention if symptoms occur. If burned by contact with hot material, cool molten material adhering to skin as quickly as possible with water, and see a physician for removal of adhering material and treatment of burn.
OSHA, PEL	mg/m^3	350/chlorobenzene
ACGIH, TLV	ppm	10/chlorobenzene
OSHA, PEL	ppm	75/chlorobenzene
UN/NA class	-	1139
ECOLOGICAL PROPERTIES		
BOC/COD ratio	%	7.3/chlorobenzene
USE & PERFORMANCE		
Manufacturer		Eastman Chemical Company
Outstanding properties		good resistance to high temperatures
Recommended for polymers		PP, TPO, non-olefin plastics, aluminum and galvanized steel
Recommended for products		automotive, auto plastics, auto refinish, general industrial coatings, graphic arts, protective coatings, trucks/buses/RVs

Eastman CP 343-1 (100% solids)

PARAMETER	UNIT	VALUE
Recommended applications		adhesion promoter for automotive OEM, coatings for automotive plastics

Eastman CP 343-3 (25% solids in xylene)

PARAMETER	UNIT	VALUE
GENERAL INFORMATION		
Name		Eastman CP 343-3 (25% solids in xylene)
CAS #		68609-36-9, 1330-20-7, 100-41-4, 108-90-7, 61789-01-3
Composition		>23% modified chlorinated polyolefin, 54.3-75% xylene, <18.8% ethylbenzene, <1.9% chlorobenzene, <2% epoxidized oil
Acronym		CP
Chemical class		chlorinated polyolefin
Mixture	-	yes
Active matter	wt%	>23
PHYSICAL PROPERTIES		
State	-	viscous liquid
Odor	-	aromatic
Color	-	amber
Color, Gardner scale	-	6-7
Boiling point	°C	138-140
Chloride content	wt%	26-32
Solubility (diluents)		soluble in esters and ketone, good solvent tolerance for a wide range of solvents.
Solubility in water at 25°C	g/l	negligible
Specific gravity at 25°C	-	0.906
HEALTH & SAFETY		
HMIS classification	Flammability	3
	Health	2
	Reactivity	0
Carcinogenicity		IARC, OSHA, NTP: no ingredient of this product present at levels greater than or equal to 0.1% is identified as probable, possible or confirmed human carcinogen
Mutagenicity		no data available
Teratogenicity		no data available
DOT class		Coating solution 3, III. Shipping descriptions may vary based on mode of transport, quantities, package size, and/or origin and destination.

Eastman CP 343-3 (25% solids in xylene)

PARAMETER	UNIT	VALUE
ICAO/IATA class		Coating solution 3, III. Shipping descriptions may vary based on mode of transport, quantities, package size, and/or origin and destination.
IMDG class		Coating solution 3, III. Shipping descriptions may vary based on mode of transport, quantities, package size, and/or origin and destination.
Autoignition temperature	°C	>578
Flash point	°C	23
Flash point method	-	TCC
Animal testing, acute toxicity, Rat oral LD50	mg/kg	3523-4000/xylene, 3500/ethylbenzene, 2262/chlorobenzene, >3200/ epoxidized oil
Animal testing, acute toxicity, Rabbit dermal LD50	mg/kg	4200/xylene, 15400/ ethylbenzene
Animal testing, acute toxicity, Guinea pig dermal LD50	mg/kg	>20000/ chlorobenzene
Animal testing, acute toxicity, Rat dermal LD50	mg/kg	>3200
Animal testing, acute toxicity, Rat inhalation, LC50	mg/m^3	6700 ppm/4H/xylene, 4000ppm/4H/ ethylbenzene, 29700/4H/ chlorobenzene
Effect of exposure, eye (human)		Causes serious eye irritation. May cause redness and pain.
Effect of exposure, inhalation (human)		May cause respiratory irritation. May cause drowsiness or dizziness. Narcotic effect.
Effect of exposure, skin (human)		Causes skin irritation.
Effect of exposure, swallowing (human)		May be fatal if swallowed and enters airways.
Effect of repeated or overexposure (human)		May cause damage to organs (auditory organ) through prolonged or repeated exposure.
Exposure, personal protection		Safety glasses, protective clothing based on chemical resistance data, chemical-resistant gloves, general and local exhaust ventilation.

Eastman CP 343-3 (25% solids in xylene)

PARAMETER	UNIT	VALUE
First aid, eye		Rinse cautiously with water for several minutes. Remove contact lenses, if present and easy to do. Continue rinsing. If molten material contacts the eye, immediately flush with plenty of water for at least 15 minutes. Get medical attention immediately.
First aid, inhalation		Remove to fresh air. If not breathing, give artificial respiration. If breathing is difficult, give oxygen. If irritation persists, obtain medical advice.
First aid, skin		Immediately flush with plenty of water for at least 15 minutes while removing contaminated clothing and shoes. Get medical attention. Wash contaminated clothing before reuse. Destroy or thoroughly clean contaminated shoes.
OSHA, PEL	mg/m^3	435/xylene, 435/ ethylbenzene, 350/ chlorobenzene
ACGIH, TLV	ppm	100/xylene, 20/ ethylbenzene,10/ chlorobenzene, STEL150/xylene
NIOSH, REL	ppm	100/ethylbenzene
OSHA, PEL	ppm	100/xylene, 100/ ethylbenzene, 75/ chlorobenzene
UN/NA class	-	1139
ECOLOGICAL PROPERTIES		
Aquatic toxicity Green algae, 96-h EC50	mg/l	2.2/72H/xylene
Aquatic toxicity Fathead minnow, 96-h LC50	mg/l	42.3-48.5/ ethylbenzene
Aquatic toxicity, Rainbow trout, 96-h LC50	mg/l	2.6/xylene
Bioconcentration factor	BCF	7.4-18.5/xylene
Biodegradation probability	readily biodegradable/xylene	
BOC/COD ratio	%	7.32/ chlorobenzene
Partition coefficient, log K$_{ow}$	-	3.12-3.30/xylene, 3.15/ethylbenzene

Eastman CP 343-3 (25% solids in xylene)

PARAMETER	UNIT	VALUE
USE & PERFORMANCE		
Manufacturer		Eastman Chemical Company
Outstanding properties		excellent compatibility, average humidity resistance, poor fuel resistance, great tolerance to esters and ketones makes them easier to incorporate into the topcoat
Recommended for polymers		PP, TPO, non-olefin plastics, aluminum, and galvanized steel
Recommended for products		automotive, auto refinish, general industrial coatings, graphic arts, non-medical housings & hardware for electronics, protective coatings, trucks/buses/RVs
Recommended applications		adhesion promoter for inks and graphic arts, automotive OEM, coatings for automotive plastics, coatings for plastics, and printing inks

Eastman CP 347W

PARAMETER	UNIT	VALUE
GENERAL INFORMATION		
Name		Eastman CP 347W
CAS #		7732-18-5, 68609-36-9, 124-68-5
Composition		>72% water, 20% modified chlorinated polyolefin, <5% alkylphenol ethoxylate, 5% 2-amino-2-methyl-1-propanol, <2.5 ethylene glycol, <1% additive
Acronym		CP
Chemical class		chlorinated polyolefin
Mixture	-	yes
PHYSICAL PROPERTIES		
State	-	liquid
Odor	-	amine-like
Color	-	tan
Boiling point	°C	100
Melting point	°C	0
Chloride content	wt%	20.5
Solids	wt%	30
pH	-	9-10
Solubility in water at 25°C	g/l	completely soluble
Specific gravity at 25°C	-	1.02
Viscosity at 25°C	mPas	10
HEALTH & SAFETY		
HMIS classification	Flammability	1
	Health	1
	Reactivity	0
Carcinogenicity	no data available	
Mutagenicity	no data available	
Teratogenicity	no data available	
DOT class	not regulated	
TDG class	not regulated	
ICAO/IATA class	not regulated	
IMDG class	not regulated	
Hazardous combustion products	Carbon monoxide, carbon dioxide	
Agency rating, listed	TSCA USA, DSL Canada, AICS Australia, MITI Japan, ECL Korea, IECSC China	

Eastman CP 347W

PARAMETER	UNIT	VALUE
Animal testing, acute toxicity, Rat oral LD50	mg/kg	1670-3250 alkyl-phenol ethoxylate, 2900/2-amino-2-methyl-1propanol
Animal testing, acute toxicity, Rabbit dermal LD50	mg/kg	non irritant/24h/product, >2000/2-amino-2-methyl-1propa-nol, 1750-4570/ alkylphenol ethox-ylate
Animal testing, acute toxicity, Rat inhalation, LC50	mg/m^3	650ppm/8H/alkyl-phenol ethoxylate
Effect of exposure, eye (human)	no data available	
Effect of exposure, inhalation (human)	no data available	
Effect of exposure, skin (human)	no data available	
Exposure, personal protection	Safety glasses, protective clothing based on chemical resistance data, chemical-resistant gloves, general and local exhaust ventilation.	
First aid, eye	Immediately flush with plenty of water for at least 15 minutes. If easy to do, remove contact lenses. Call a physician or poison control center immediately. In case of irritation from airborne exposure, move to fresh air. Get medical attention if symptoms persist.	
First aid, inhalation	Remove to fresh air. If not breathing, give artificial respiration. If breathing is difficult, give oxygen. If irritation persists, obtain medical advice.	
First aid, skin	Wash with soap and water. Get medical attention if symptoms occur.	
ACGIH, TLV	mg/m^3	100/ethylene glycol
OSHA, PEL	mg/m^3	125/ethylene glycol
OSHA, PEL	ppm	50/ethylene glycol
ECOLOGICAL PROPERTIES		
Aquatic toxicity *Fathead minnow*, 96-h LC50	mg/l	6.9-8.6/alkylphenol ethoxylate
Biodegradation probability	58.7/28d/ alkylphenol ethoxylate	
Biological oxygen demand, 5 days	g/g	0.030/ chlorobenzene
Chemical oxygen demand	g/g	0.41/ chlorobenzene

Eastman CP 347W

PARAMETER	UNIT	VALUE
Theoretical oxygen demand	g/g	2.060/ chlorobenzene
Partition coefficient, log K_{oc}	-	2.4/chlorobenzene
USE & PERFORMANCE		
Manufacturer		Eastman Chemical Company
Outstanding properties		compatible with other waterborne resins and therefore more useful as an additive and with many amine-neutralizable solution resins.
Recommended for products		auto OEM, auto refinish, general industrial coatings, graphic arts, trucks/busses/RVs
Recommended applications		adhesion promoter for paints, compatible with other waterborne resins and therefore more useful as an additive

Eastman CP 515-2 (40% solids in aromatic 100)

PARAMETER	UNIT	VALUE
GENERAL INFORMATION		
Name		Eastman CP 515-2 (40% solids in aromatic 100)
CAS #		68442-33-1, 64742-95-6, 95-63-6, 108-90-7, 61789-01-3, 94-96-2
Composition		>35% chlorinated polyolefin, >38% solvent naphtha (petroleum), light aromatic, 19.2% 1,2,4-trimethylbenzene, <4% chlorobenzene, <4% epoxidized oil, <1.4% xylene, <0.9% cumene, <0.5 ethylbenzene
Acronym		CP
Chemical class		chlorinated polyolefin
Mixture	-	yes
PHYSICAL PROPERTIES		
State		viscous liquid
Odor	-	odorless
Color	-	amber
Color, Gardner scale	-	7
Boiling point	°C	155
Decomposition temperature	°C	300
Decomposition energy	J/g	134
Chloride content	wt%	26-32
Solubility (diluents)		xylene, toluene, selected ketones, and ester solvents
Solubility in water at 25°C	g/l	negligible
Specific gravity at 25°C	-	0.944
HEALTH & SAFETY		
NFPA classification	Flammability	2
	Health	2
	Reactivity	0
HMIS classification	Flammability	2
	Health	3
	Reactivity	0
Carcinogenicity		IARC, OSHA, NTP: no ingredient of this product present at levels greater than or equal to 0.1% is identified as probable, possible or confirmed human carcinogen
Mutagenicity		no evidence
Teratogenicity		no data available

Eastman CP 515-2 (40% solids in aromatic 100)

PARAMETER	UNIT	VALUE
ICAO/IATA class		Coating solution 3, III. Shipping descriptions may vary based on mode of transport, quantities, package size, and/or origin and destination.
IMDG class		Coating solution 3, III. Shipping descriptions may vary based on mode of transport, quantities, package size, and/or origin and destination.
Autoignition temperature	°C	471
Flash point	°C	42
Flash point method	-	TCC
Hazardous ingredients, labelling	COMBUSTIBLE LIQUID, N.O.S.	
Animal testing, acute toxicity, Rat oral LD50	mg/kg	>5000/solvent naphtha (petroleum) light arom, 6000/1,2,4-trimethylbenzene, 3500/ethylbenzene, 2262/chlorobenzene, 2910/cumene
Animal testing, acute toxicity, Rabbit dermal LD50	mg/kg	>2000/Solvent naphtha (petroleum) light arom, 15400/ethylbenzene, 10000/cumene
Animal testing, acute toxicity, Rat dermal LD50	mg/kg	>3200
Animal testing, acute toxicity, Rat inhalation, LC50	mg/m³	763000/4h/solvent naphtha (petroleum), light arom, 29700/4H/chlorobenzene, 41600/4H/cumene
Effect of exposure, eye (human)	Causes serious eye irritation.	
Effect of exposure, inhalation (human)	May cause respiratory irritation. Avoid inhalation of vapor or mist.	
Effect of exposure, skin (human)	Causes skin irritation.	
Effect of exposure, swallowing (human)	May be fatal if swallowed and enters airways.	
Exposure, personal protection	Safety glasses, protective clothing based on chemical resistance data, chemical-resistant gloves, general and local exhaust ventilation.	

Eastman CP 515-2 (40% solids in aromatic 100)

PARAMETER	UNIT	VALUE
First aid, eye		Rinse cautiously with water for several minutes. Remove contact lenses, if present and easy to do. Continue rinsing. If molten material contacts the eye, immediately flush with plenty of water for at least 15 minutes. Get medical attention immediately.
First aid, inhalation		Remove to fresh air. If not breathing, give artificial respiration. If breathing is difficult, give oxygen. If irritation persists, obtain medical advice.
First aid, skin		Immediately flush with plenty of water for at least 15 minutes while removing contaminated clothing and shoes. Get medical attention. Wash contaminated clothing before reuse. Destroy or thoroughly clean contaminated shoes.
ACGIH, TLV	mg/m^3	435/xylene
NIOSH, REL	mg/m^3	125/1,2,4-trimethylbenzene, 245/cumene
OSHA, PEL	mg/m^3	435/xylene, 435/ethylbenzene, 350/chlorobenzene, STEL655/xylene, 245,cumene
ACGIH, TLV	ppm	100/xylene, 20/ethylbenzene,10/chlorobenzene, 50/cumene
NIOSH, REL	ppm	20/1,2,4-trimethylbenzene, 100/ethylbenzene, 50/cumene
OSHA, PEL	ppm	100/xylene,100/ethylbenzene,75/chlorobenzene, STEL150/xylene, 50/cumene
UN/NA class	-	1139

Eastman CP 515-2 (40% solids in aromatic 100)

PARAMETER	UNIT	VALUE
ECOLOGICAL PROPERTIES		
Aquatic toxicity, *Green algae*, 96-h EC50	mg/l	2.356/72H/1,2,4-trimethylbenzen, 3.1/72H/solvent naphtha (petroleum) light arom, 2.1/72H/cumene
Aquatic toxicity, *Daphnia magna*, 48-h LC50	mg/l	4.5/solvent naphtha (petroleum) light arom, 3.6/1,2,4-Trimethylbenzene, 4.3/chlorobenzene
Aquatic toxicity *Fathead minnow*, 96-h LC50	mg/l	8.2/solvent naphtha (petroleum) light arom,7.72/1,2,4-trimethylbenzene, 42.3-48.5/ethylbenzene
Biodegradation probability		74.0%/28d/solvent naphtha (petroleum) light arom/inherently biodegradable, 8.0-14.0%/28d/ (conc100mg/l)/1,2,4-trimethylbenzene
BOC/COD ratio	%	7.32/chlorobenzene
Biological oxygen demand, 5 days	g/g	0.030/chlorobenzene
Theoretical oxygen demand	g/g	0.410/chlorobenzene
Partition coefficient, log K_{oc}	-	2.4/chlorobenzene
USE & PERFORMANCE		
Manufacturer	Eastman Chemical Company	
Outstanding properties	excellent compatibility with a variety of resins	
Recommended for polymers	PP, TPO, aluminum, and galvanized steel	
Recommended for products	automotive, commercial printing inks, graphic arts, protective coatings	
Recommended applications	adhesion promoter for inks and graphic arts, automotive OEM, coatings for automotive plastics, coatings for plastics, and printing inks	

Eastman CP 515-2 (40% solids in toluene)

PARAMETER	UNIT	VALUE
GENERAL INFORMATION		
Name		Eastman CP 515-2 (40% solids in toluene)
CAS #		68442-33-1, 108-88-3, 108-90-7, 61789-01-3
Composition		>38% chlorinated polyolefin, <59% toluene, <6% chlorobenzene, <3% epoxidized oil
Acronym		CP
Chemical class		chlorinated polyolefin
Mixture	-	yes
PHYSICAL PROPERTIES		
State	-	viscous liquid
Odor	-	aromatic
Odor threshold	ppm	2.9
Color	-	yellow
Color, Gardner scale	-	6
Boiling point	°C	110
Chloride content	wt%	26.5-31.5
Kinematic viscosity at 25°C	cSt	333
Solubility (diluents)		xylene, toluene, selected ketones, and ester solvents
Solubility in water at 25°C	g/l	negligible
Specific gravity at 25°C	-	0.96
Vapor pressure at 20°C	kPa	5.05
Viscosity at 25°C	mPas	320
HEALTH & SAFETY		
HMIS classification	Flammability	3
	Health	2
	Reactivity	0
Carcinogenicity		IARC, OSHA, NTP: no ingredient of this product present at levels greater than or equal to 0.1% is identified as probable, possible or confirmed human carcinogen
Mutagenicity		no data available
Teratogenicity		no data available
DOT class		Coating solution 3, III. Shipping descriptions may vary based on mode of transport, quantities, package size, and/or origin and destination.

Eastman CP 515-2 (40% solids in toluene)

PARAMETER	UNIT	VALUE
ICAO/IATA class		Coating solution 3, III. Shipping descriptions may vary based on mode of transport, quantities, package size, and/or origin and destination.
IMDG class		Coating solution 3, III. Shipping descriptions may vary based on mode of transport, quantities, package size, and/or origin and destination.
Flash point	°C	5
Flash point method	-	TCC
Animal testing, acute toxicity, Rat oral LD50	mg/kg	>5000/toluene, 2262/chlorobenzene, >3200/ epoxidized oil
Animal testing, acute toxicity, Rabbit dermal LD50	mg/kg	>5000/toluene
Animal testing, acute toxicity, Guinea pig dermal LD50	mg/kg	>20000/ chlorobenzene
Animal testing, acute toxicity, Rat dermal LD50	mg/kg	>3200
Animal testing, acute toxicity, Rat inhalation, LC50	mg/m^3	20000/4H/toluene, 29700/4H/ chlorobenzene
Effect of exposure, eye (human)		Causes serious eye irritation. May cause redness and pain.
Effect of exposure, inhalation (human)		May cause respiratory irritation. May cause drowsiness or dizziness. Narcotic effect.
Effect of exposure, skin (human)		Causes skin irritation.
Effect of exposure, swallowing (human)		May be fatal if swallowed and enters airways.
Effect of repeated or overexposure (human)		May cause damage to organs (auditory organ) through prolonged or repeated exposure.
Exposure, personal protection		Safety glasses, protective clothing based on chemical resistance data, chemical-resistant gloves, general and local exhaust ventilation.
First aid, eye		Rinse cautiously with water for several minutes. Remove contact lenses, if present and easy to do. Continue rinsing. If molten material contacts the eye, immediately flush with plenty of water for at least 15 minutes. Get medical attention immediately.

Eastman CP 515-2 (40% solids in toluene)

PARAMETER	UNIT	VALUE
First aid, inhalation		Remove to fresh air. If not breathing, give artificial respiration. If breathing is difficult, give oxygen. If irritation persists, obtain medical advice.
First aid, skin		Immediately flush with plenty of water for at least 15 minutes while removing contaminated clothing and shoes. Get medical attention. Wash contaminated clothing before reuse. Destroy or thoroughly clean contaminated shoes.
OSHA, PEL	mg/m^3	435/ethylbenzene, 350/chlorobenzene
ACGIH, TLV	ppm	20/toluene, 20/ethylbenzene,10/chlorobenzene
OSHA, PEL	ppm	toluene: TWA300, Celling300, Max Conc500, 100/ethylbenzene, 75/chlorobenzene
UN/NA class	-	1139
ECOLOGICAL PROPERTIES		
Aquatic toxicity, *Green algae*, 96-h EC50	mg/l	2.2/72H/xylene
Aquatic toxicity, *Daphnia magna*, 48-h LC50	mg/l	4.3/chlorobenzene
Bioconcentration factor	BCF	7.4-18.5/xylene
Biodegradation probability		readily biodegradable (according to OECD criteria)/toluene
Partition coefficient, log K_{ow}	-	2.69/toluene
USE & PERFORMANCE		
Manufacturer		Eastman Chemical Company
Outstanding properties		good resistance to high temperature
Recommended for polymers		PP, TPO, aluminum, and galvanized steel
Recommended for products		automotive, inks, and paints
Recommended applications		adhesion promoter for inks and graphic arts, automotive OEM, coatings for automotive plastics, coatings for plastics, and printing inks

Eastman CP 515-2 (40% solids in xylene)

PARAMETER	UNIT	VALUE
GENERAL INFORMATION		
Name		Eastman CP 515-2 (40% solids in xylene)
CAS #		68442-33-1, 1330-20-7, 100-41-4, 108-90-7, 61789-01-3
Composition		>38% chlorinated polyolefin, <60% xylene, <15% ethylbenzene, <5% chlorobenzene, <2% epoxidized oil
Acronym		CP
Chemical class		chlorinated polyolefin
Mixture	-	yes
PHYSICAL PROPERTIES		
State	-	liquid
Odor	-	aromatic
Color	-	amber
Color, Gardner scale	-	3
Boiling point	°C	138
Chloride content	wt%	26-32
Solubility (diluents)		xylene, toluene, selected ketones, and ester solvents
Solubility in water at 25°C	g/l	negligible
Specific gravity at 25°C	-	0.955
HEALTH & SAFETY		
HMIS classification	Flammability	3
	Health	2
	Reactivity	0
Carcinogenicity		IARC, OSHA, NTP: no ingredient of this product present at levels greater than or equal to 0.1% is identified as probable, possible or confirmed human carcinogen
Mutagenicity		no data available
Teratogenicity		no data available
DOT class		Coating solution 3, III. Shipping descriptions may vary based on mode of transport, quantities, package size, and/or origin and destination.
ICAO/IATA class		Coating solution 3, III. Shipping descriptions may vary based on mode of transport, quantities, package size, and/or origin and destination.

Eastman CP 515-2 (40% solids in xylene)

PARAMETER	UNIT	VALUE
IMDG class		Coating solution 3, III. Shipping descriptions may vary based on mode of transport, quantities, package size, and/or origin and destination.
Autoignition temperature	°C	525
Flash point	°C	28
Flash point method	-	TCC
Animal testing, acute toxicity, Rat oral LD50	mg/kg	3523-4000/xylene, 3500/ethylbenzene, 2262/chlorobenzene, >3200/ epoxidized oil
Animal testing, acute toxicity, Rabbit dermal LD50	mg/kg	4200/xylene, 15400/ ethylbenzene
Animal testing, acute toxicity, Guinea pig dermal LD50	mg/kg	>20000/ chlorobenzene
Animal testing, acute toxicity, Rat dermal LD50	mg/kg	>3200
Animal testing, acute toxicity, Rat inhalation, LC50	mg/m^3	6700 ppm/4H/xylene, 4000ppm/4H/ ethylbenzene, 29700/4H/ chlorobenzene
Effect of exposure, eye (human)		Causes serious eye irritation. May cause redness and pain.
Effect of exposure, inhalation (human)		May cause respiratory irritation. May cause drowsiness or dizziness. Narcotic effect.
Effect of exposure, skin (human)		Causes skin irritation.
Effect of exposure, swallowing (human)		May be fatal if swallowed and enters airways.
Effect of repeated or overexposure (human)		May cause damage to organs (auditory organ) through prolonged or repeated exposure.
Exposure, personal protection		Safety glasses, protective clothing based on chemical resistance data, chemical-resistant gloves, general and local exhaust ventilation.
First aid, eye		Rinse cautiously with water for several minutes. Remove contact lenses, if present and easy to do. Continue rinsing. If molten material contacts the eye, immediately flush with plenty of water for at least 15 minutes. Get medical attention immediately.

Eastman CP 515-2 (40% solids in xylene)

PARAMETER	UNIT	VALUE
First aid, inhalation		Remove to fresh air. If not breathing, give artificial respiration. If breathing is difficult, give oxygen. If irritation persists, obtain medical advice.
First aid, skin		Immediately flush with plenty of water for at least 15 minutes while removing contaminated clothing and shoes. Get medical attention. Wash contaminated clothing before reuse. Destroy or thoroughly clean contaminated shoes.
OSHA, PEL	mg/m^3	435/xylene, 435/ethylbenzene, 350/chlorobenzene
ACGIH, TLV	ppm	100/xylene, 20/ethylbenzene,10/chlorobenzene, STEL150/xylene
NIOSH, REL	ppm	100/ethylbenzene
OSHA, PEL	ppm	100/xylene, 100/ethylbenzene, 75/chlorobenzene
UN/NA class	-	1139
ECOLOGICAL PROPERTIES		
Aquatic toxicity, *Green algae*, 96-h EC50	mg/l	2.2/72H/xylene
Aquatic toxicity *Fathead minnow*, 96-h LC50	mg/l	42.3-48.5/ethylbenzene
Aquatic toxicity, *Rainbow trout*, 96-h LC50	mg/l	2.6/xylene
Bioconcentration factor	BCF	7.4-18.5/xylene
Biodegradation probability		readily biodegradable/xylene
BOC/COD ratio	%	7.32/chlorobenzene
Partition coefficient, log K_{ow}	-	3.12-3.30/xylene, 3.15/ethylbenzene
USE & PERFORMANCE		
Manufacturer		Eastman Chemical Company
Outstanding properties		excellent compatibility with a variety of resins, good heat resistance.
Recommended for polymers		PP, TPO, aluminum, and galvanized steel
Recommended for products		automotive, auto refinish, general industrial coatings,graphic arts, protective coatings

Eastman CP 515-2 (40% solids in xylene)

PARAMETER	UNIT	VALUE
Recommended applications		adhesion promoter for inks and graphic arts, automotive OEM, coatings for automotive plastics, coatings for plastics, and printing inks

Eastman CP 730-1 (20% solids in xylene)

PARAMETER	UNIT	VALUE
GENERAL INFORMATION		
Name		Eastman CP 730-1 (20% solids in xylene)
CAS #		68609-36-9, 1330-20-7, 100-41-4, 108-90-7, 61789-01-3
Composition		<20% modified chlorinated polyolefin, <80% xylene, <20% ethylbenzene, <2.5% chlorobenzene, <3% epoxidized oil,
Acronym		CP
Chemical class		chlorinated polyolefin
Mixture	-	yes
Active matter	wt%	20
PHYSICAL PROPERTIES		
State	-	liquid
Odor	-	slight aromatic
Color	-	yellow
Color, Gardner scale	-	4
Boiling point	°C	135
Chloride content	wt%	20.50-23.50
Kinematic viscosity at 20°C	cSt	157-210
Solubility (diluents)		soluble in aromatic hydrocarbons/xylene, toluene, cyclic hydrocarbons such as methylcyclohexane and ethylcyclohexane can be used to dilute CP 730-1. It is not soluble in aliphatic hydrocarbons, esters, ketones, or alcohols but can be diluted, provided there are long chain ketones, esters like methyl amyl ketone.
Solubility in water at 25°C	g/l	negligible
Specific gravity at 25°C	-	<1
Viscosity at 25°C	mPas	300-400
HEALTH & SAFETY		
HMIS classification	Flammability	3
	Health	3
	Reactivity	0

Eastman CP 730-1 (20% solids in xylene)

PARAMETER	UNIT	VALUE
Carcinogenicity		ethylbenzene IARC 2B: possibly carcinogenic to humans. NTP Not Listed. OSHA Not Listed. Expert judgment and weight of evidence determination: Not classified
Mutagenicity		no data available
Teratogenicity		no data available
DOT class		Coating solution 3, III. Shipping descriptions may vary based on mode of transport, quantities, package size, and/or origin and destination.
ICAO/IATA class		Coating solution 3, III
IMDG class		Coating solution 3, III
Autoignition temperature	°C	450
Flash point	°C	26
Flash point method	-	PMCC
Animal testing, acute toxicity, Rat oral LD50	mg/kg	3523-4000/xylene, 3500/ethylbenzene, 2262/chlorobenzene, >3200/ epoxidized oil
Animal testing, acute toxicity, Rabbit dermal LD50	mg/kg	4200/xylene, 5400/ ethylbenzene
Animal testing, acute toxicity, Guinea pig dermal LD50	mg/kg	>20000/ chlorobenzene
Animal testing, acute toxicity, Rat inhalation, LC50	mg/m³	6700 ppm/4H/xylene, 4000ppm/4H/ ethylbenzene, 29700/4H/ chlorobenzene
Effect of exposure, eye (human)		Causes serious eye irritation. May cause redness and pain.
Effect of exposure, inhalation (human)		May cause respiratory irritation. May cause drowsiness or dizziness. Narcotic effect. Symptoms may be delayed.
Effect of exposure, skin (human)		Causes skin irritation. May cause redness and pain.
Effect of exposure, swallowing (human)		May be fatal if swallowed and enters airways.
Effect of repeated or overexposure (human)		Contains ethylbenzene. May cause damage to organs (auditory organ) through prolonged or repeated exposure.

Eastman CP 730-1 (20% solids in xylene)

PARAMETER	UNIT	VALUE
Exposure, personal protection		Safety glasses, protective clothing based on chemical resistance data, chemical-resistant gloves, general and local exhaust ventilation.
First aid, eye		Rinse cautiously with water for several minutes. Remove contact lenses, if present and easy to do. Continue rinsing. If molten material contacts the eye, immediately flush with plenty of water for at least 15 minutes. Get medical attention immediately.
First aid, inhalation		Remove to fresh air. If not breathing, give artificial respiration. If breathing is difficult, give oxygen. If irritation persists, obtain medical advice.
First aid, skin		Wash with soap and water. Get medical attention if symptoms occur. If burned by contact with hot material, cool molten material adhering to skin as quickly as possible with water, and see a physician for removal of adhering material and treatment of burn.
OSHA, PEL	mg/m^3	435/xylene, 435/ethylbenzene, 350/chlorobenzene
ACGIH, TLV	ppm	100/xylene, 20/ethylbenzene, 10/chlorobenzene, STEL150/xylene
NIOSH, REL	ppm	100/ethylbenzene
OSHA, PEL	ppm	100/xylene, 100/ethylbenzene, 75/chlorobenzene
UN/NA class	-	1139
ECOLOGICAL PROPERTIES		
Aquatic toxicity, *Daphnia magna*, 48-h LC50	mg/l	4.3/chlorobenzene
Aquatic toxicity *Fethead minnow*, 96-h LC50	mg/l	42.3-48.5/ethylbenzene
Aquatic toxicity, *Rainbow trout*, 96-h LC50	mg/l	2.6/xylene
Biodegradation probability		readily biodegradable/xylene
BOC/COD ratio	%	7.32/chlorobenzene
Partition coefficient, log K_{ow}	-	3.12-3.30/xylene, 3.15/ethylbenzene

Eastman CP 730-1 (20% solids in xylene)

PARAMETER	UNIT	VALUE
USE & PERFORMANCE		
Manufacturer		Eastman Chemical Company
Outstanding properties		provides excellent adhesion properties for all typical basecoat chemistries, excellent gasoline resistance, and humidity resistance. Excellent redissolve resistance.
Recommended for polymers		PP, TPO, non-olefin plastics
Recommended for products		auto OEM, auto plastics, auto refinish, general industrial coatings, graphic arts, protective coatings, trucks/buses/RVs
Recommended applications		adhesion promoter designed to be an active component in adhesion promoter primers to ensure adhesion of color coats and topcoats to polypropylene (PP) and thermoplastic olefin (TPO)

Eastman CP 730-1 100% solids)

PARAMETER	UNIT	VALUE
GENERAL INFORMATION		
Name		Eastman CP 730-1 (100% solids)
CAS #		68609-36-9, 108-90-7, 61789-01-3
Composition		97% modified chlorinated polyolefin, 1-<5% chlorobenzene, 1% epoxidized oil
Acronym		CP
Chemical class		chlorinated polyolefin
Mixture	-	yes
PHYSICAL PROPERTIES		
State	-	solid/granules
Odor	-	odorless
Color	-	beige
Softening po nt	°C	>100
Chlorine content	%	20.5-23.5
Solubility (diluents)		soluble in aromatic hydrocarbons/xylene, toluene, cyclic hydrocarbons such as methylcyclohexane and ethylcyclohexane can be used to dilute CP 730-1. It is not soluble in aliphatic hydrocarbons, esters, ketones, or alcohols but can be diluted, provided there are long-chain ketones, esters like methyl amyl ketone
Specific gravity	-	>1
HEALTH & SAFETY		
HMIS classification	Flammability	1
	Health	1
	Reactivity	0
Carcinogenicity		IARC, OSHA, NTP: no ingredient of this product present at levels greater than or equal to 0.1% is identified as probable, possible or confirmed human carcinogen
DOT class		Class 9, Packing Group III when material is shipped in quantities in one package at or above the Reportable Quantity and when no other hazard class applies; otherwise, not regulated.
ICAO/IATA class		not regulated
IMDG class		not regulated

Eastman CP 730-1 100% solids)

PARAMETER	UNIT	VALUE
Animal testing, acute toxicity, Rat oral LD50	mg/kg	2262/chlorobenzene, >3200/ epoxidized oil
Animal testing, acute toxicity, Guinea pig dermal LD50	mg/kg	>20000/ chlorobenzene
Animal testing, acute toxicity, Rat inhalation, LC50	mg/m^3	29700/4H/ chlorobenzene
Effect of exposure, eye (human)	May cause eye irritation and redness.	
Effect of exposure, skin (human)	Causes skin irritation.	
Exposure, personal protection	Safety glasses, protective clothing based on chemical resistance data, chemical-resistant gloves, general and local exhaust ventilation.	
First aid, eye	Rinse cautiously with water for several minutes. Remove contact lenses, if present and easy to do. Continue rinsing. If molten material contacts the eye, immediately flush with plenty of water for at least 15 minutes. Get medical attention immediately.	
First aid, inhalation	Remove to fresh air. If not breathing, give artificial respiration. If breathing is difficult, give oxygen. If irritation persists, obtain medical advice.	
First aid, skin	Wash with soap and water. Get medical attention if symptoms occur. If burned by contact with hot material, cool molten material adhering to skin as quickly as possible with water, and see a physician for removal of adhering material and treatment of burn.	
OSHA, PEL	mg/m^3	350/chlorobenzene
ACGIH, TLV	ppm	10/chlorobenzene
OSHA, PEL	ppm	75/chlorobenzene
ECOLOGICAL PROPERTIES		
Aquatic toxicity, *Daphnia magna*, 48-h LC50	mg/l	4.3/chlorobenzene
BOC/COD ratio	%	7.32/ chlorobenzene
USE & PERFORMANCE		
Manufacturer	Eastman Chemical Company	

Eastman CP 730-1 100% solids)

PARAMETER	UNIT	VALUE
Outstanding properties		provides excellent adhesion properties for all typical basecoat chemistries, excellent gasoline resistance, and humidity resistance. Excellent redissolve resistance.
Recommended for polymers		PP, TPO
Recommended for products		automotive OEM, refinish, coatings for automotive plastics, general industrial coatings, graphic arts, protective coatings, trucks/buses/RVs
Recommended applications		adhesion promoter designed to be an active component in adhesion promoter primers to ensure adhesion of color coats and topcoats to polypropylene (PP) and thermoplastic olefin (TPO) plastics. Used for automotive OEM, refinish, coatings for automotive plastics

3.7 Crosslinkers
Visiomer TMPTMA

PARAMETER	UNIT	VALUE
GENERAL INFORMATION		
Name		Visiomer TMPTMA
CAS #	-	3290-92-4
EC number	-	221-950-4
Common synonym		propylidynetrimethyl trimethacrylate
IUPAC name		propylidynetrimethyl trimethacrylate
Empirical formula		C18H26O6
Formula		
Molecular mass	daltons	338.3
Chemical class	crosslinker	
Active matter	wt%	98
PHYSICAL PROPERTIES		
State	-	liquid
Color	-	yellowish
Color, Platinum-cobalt scale	-	100
Boiling point	°C	200 (1013 hPa)
Melting point	°C	-25
Density at 20°C	kg/m³	1090
Refractive index at 20°C	-	1.473
Solubility in water at 25°C	wt%	0.3
Viscosity at 20°C	mPas	64
HEALTH & SAFETY		
Flash point	°C	187
USE & PERFORMANCE		
Manufacturer		Evonik
Outstanding features		trifunctional methacrylate monomer. It is ideal for use as a crosslinking agent in free radical polymerization, as an adhesion promoter and hardener for PVC plastisol, and coagent for peroxide crosslinking of elastomers. Cohesive strength (dense networks) and high-temperature resistance to polymers and elastomers
Recommended for polymers		PVC
Recommended for products		reactive systems, adhesives, sealants, rubber, elastomers, composites

Visiomer TRGDMA

PARAMETER	UNIT	VALUE
GENERAL INFORMATION		
Name	Visiomer TRGDMA	
CAS #	-	109-16-0
EC number	-	203-652-6
Common synonym	triethyleneglycol dimethacrylate	
IUPAC name	2,2'-ethylenedioxydiethyl dimethacrylate	
Empirical formula	C14H22O6	
Formula		
Molecular mass	daltons	286.3
Chemical class	crosslinker	
Active matter	wt%	98
Purity	wt%	95
PHYSICAL PROPERTIES		
State	-	liquid
Color	clear, colorless, slightly yellow	
Color, Platinum-cobalt scale	-	50
Boiling point	°C	250 (1013 hPa)
Density at 20°C	kg/m^3	1075
Refractive index at 20°C	-	1.46
Solubility in water at 25°C	wt%	3.1
Vapor pressure at 100°C	kPa	13
Viscosity at 20°C	mPas	10
HEALTH & SAFETY		
Autoignition temperature	°C	255
Flash point	°C	150-169
USE & PERFORMANCE		
Manufacturer	Evonik	
Outstanding properties	low volatility, high reactivity	
Recommended for polymers	PVC	
Recommended for products	plastisols	

3.8 Epoxides
Isopropyl glycidyl ether

PARAMETER	UNIT	VALUE
GENERAL INFORMATION		
Name	Isopropyl glycidyl ether	
CAS #	-	4016-14-2
EC number	-	223-672-9
Composition	isopropyl glycidyl ether	
IUPAC name	2-(propan-2-yloxymethyl)oxirane	
Acronym		IPGE
Empirical formula	C6H12O2	
Formula		
Molecular mass	daltons	116.16
RTECS number	-	TZ3500000
Chemical class	epoxides	
Active matter	wt%	98
PHYSICAL PROPERTIES		
State	-	liquid
Color	-	colorless
Boiling point	°C	131-137
Density at 25°C	kg/m^3	92
Refractive index at 20°C	-	1.41
Solubility in water at 25°C	wt%	18.8
Solubility in solvents	ketons, alcohols	
Vapor density	-	4.15
Vapor pressure at 25°C	kPa	1.25
UN #	-	1993, 3271
HEALTH & SAFETY		
Flash point	°C	33
Carcinogenicity	-	not listed
Animal testing, acute toxicity, Rat oral LD50	mg/kg	4200
Animal testing, acute toxicity, Mouse oral LD50	mg/kg	1300
Animal testing, acute toxicity, Rabbit dermal LD50	mg/kg	9650
Animal testing, acute toxicity, Rat inhalation, LC50	ppm/8H	1100

3.8 Epoxides
Isopropyl glycidyl ether

PARAMETER	UNIT	VALUE
First aid, eye	First rinse with plenty of water for several minutes (remove contact lenses if easily possible), then refer for medical attention.	
First aid, inhalation	Fresh air, rest. Half-upright position. Refer for medical attention	
First aid, skin	Remove contaminated clothes. Rinse and then wash shin with water and soap	
NIOSH, REL	mg/m^3	240
OSHA, PEL	mg/m^3	240
NIOSH, REL	ppm	50
OSHA, PEL	ppm	50
UN/NA class	-	3271/1993
USE & PERFORMANCE		
Manufacturer	generic	
Recommended for polymers	PVC	
Recommended applications	adhesion promoter for PVC	

Phenyl glycidyl ether

PARAMETER	UNIT	VALUE
GENERAL INFORMATION		
Name		Phenyl glycidyl ether
CAS #	-	122-60-1
EC number	-	204-557-2
Common synonym		1,2-epoxy-3-phenoxypropane
Empirical formula		C9H10O2
Formula		
Molecular mass	daltons	150.177
RTECS number	-	TZ3675000
Chemical class	-	epoxides
Active matter	wt%	85
PHYSICAL PROPERTIES		
State	-	liquid
Odor	-	characteristic
Boiling point	°C	243
Melting point	°C	3.5
Density at 20°C	kg/m³	1,107-1,109
Refractive index at 20°C	-	1.531
Solubility in water at 25°C	mg/l	2400
Vapor density	-	4.37
Vapor pressure at 25°C	kPa	0.0013
Henry law constant at 25°C	atm-m³/mole	8.23E-07
HEALTH & SAFETY		
Carcinogenicity		2B, possible carcinogen
DOT class		Poison
Flash point	°C	120
Flash point method	-	CC
Animal testing, acute toxicity, Rat oral LD50	mg/kg	3850
Animal testing, acute toxicity, Mouse oral LD50	mg/kg	1400
Animal testing, acute toxicity, Rabbit dermal LD50	mg/kg	1500
Animal testing, acute toxicity, Rat inhalation, LC50	ppm	>100/8H
NIOSH, REL	mg/m³	6

Phenyl glycidyl ether

PARAMETER	UNIT	VALUE
ACGIH, TLV	ppm	0.1
NIOSH, REL	ppm	1
OSHA, PEL	ppm	10
UN/NA class	-	2810
ECOLOGICAL PROPERTIES		
Partition coefficient, log K_{ow}	-	1.61
USE & PERFORMANCE		
Manufacturers	Acros Organics, Epotec	

3.9 Esters
Phthalate diethylene glycol diacrylate

PARAMETER	UNIT	VALUE
GENERAL INFORMATION		
Name		Phthalate diethylene glycol diacrylate
Acronym		PDDA
Molecular mass	daltons	450
Chemical class	-	esters
Active matter	wt%	97
PHYSICAL PROPERTIES		
State	-	liquid
Color	-	clear
Color, Platinum-cobalt scale	-	60
Acid number	mg KOH/g	1
Viscosity at 25°C	mPas	100-300
USE & PERFORMANCE		
Manufacturer		HUPC Chemical
Outstanding properties		good flexibility and adhesion
Recommended for polymers		PVC
Recommended for products		wood coatings, paper coatings

Radcure ODA

PARAMETER	UNIT	VALUE
GENERAL INFORMATION		
Name		Radcure ODA
CAS #		2499-59-4, 2156-96-6
Composition		mixture of octyl and decyl acrylate
Empirical formula		C21H40O2
Formula		
Molecular mass	daltons	324.54
Mixture	-	yes
PHYSICAL PROPERTIES		
State	-	liquid
Color, Gardner scale	-	3
Acid number	mg KOH/g	1
Density at 25°C	kg/m³	870
Viscosity at 25°C	mPas	3
USE & PERFORMANCE		
Manufacturer		Allnex
Recommended applications		adhesion promoter to nonpolar substrates, good water- and moderate chemical-resistance. It reduces viscosity for better processing and improves crosslinking. Used in industrial coatings.

Uniplex 260

PARAMETER	UNIT	VALUE
GENERAL INFORMATION		
Name	Uniplex 260	
CAS #	-	614-33-5
Composition	glyceryl tribenzoate	
Formula		
Molecular mass	daltons	404.41
Chemical class	ester	
Active matter	wt%	99
Moisture content	wt%	0.1
PHYSICAL PROPERTIES		
State	-	solid/crystalline particles
Color	-	white
Melting point	°C	70-73
Acid number	mg KOH/g	0.28
Density at 25°C	kg/m^3	1,260
Refractive index at 20°C	-	1.565-1.570
Specific gravity at 30°C	-	1.262
HEALTH & SAFETY		
Exposure, personal protection		Safety glasses, protective clothing based on chemical resistance data, chemical-resistant gloves, general and local exhaust ventilation.
First aid, eye		Immediately flush eyes with plenty of water, occasionally lifting the upper and lower eyelids. Check for and remove any contact lenses. Rinse opened eye for several minutes under running water. Obtain medical attention if irritation develops.
First aid, inhalation		Move to fresh air. Get medical attention if nasal, throat or lung irritation develops.
USE & PERFORMANCE		
Manufacturer		Lanxess
Recommended for polymers		polyester, acrylic resins, PVAc, cellophane, nitrocellulose
Recommended for products		nail lacquer, printing inks, adhesives

Uniplex 260

PARAMETER	UNIT	VALUE
Recommended applications		adhesion-promoting plasticizer, particularly suitable for use in hot melt adhesives. Recommend for use in heat seal applications and coatings. Improves the heat seal properties of cellophane and nitrocellulose coatings.

3.10 Inorganic compounds
Markoba CB20

PARAMETER	UNIT	VALUE
GENERAL INFORMATION		
Name	Markoba CB20	
CAS #	-	10139-54-5
Composition	cobalt neodecanoate	
Empirical formula	C10H19CoO2	
Molecular mass	daltons	230.19
Chemical class	inorganic compounds	
PHYSICAL PROPERTIES		
State	-	solid/pastilles
Color	-	purple to blue
USE & PERFORMANCE		
Manufacturer	Wholemark Fine Chemical	
Recommended for polymers	rubber	
Recommended applications	adhesion promoter	

Markoba CB-S

PARAMETER	UNIT	VALUE
GENERAL INFORMATION		
Name	Markoba CB-S	
CAS #	-	1002-88-6
EC number	-	213-694-7
Composition	cobalt stearate	
Empirical formula	C18H36CoO2	
Molecular mass	daltons	284.48
Chemical class	inorganic compounds	
PHYSICAL PROPERTIES		
State	-	solid/pastilles
Color	-	purple
USE & PERFORMANCE		
Manufacturer	Wholemark Fine Chemical	
Recommended for polymers	rubber	
Recommended applications	adhesion promoter	

Markoba CB23

PARAMETER	UNIT	VALUE
GENERAL INFORMATION		
Name	Markoba CB23	
CAS #	-	72432-84-9
Common name	cobalt carboxy-boroaceglate	
Empirical formula	C20H38CoO4	
Molecular mass	daltons	230.19
Chemical class	inorganic compounds	
PHYSICAL PROPERTIES		
State	-	solid/pastilles
Color	-	purple to blue
USE & PERFORMANCE		
Manufacturer	Wholemark Fine Chemical	
Recommended for polymers	rubber	
Recommended applications	adhesion promoter	
Food approval (FDA)	U.S. FDA FAR 21 CFR 175.300	

3.11 Ionomers
Loxanol MI 6721

PARAMETER	UNIT	VALUE
GENERAL INFORMATION		
Name	Loxanol MI 6721	
CAS #	-	9002-98-6
General description	75-100% branched polyethyleneimine	
Synonym	aziridine homopolymer	
Molecular mass	daltons	1,300
Chemical class	-	ionomer
Active matter	wt%	99
PHYSICAL PROPERTIES		
State	-	liquid
Color	colorless to yellowish	
Boiling point	°C	>200
Freezing point	°C	-20
Pour point	°C	-16
Solidification temperature	°C	-20
Density at 20°C	kg/m^3	1,030
pH	-	11
Refractive index at 20°C	-	1.526
Viscosity at 20°C	mPas	8,000
HEALTH & SAFETY		
DOT class	-	9 III
ICAO/IATA class	-	9 III
Autoignition temperature	°C	>200
Flash point	°C	>200
Animal testing, acute toxicity, Rat oral LD50	mg/kg	680
First aid, eye	Immediately wash affected eyes for at least 15 minutes under running water with eyelids held open, consult an eye specialist.	
First aid, inhalation	Keep patient calm, remove to fresh air, seek medical attention. Immediately administer a corticosteroid from a controlled/metered dose inhaler	
First aid, skin	Immediately wash thoroughly with plenty of water, apply sterile dressings, consult a skin specialist.	
UN/NA class	-	3082

3.11 Ionomers
Loxanol MI 6721

PARAMETER	UNIT	VALUE
USE & PERFORMANCE		
Manufacturer	BASF	
Recommended for products	inks and coatings	
Concentrations used	wt%	0.1-2

Lupasol SC 61 B

PARAMETER	UNIT	VALUE
GENERAL INFORMATION		
Name		Lupasol SC 61 B
Composition		37% aqueous solution of hydroxyethylated polyethyleneimine
Chemical class		ionomer
Active matter	wt%	37
PHYSICAL PROPERTIES		
State	-	liquid
Odor	-	mild, amine-like
Color	-	slightly yellow
Boiling point	°C	100 (1013 hPa)
Freezing pcint	°C	-20
Pour point	°C	-16
Decomposition temperature	°C	>250
Density at 20°C	kg/m^3	1,080
pH	-	11
Solubility		distilled water, alcohols, methanol, ethanol, 1-propanol, 2-propanol
Vapor pressure at 20°C	mbar	24
HEALTH & SAFETY		
NFPA classification	Flammability	1
	Health	1
	Reactivity	0
HMIS classification	Flammability	1
	Health	1
	Reactivity	0
DOT class		not classified as dangerous
TDG class		not classified as dangerous
ICAO/IATA class		not clasiffied as dangerous
IMDG class		not clasiffied as dangerous
Flash point	°C	100
Autoignition	°C	>200
Animal testing, acute toxicity, Rat oral LD50	mg/kg	>300-2000
First aid, eye		Wash affected eyes for at least 15 minutes under running water with eyelids held open.

Lupasol SC 61 B

PARAMETER	UNIT	VALUE
First aid, inhalation		Keep patient calm, remove to fresh air. Assist in breathing if necessary. Seek medical attention if necessary.
First aid, skin		Wash thoroughly with soap and water.
UN #	-	3082
ECOLOGICAL PROPERTIES		
Aquatic toxicity, *Daphnia magna*, 48-h LC50	mg/l	100
USE & PERFORMANCE		
Manufacturer		BASF
Outstanding properties		improves dye acceptance, paintability, and barrier properties
Recommended for products		coatings, inks
Recommended applications		ideal adhesion promoter between different types of plastics or between plastics and polar substrates, such as polyolefin films and paper

Lupasol SK

PARAMETER	UNIT	VALUE
GENERAL INFORMATION		
Name		Lupasol SK
Composition		24% aqeous solution of a modified polyethylenimine with molecular weight of 2,000,000
Molecular mass	daltons	2,000,000
Chemical class	ionomer	
Active matter	wt%	24
PHYSICAL PROPERTIES		
State	-	liquid
Odor	-	ether-like
Color	-	yellowish
Boiling point	°C	100
Solidification temperature	°C	-5
Density at 20°C	kg/m³	1,060
pH	-	7.8-8.7
Solubility in water at 25°C	g/l	soluble
Vapor pressure at 50°C	kPa	12
Viscosity at 20°C	mPas	500-1000
HEALTH & SAFETY		
NFPA classification	Flammability	1
	Health	1
	Reactivity	0
HMIS classification	Flammability	0
	Health	1
	Reactivity	0
DOT class	not classified as dangerous	
TDG class	not classified as dangerous	
ICAO/IATA class	not clasiffied as dangerous	
IMDG class	not clasiffied as dangerous	
Autoignition temperature	°C	380
Flash point	°C	>100
Animal testing, acute toxicity, Rat oral LD50	mg/kg	13400
First aid, eye	Flush with copious amounts of water for at least 15 minutes. Hold eyelids open to facilitate rinsing. If irritation develops, seek medical attention.	

Lupasol SK

PARAMETER	UNIT	VALUE
First aid, inhalation		Keep patient calm, remove to fresh air. Assist in breathing if necessary. Seek medical attention if necessary.
First aid, skin		Wash affected areas thoroughly with soap and water. If irritation develops, seek medical attention.
ECOLOGICAL PROPERTIES		
Aquatic toxicity, *Daphnia magna*, 48-h LC50	mg/l	19.8
USE & PERFORMANCE		
Manufacturer	BASF	

3.12 Isocyanates
Desmodur® BL 2078/2

PARAMETER	UNIT	VALUE
GENERAL INFORMATION		
Name		Desmodur® BL 2078/2
CAS #		127184-53-6
Composition		blocked aliphatic polyisocyanate based on IPDI in Solvesso 100.
Chemical class	-	isocyanate
Active matter	wt%	60
Free NCO content	%	<=0.2
Blocked NCO content	%	7
Equivalent weight	-	600
PHYSICAL PROPERTIES		
State	-	liquid
Color		colorless to light yellow
Hazen color value	-	<=100
Odor	-	solvent-like
Boiling point	°C	170.5
Pour point	°C	-15
Unblocking temperature	°C	140
Density at 25°C	kg/m^3	1040
Vapor pressure at 20°C	kPa	2.0
Viscosity at 23°C	mPas	1750
HEALTH & SAFETY		
Carcinogenicity	-	2B (IARC)
Autoignition temperature	°C	430
Flash point	°C	4
Explosive LEL	vol%	1 (solvent)
Explosive UEL	vol%	7.5 (solvent
Animal testing, acute toxicity, Rat oral LD50	mg/kg	>10,000
Animal testing, acute toxicity, Rat inhalation LC50, 4 h	mg/l	5.3
Effect of exposure, eye (human)		In case of contact, immediately flush eyes with plenty of water for at least 15 minutes. Use lukewarm water. Use fingers to ensure that eyelids are separated and that the eye is being irrigated. Then remove contact lenses, if easily removable, and continue eye irrigation for not less than 15 minutes. Get medical attention if irritation develops.

3.12 Isocyanates
Desmodur® BL 2078/2

PARAMETER	UNIT	VALUE
Effect of exposure, inhalation (human)		Move to an area free from further exposure. Extreme asthmatic reactions that may occur in sensitized persons can be life threatening. Get medical attention immediately.
Effect of exposure, skin (human)		If direct skin contact with isocyanates occurs, immediately remove contaminated clothing and shoes. Wipe off the isocyanate product from the skin using dry towels or other similar absorbent fabric. Discard or wash contaminated clothing before reuse.
Effect of exposure, swallowing (human)		Do NOT induce vomiting. Wash mouth out with water. Do not give anything by mouth to an unconscious person. Get medical attention.
Exposure, personal protection		Safety glasses, protective clothing based on chemical resistance data, chemical-resistant gloves, general and local exhaust ventilation.
ACGIH, TLV	ppm	25 (trimethylbenzene), 0.005 IPDI
UN/NA number	-	1866
ECOLOGICAL PROPERTIES		
Aquatic toxicity, *Zebra fish*, 96-h LC50	mg/l	no toxic effect
Aquatic toxicity, *Daphnia magna*, 48-h EC50	mg/l	no toxic effect
Biodegradation probability	1%, exposure time: 28 d, i.e., not readily degradable	
USE & PERFORMANCE		
Manufacturer	Covestro	
Outstanding properties	additive to conventional baking systems to improve adhesion and elasticity	
Recommended for products	adhesives	
Recommended applications	highly effective crosslinker for adhesives based on Desmocoll, natural or synthetic rubber with special adhesion on rubber materials. It is used in production of contact and heat-activated adhesives, as well as reactive and hot-melt adhesives for the transportation, furniture, footwear, packaging, and construction markets.	

Desmodur® BL 3175A

PARAMETER	UNIT	VALUE
GENERAL INFORMATION		
Name		Desmodur® BL 3175A
CAS #		85940-94-9
Composition		60-80% MEKO-blocked HDI polymer, 10-30% petroleum solvent, 6-15% tri-methylbenzene, 1-5 N-propylbenzene
Chemical class		isocyanate
Active matter	wt%	73-77
Free NCO content	%	<=0.2
Blocked NCO content	%	11.1
PHYSICAL PROPERTIES		
State	-	liquid
Color	-	light yellow
Hazen color value	-	<=60
Odor	-	solvent-like
Unblocking temperature	°C	130
Density at 25°C	kg/m^3	1060
Vapor pressure at 25°C	kPa	<1.33
Viscosity at 20°C	mPas	2000-4000
HEALTH & SAFETY		
Carcinogenicity	2B (IARC)	
Flash point	°C	51
Explosive LEL	vol%	0.9 (solvent)
Explosive UEL	vol%	6 (solvent
Animal testing, acute toxicity, Rat oral LD50	mg/kg	>2000 to >2757
Animal testing, acute toxicity, Rat dermal LD50	mg/kg	2667
Effect of exposure, eye (human)		In case of contact, immediately flush eyes with plenty of water for at least 15 minutes. Use lukewarm water. Use fingers to ensure that eyelids are separated and that the eye is being irrigated. Then remove contact lenses, if easily removable, and continue eye irrigation for not less than 15 minutes. Get medical attention if irritation develops.
Effect of exposure, inhalation (human)		Move to an area free from further exposure. Extreme asthmatic reactions that may occur in sensitized persons can be life threatening. Get medical attention immediately.

Desmodur® BL 3175A

PARAMETER	UNIT	VALUE
Effect of exposure, skin (human)		If direct skin contact with isocyanates occurs, immediately remove contaminated clothing and shoes. Wipe off the isocyanate product from the skin using dry towels or other similar absorbent fabric. Discard or wash contaminated clothing before reuse.
Effect of exposure, swallowing (human)		Do NOT induce vomiting. Wash mouth out with water. Do not give anything by mouth to an unconscious person. Get medical attention.
Exposure, personal protection		Safety glasses, protective clothing based on chemical resistance data, chemical-resistant gloves, general and local exhaust ventilation.
ACGIH, TLV	ppm	25 (trimethylbenzene), 0.005 HDI
UN/NA number	-	1866
ECOLOGICAL PROPERTIES		
Aquatic toxicity, *Zebra fish*, 96-h LC50	mg/l	141
Aquatic toxicity, *Daphnia magna*, 48-h LC50	mg/l	4
Biodegradation probability		aerobic 9%, exposure time: 28 d, i.e., not readily degradable
USE & PERFORMANCE		
Manufacturer		Covestro
Outstanding properties		improves flexibility and adhesion
Recommended for products		light-stable one-component baking systems, industrial finishing of appliances, coil coating, and can coating, and in automotive industry for chip-resistant primer-surfacers and topcoats
Recommended applications		highly effective crosslinker for adhesives based on Desmocoll, natural or synthetic rubber with special adhesion on rubber materials. It is used in the production of contact and heat-activated adhesives, as well as reactive and hot-melt adhesives for the transportation, furniture, footwear, packaging, and construction markets.

Desmodur RFE

PARAMETER	UNIT	VALUE
GENERAL INFORMATION		
Name		Desmodur RFE
CAS #		4151-51-3, 141-78-6
EC number		223-981-9, 205-500-4
Composition		28% tris(p-isocyanatophenyl) thiophosphate in 71% ethyl acetate, impurity: <1% chlorobenzene
Formula		
Chemical class		isocyanate
Active matter	wt%	28
PHYSICAL PROPERTIES		
State	-	liquid
Color	-	yellow to brown
Density at 25°C	kg/m³	1000
Vapor pressure at 25°C	kPa	0.097
Viscosity at 20°C	mPas	3
HEALTH & SAFETY		
Carcinogenicity		no data available
ICAO/IATA class		FLAMMABLE LIQUID, N.O.S. (Ethyl Acetate, Monochlorobenzene) 3, II
IMDG class		FLAMMABLE LIQUID, N.O.S. (Ethyl Acetate, Monochlorobenzene) 3, II
Autoignition temperature	°C	460
Flash point	°C	-4.00
Explosive LEL	wt%	2.2
Explosive UEL	wt%	11.5
Animal testing, acute toxicity, Rat oral LD50	mg/kg	>2000
Animal testing, acute toxicity, Rabbit dermal LD50	mg/kg	causes skin irritation
Animal testing, acute toxicity, Rat dermal LD50	mg/kg	18000/ethyl acetate
Animal testing, acute toxicity, Rat inhalation, LC50	mg/m³	225000/6H/ethyl acetate
Effect of exposure, eye (human)		Causes serious eye irritation

Desmodur RFE

PARAMETER	UNIT	VALUE
Effect of exposure, inhalation (human)		May cause respiratory irritation. May cause drowsiness or dizziness. May cause allergy or asthma symptoms or breathing difficulties if inhaled.
Effect of exposure, skin (human)		Causes skin irritation. May cause allergic skin irritation.
Effect of exposure, swallowing (human)		Harmful if swallowed.
Effect of repeated or overexposure (human)		Repeated exposure may cause skin dryness or cracking.
Exposure, personal protection		Safety glasses, protective clothing based on chemical resistance data, chemical-resistant gloves, general and local exhaust ventilation.
First aid, eye		Immediately flush eyes with plenty of water, occasionally lifting the upper and lower eyelids. Check for and remove any contact lenses. Rinse opened eye for several minutes under running water. Obtain medical attention if irritation develops.
First aid, inhalation		Move person into the fresh air and keep him warm, let him rest; if there is difficulty in breathing, medical advice is required.
First aid, skin		Wash with a cleanser based on polyethylene glycol or with plenty of warm water and soap. Consult a doctor in the event of a skin reaction
ACGIH, TLV	mg/m^3	720/AU NOE/ethyl acetate
ACGIH, TLV	ppm	200/AU NOE/ethyl acetate
UN risk phrases, R	-	R20,R52/53
UN/NA class	-	1993
ECOLOGICAL PROPERTIES		
Aquatic toxicity, *Zebra fish*, 96-h LC50	mg/l	no toxic effects with saturated solution.
Bioconcentration factor	BCF	30/3d (Golden orfe)/ethyl acetate
Biodegradation probability		69%/20d/acetyl acetate/readily biodegradable
USE & PERFORMANCE		
Manufacturer		Covestro

Desmodur RFE

PARAMETER	UNIT	VALUE
Recommended for products		furniture, footwear, packaging, construction
Recommended applications		highly effective crosslinker for adhesives based on Desmocoll, natural or synthetic rubber with special adhesion on rubber materials. It is used in the production of contact and heat-activated adhesives, as well as reactive and hot-melt adhesives for the transportation, furniture, footwear, packaging, and construction markets.

Nourybond 289

PARAMETER	UNIT	VALUE
GENERAL INFORMATION		
Name		Nourybond 289
Composition		blocked isocyanate (1.8-2.1%)
Chemical class		isocyanate (toluene diisocyanate)
PHYSICAL PROPERTIES		
State	-	liquid
Color, Gardner scale	-	2
Viscosity at 25°C	mPas	30,000-50,000
HEALTH & SAFETY		
Flash point	°C	>100
Flash point method	-	PMCC
USE & PERFORMANCE		
Manufacturer		Evonik
Outstanding properties		provides excellent color stability and superior plastisol rheological performance.
Recommended for polymers		low bake PVC plastisols
Recommended for products		automobile underbody coating and seam sealant
Recommended applications		adhesion promoter, it cures at 120°C with an addition of zinc octoate catalyst
Concentrations used	wt%	2-5
Guidelines for use		they are applied on electrodeposition primed (E-coat primed) metals and exposed to high temperatures in a body shop or paint shop, usually ranging between 120°C and 160°C. During this exposure, the diffusion of PVC or acrylic resin in a plasticizer forms a continuous adhesive film.

Nourybond 290

PARAMETER	UNIT	VALUE
GENERAL INFORMATION		
Name	Nourybond 290	
Composition	blocked isocyanate	
Chemical class	isocyanate	
HEALTH & SAFETY		
Flash point	°C	>100
USE & PERFORMANCE		
Manufacturer	Evonik	
Outstanding properties	provides excellent plastisol rheological performance, high tensile strength, and 130°C cure.	
Recommended for polymers	low bake PVC plastisols	
Recommended applications	adhesion promoters designed to provide adhesion to electrodeposition primers used in the manufacture of automobiles, trucks, and buses. Used in conjunction with a small amount of polyamidoamine adhesion promoter.	
Concentrations used	parts	3-4

3.13 Isocyanurates
Dynasylan VPS 7163

PARAMETER	UNIT	VALUE
GENERAL INFORMATION		
Name		Dynasylan VPS 7163
CAS #	-	26115-70-8
IUPAC name		1,3,5-tris[3-(trimethoxysilyl)propyl]-1,3,5-triazine-2,4,6(1H,3H,5H)-trione
Empirical formula	-	C21H45N3O12Si3
Formula		
Molecular mass	daltons	615.85
Active matter	wt%	>90
PHYSICAL PROPERTIES		
State	-	liquid
Odor	-	characteristic
Color		colorless to light yellow
Boiling point at 35 hPa	°C	237-247
Freezing point	°C	-20
Density at 20°C	kg/m³	1181
Viscosity at 20°C	mPas	440-580
HEALTH & SAFETY		
HMIS classification	Flammability	1
	Health	2
	Physical hazard	0
Carcinogenicity		no carcinogens present or none present in regulated quantities
Flash point	°C	>95
Flash point method	-	PMCC
Animal testing, acute toxicity, Rat oral LD50	mg/kg	1717
First aid, eye		Rinse thoroughly with plenty of water keeping eyelid open. In case of persistent discomfort: Consult an ophthalmologist.

3.13 Isocyanurates
Dynasylan VPS 7163

PARAMETER	UNIT	VALUE
First aid, inhalation		If aerosol or mists are inhaled, take affected persons out into the fresh air. In case of persistent discomfort or other symptoms, consult a physician immediately.
First aid, skin		Immediately wash skin with soap and plenty of water. Remove contaminated clothing. Obtain medical attention immediately if symptoms occur. Wash clothing before reuse.
USE & PERFORMANCE		
Manufacturer		Evonik
Outstanding properties		strong crosslinker. It can be used as an adhesion promoter on different types of substrates like metals and plastics. The molecule with the heterocyclic planar six-membered ring system imparts good wetting and thermal resistance, thus leading to good adhesion even at high temperatures on metal substrates.
Recommended for polymers		epoxy, silane modified polymers. PU, silicones, EVA, PA, polyester, polyolefins
Recommended for products		adhesives and sealants, coatings, primers, hotmelts

Vulcabond MDX

PARAMETER	UNIT	VALUE
GENERAL INFORMATION		
Name	Vulcabond MDX	
Composition	25% solution of isocyanurate trimer in diisononyl phthalate	
Chemical class	isocyanurate	
Active matter	wt%	25
PHYSICAL PROPERTIES		
State	-	liquid
Specific gravity at 25°C	-	1.02
Viscosity at 25°C	mPas	3,000
HEALTH & SAFETY		
Flash point	°C	150
USE & PERFORMANCE		
Manufacturer	Valtris	
Outstanding properties	minimal influence on initial plastisol viscosity. Low odor, no flammable solvents, high bond strengths.	
Recommended for polymers	PVC plastisols	
Recommended for products	coated fabrics, conveyor belts, certain types of floor covering	
Recommended applications	highly effective as a bonding agent to increase adhesion between a PVC plastisol and synthetic fabrics such as polyesters and polyamides. Areas of application include tarpaulins, protective clothing, marquees, conveyor belts, and certain types of floor covering.	

Vulcabond VP

PARAMETER	UNIT	VALUE
GENERAL INFORMATION		
Name		Vulcabond VP
CAS #		9017-01-0, 84-74-2, 26471-62-5
Composition		25% solution of isocyanurate trimer 78-80% dibutyl phthalate, <1% toluene diisocyanate
Chemical class		isocyanurate
Active matter	wt%	25
PHYSICAL PROPERTIES		
State	-	liquid
Color	-	pale yellow
Density at 25°C	kg/m^3	1,130
Vapor pressure at 20°C	kPa	0.01
Viscosity at 25°C	mPas	3,000
HEALTH & SAFETY		
Teratogenicity		May cause harm to the unborn child. Possible risk of impaired fertility.
DOT class		Environmentally Hazardous Substance, Liquid, n.o.s. (contains dibutyl phthalate) 9, III
ICAO/IATA class		Environmentally Hazardous Substance, Liquid, n.o.s. (contains dibutyl phthalate) 9, III
IMDG class		Environmentally Hazardous Substance, Liquid, n.o.s. (contains dibutyl phthalate) 9, III
Autoignition temperature	°C	400.00 dibutyl phthalate
Flash point	°C	145
Flash point method	-	CC
Animal testing, acute toxicity, Rat oral LD50	mg/kg	8000/ dibutyl phthalate
Animal testing, acute toxicity, Rat dermal LD50	mg/kg	>2000/ dibutyl phthalate
Animal testing, acute toxicity, Rat inhalation, LC50	mg/m^3	4250/ dibutyl phthalate
Effect of exposure, eye (human)		Causes serious eye damage.
Effect of exposure, inhalation (human)		May cause sensitization by inhalation.
Effect of exposure, skin (human)		Causes skin irritation. May cause sensitization by skin contact.
Effect of exposure, swallowing (human)		Harmful if swallowed.

Vulcabond VP

PARAMETER	UNIT	VALUE
Exposure, personal protection		Safety glasses, protective clothing based on chemical resistance data, chemical-resistant gloves, general and local exhaust ventilation.
First aid, eye		Immediately flush eyes with plenty of water, occasionally lifting the upper and lower eyelids. Check for and remove any contact lenses. Rinse opened eye for several minutes under running water. Obtain medical attention if irritation develops.
First aid, inhalation		Move person into the fresh air and keep him warm, let him rest; if there is difficulty in breathing, medical advice is required.
First aid, skin		Immediately flush skin with plenty of water for at least 15 minutes while removing contaminated clothing. If symptoms persist, call a physician.
ACGIH, TLV	mg/m^3	5/dibutyl phthalate, 0.02/toluene diisocyanate
UN risk phrases, R		R36/38,R42/43,R50,R61,R62,R26,R36/37/38,R40,R52/53
US safety phrases, S	-	S45,S53,S61
UN/NA class	-	3082
USE & PERFORMANCE		
Manufacturer	Valtris	
Outstanding properties		minimal influence on initial plastisol viscosity. Low odor, no flammable solvents, high bond strengths.
Recommended for polymers	PVC, rigid PVC	
Recommended for products		coated fabrics, conveyor belts, tarpaulins
Recommended applications		adhesion promoters for coated fabrics. Bonding agent for PVC plastisols.

3.14 Lignin
Lignin

PARAMETER	UNIT	VALUE
GENERAL INFORMATION		
Name	Lignin	
CAS #	-	8068-05-1; 8068-03-9 9005-53-2
Molecular mass	daltons	2000-50000
Chemical class	lignin	
Molecular formula	C81H92O28	
Molecular weight	100-3000 (kraft lignins), 800-3000 (soda lignins), 500-4000 (organosolv lignins), 20000-50000 (lignosulfonates)	
Polydispersity, M_n/M_w	2.5-3.5 (kraft lignins), 2.5-3.5 (soda lignins), 1.3-4 (organosolv lignins), 6-8 (lignosulfonates)	
Functional organic groups	COOH, OH, OCH_3	
PHYSICAL PROPERTIES		
Color	yellow, light to dark brown	
Glass transition temperature	°C	97-162
Ash content	wt%	2-6
Density at 20°C	kg/m³	1,350-1,500
Hildebrand solubility parameter	$(MPa)^{1/2}$	22.7-33.1
USE & PERFORMANCE		
Outstanding properties	contains both hydrophobic and hydrophilic groups	
Recommended for polymers	phenolic resin, poly(lactic acid), poly-urethane, epoxy, rubber, polyesters, polyolefins	
Recommended for products	adhesives and sealants, coatings, com-posites, agricultural chemicals, natural binders, adhesives	

3.15 Maleic anhydride modified polymers
Amplify TY 1451B

PARAMETER	UNIT	VALUE
GENERAL INFORMATION		
Name		Amplify™ TY 1451B
General description		tie layer resin
Composition		base polymer LLDPE
Graft content	wt%	<0.2
PHYSICAL PROPERTIES		
State	-	pellets
Density at 20°C	kg/m³	910
Melt flow rate at 190°C/2.16 kg	dg/min	1.7
USE & PERFORMANCE		
Manufacturer		Dow
Recommended for polymers		PE, PA, EVOH, ionomer
Recommended for products		multi-layer films, bottles, sheets and tube structures
Recommended applications		food & specialty packaging, agricultural, pharmaceuticals, pipe, flooring
Processing methods		high-performance thermoforming, blown & cast film

Amplify TY 4817

PARAMETER	UNIT	VALUE
GENERAL INFORMATION		
Name	Amplify™ TY 4817	
General description	PET tie layer resin	
Graft content	wt%	<0.2
PHYSICAL PROPERTIES		
State	-	pellets
Density at 20°C	kg/m³	917
Melt flow rate at 190°C/2.16 kg	g/min	0.8
USE & PERFORMANCE		
Manufacturer	Dow	
Recommended for polymers	OPET, OPP, PE, PP, PA, EVOH	
Recommended for products	multilayer films, sheets and tube structures	
Recommended applications	food, pharmaceuticals	
Processing methods	extrusion coating & lamination	

Bynel™21E4817

PARAMETER	UNIT	VALUE
GENERAL INFORMATION		
Name		Bynel™21E4817
General description		modified polyolefin blend
Graft content	wt%	0.09
PHYSICAL PROPERTIES		
State	-	pellets
Odor	-	odorless to mild
Color	-	white
Melting point	°C	105
Freezing point	°C	92
Vicat softening point	°C	53
Maximum processing temperature	°C	260
Density at 20°C	kg/m³	917
Melt flow rate at 190°C/2.16 kg	g/10 min	10
HEALTH & SAFETY		
DOT class		not regulated for transport
TDG class		not regulated for transport
ICAO/IATA class		not regulated for transport
Hazardous combustion products		carbon monoxide, carbon dioxide
Animal testing, acute toxicity, Rat oral LD50	mg/kg	>5000
Animal testing, acute toxicity, Rabbit dermal LD50	mg/kg	>2000
First aid, eye		Flush eyes thoroughly with water for several minutes.
First aid, inhalation		Move person to fresh air and keep comfortable for breathing; consult a physician.
First aid, skin		Wash off with plenty of water. Seek first aid or medical attention as needed.
ACGIH, TLV	mg/m³	0.01 (MAH)
OSHA, PEL	ppm	0.25 (MAH)
ECOLOGICAL PROPERTIES		
Bioaccumulative potential		no bioconcentration is expected because of the relatively high molecular weight (MW greater than 1000).
Biodegradation probability		this water-insoluble polymeric solid is expected to be inert in the environment. Surface photodegradation is expected with exposure to sunlight. No appreciable biodegradation is expected.

Bynel™21E4817

PARAMETER	UNIT	VALUE
USE & PERFORMANCE		
Manufacturer		Dow
Outstanding properties		contains a temperature stable ester which makes it functional in high-temperature coextrusion.
Recommended for polymers		PET to EVOH or PA, and o PE, PP, and ethylene copolymers
Processing methods		conventional extrusion and coextrusion equipment designed to process polyethylene, coextrusion coating and laminating applications

Bynel™ 41E1352

PARAMETER	UNIT	VALUE
GENERAL INFORMATION		
Name		Bynel™ 41E1352
General description		anhydride-modified, linear low-density polyethylene
Chemical class		maleic anhydride modified polymer
Graft content	wt%	<0.2
PHYSICAL PROPERTIES		
State	-	pellets
Odor	-	acidic
Color	-	white to off-white
Melting point	°C	125
Vicat softening point	°C	104
Maximum processing temperature	°C	260
Density at 20°C	kg/m³	922
Melt flow rate at 190°C/2.16 kg	g/10 min	1
HEALTH & SAFETY		
DOT class		not regulated for transport
TDG class		not regulated for transport
ICAO/IATA class		not regulated for transport
Animal testing, acute toxicity, Rat oral LD50	mg/kg	>5000
Animal testing, acute toxicity, Rabbit dermal LD50	mg/kg	>2000
First aid, eye		Flush eyes thoroughly with water for several minutes
First aid, inhalation		Move person to fresh air and keep comfortable for breathing; consult a physician.
First aid, skin		Wash off with plenty of water. Seek first aid or medical attention as needed.
USE & PERFORMANCE		
Manufacturer		Dow
Recommended for polymers		EVOH, polyamide, PE, and ethylene copolymers
Recommended for products		boil-in-bag structures, blow molded containers in which drop strength is important, bag-in-box films where LLDPE is the heat seal layer
Recommended applications		blown film, cast film/sheet, blow molding melt and solid phase thermoforming sheet, tubing

Bynel™ 41E1352

PARAMETER	UNIT	VALUE
Processing methods		conventional extrusion and coextrusion equipment designed to process polyethylene resins
Food approval (FDA)		complies with Food and Drug Administration Regulation 21 CFR 175.105 - - Adhesives

Bynel™ 41E3351B

PARAMETER	UNIT	VALUE
GENERAL INFORMATION		
Name		Bynel™ 41E3351B
General description		ethylene copolymer
Mixture	-	yes
Graft content	wt%	0.2-0.5
PHYSICAL PROPERTIES		
State	-	pellets
Odor	-	acidic
Color	-	white to off-white
Melting point	°C	126
Vicat softening point	°C	90
Maximum processing temperature	°C	250
Density at 20°C	kg/m³	940
Melt flow rate at 190°C/2.16 kg	g/10 min	3
HEALTH & SAFETY		
DOT class		not regulated for transport
TDG class		not regulated for transport
ICAO/IATA class		not regulated for transport
Animal testing, acute toxicity, Rat oral LD50	mg/kg	>5000
Animal testing, acute toxicity, Rabbit dermal LD50	mg/kg	>2000
ECOLOGICAL PROPERTIES		
Bioaccumulative potential		No bioconcentration is expected because of the relatively high molecular weight (M_W greater than 1000).
Biodegradation probability		This water-insoluble polymeric solid is expected to be inert in the environment
USE & PERFORMANCE		
Manufacturer		Dow
Outstanding properties		tie layer designed for cast sheet adhesion between EVOH, polyolefins and polystyrene, exceptional thermal stability, direct adhesio, no curing needed
Recommended for polymers		EVOH, PA, PE, PP, HIPS and PS
Recommended applications		coffee capsules, thermoformed food containers produced via cast sheet
Processing methods		coextrusion, cast film, sheet/cast extrusion
Food approval (FDA)		food use

Eastman G-3003

PARAMETER	UNIT	VALUE
GENERAL INFORMATION		
Name	Eastman G-3003	
CAS #	25722-45-6, 108-31-6	
Composition	>99% maleated polypropylene (maleic anhydride grafted polypropylene), <1% maleic anhydride	
Acronym		MAPP
Empirical formula	$(C_4H_2O_3 \cdot C_3H_6)x$	
Molecular mass	daltons	52,000
Chemical class	maleic anhydride modified polymer	
PHYSICAL PROPERTIES		
State	-	solid/flake or pellet
Odor	slight /0.32ppm odor threshold	
Melting point	°C	150-170/ softening point
Acid number	mg KOH/g	9
Specific gravity at 25°C	-	<1
Viscosity at 190°C	mPas	60,000
HEALTH & SAFETY		
NFPA classification	Flammability	1
	Health	1
	Reactivity	0
HMIS classification	Flammability	1
	Health	1
	Reactivity	0
Carcinogenicity	no data available	
Mutagenicity	no data available	
Teratogenicity	no data available	
DOT class	no data available	
TDG class	not regulated	
ICAO/IATA class	not regulated	
IMDG class	not regulated	
Animal testing, acute toxicity, Rat oral LD50	mg/kg	>2000/maleated polypropylene, 400-800/maleic anhydride

Eastman G-3003

PARAMETER	UNIT	VALUE
Animal testing, acute toxicity, Rabbit dermal LD50	mg/kg	slight/maleated polypropylene, severe irritation/ maleic anhydride
Animal testing, acute toxicity, Rat inhalation, LC50	mg/m^3	>4350/1H/ maleic anhydride
Effect of exposure, eye (human)		Contact with molten substance/product may cause severe eyes damage.
Effect of exposure, inhalation (human)		Avoid inhalation of vapors from heated material. At elevated temperatures, vapor may cause allergic respiratory reaction.
Effect of exposure, skin (human)		Contact with molten substance/product may cause severe burns to skin.
Exposure, personal protection		Safety glasses, protective clothing based on chemical resistance data, chemical-resistant gloves, general and local exhaust ventilation.
First aid, eye		Rinse cautiously with water for several minutes. Remove contact lenses, if present and easy to do. Continue rinsing. If molten material contacts the eye, immediately flush with plenty of water for at least 15 minutes. Get medical attention immediately.
First aid, inhalation		Remove to fresh air. If not breathing, give artificial respiration. If breathing is difficult, give oxygen. If irritation persists, obtain medical advice.
First aid, skin		Wash with soap and water. Get medical attention if symptoms occur. If burned by contact with hot material, cool molten material adhering to skin as quickly as possible with water, and see a physician for removal of adhering material and treatment of burn. Get medical attention.
ACGIH, TLV	mg/m^3	0.01/ maleic anhydride
OSHA, PEL	mg/m^3	1.00/ maleic anhydride
OSHA, PEL	ppm	0.25/ maleic anhydride
ECOLOGICAL PROPERTIES		
Aquatic toxicity, *Green algae*, 96-h EC50	mg/l	>150/72H/ maleic anhydride

Eastman G-3003

PARAMETER	UNIT	VALUE
Aquatic toxicity, *Bluegill sunfish*, 96-h LC50	mg/l	>75
Aquatic toxicity, *Daphnia magna*, 48-h LC50	mg/l	330/ maleic anhydride
Aquatic toxicity *Fathead minnow*, 96-h LC50	mg/l	>100
USE & PERFORMANCE		
Manufacturer	Eastman Chemical Company	
Outstanding properties	balanced attraction between polar fillers/ reinforcements and non polar polymers. Increased most physical properties of fiberglass-reinforced nylon composites. Increased the strength of composites that utilize reinforcements /fillers such as glass, talc, calcium carbonate, and metals.	
Recommended for polymers	PA, PP, rubber	
Recommended for products	reinforced plastics composites in automotive, building and construction, and wood plastics composite	
Recommended applications	Eastman G functionalized polyolefins can be used as an adhesive promoter to olefinic plastic and rubber to increase the polarity and surface energy of material which enhance the adhesion of plastics and rubber compounds to any polar surface like metal, nylon or natural fiber. Eastman G-3003 is useful for laminating as a compatibilizer for PP blends with nylon or EVOH. It is superior coupling agent for polypropylene mica/ PP composites.	
Processing methods	extrusion coating, injection molding	

Eastman G-3015

PARAMETER	UNIT	VALUE
GENERAL INFORMATION		
Name		Eastman G-3015
CAS #		25722-45-6, 108-31-6
Composition		>99% maleated polypropylene (maleic anhydride grafted polypropylene), <1% maleic anhydride
Acronym		MAPP
Molecular mass	daltons	47,000
Chemical class		maleic anhydride modified polymer
PHYSICAL PROPERTIES		
State	-	solid/flake or pellet
Odor	-	slight
Melting point	°C	150-170/ softening point
Acid number	mg KOH/g	15
Specific gravity at 25°C	-	<1.00
Viscosity at 190°C	mPas	18,000
HEALTH & SAFETY		
NFPA classification	Flammability	1
	Health	1
	Reactivity	0
HMIS classification	Flammability	1
	Health	2
	Reactivity	0
Carcinogenicity	no data available	
Mutagenicity	no data available	
Teratogenicity	no data available	
DOT class	not regulated	
TDG class	not regulated	
ICAO/IATA class	not regulated	
IMDG class	not regulated	
Animal testing, acute toxicity, Rat oral LD50	mg/kg	>2000/maleated polypropylene, 400-800/maleic anhydride
Animal testing, acute toxicity, Rabbit dermal LD50	mg/kg	slight/maleated polypropylene, severe irritation/ maleic anhydride

Eastman G-3015

PARAMETER	UNIT	VALUE
Animal testing, acute toxicity, Rat inhalation, LC50	mg/m^3	>4350/1H/ maleic anhydride
Effect of exposure, eye (human)		Contact with molten substance/product may cause severe eyes damage.
Effect of exposure, inhalation (human)		Avoid inhalation of vapors from heated material. At elevated temperatures, vapor may cause allergic respiratory reaction.
Effect of exposure, skin (human)		Contact with molten substance/product may cause severe burns to skin.
Exposure, personal protection		Safety glasses, protective clothing based on chemical resistance data, chemical-resistant gloves, general and local exhaust ventilation.
First aid, eye		Rinse cautiously with water for several minutes. Remove contact lenses, if present and easy to do. Continue rinsing. If molten material contacts the eye, immediately flush with plenty of water for at least 15 minutes. Get medical attention immediately.
First aid, inhalation		Remove to fresh air. If not breathing, give artificial respiration. If breathing is difficult, give oxygen. If irritation persists, obtain medical advice.
First aid, skin		Wash with soap and water. Get medical attention if symptoms occur. If burned by contact with hot material, cool molten material adhering to skin as quickly as possible with water, and see a physician for removal of adhering material and treatment of burn. Get medical attention.
ACGIH, TLV	mg/m^3	0.01/ maleic anhydride
OSHA, PEL	mg/m^3	1.00/ maleic anhydride
OSHA, PEL	ppm	0.25/ maleic anhydride
USE & PERFORMANCE		
Manufacturer		Eastman Chemical Company

Eastman G-3015

PARAMETER	UNIT	VALUE
Outstanding properties		balanced attraction between polar fillers/ reinforcements and non polar polymers. Increased most physical properties of fiberglass-reinforced nylon composites. Increased the strength of composites that utilize reinforcements /fillers such as glass, talc, calcium carbonate, and metals.
Recommended for polymers		PA, rubber, cellulose
Recommended for products		reinforced plastics composites in automotive, building and construction, and wood plastics composite
Recommended applications		Eastman G functionalized polyolefins can be used as an adhesive promoter to olefinic plastic and rubber to increase the polarity and surface energy of materials which enhance the adhesion of plastics and rubber compounds to any polar surface like metal, nylon, or natural fiber. Eastman G-3003 is useful for laminating as a compatibilizer for PP blends with nylon, or EVOH. It is superior coupling agent for polypropylene mica/PP composites.
Processing methods		extrusion coating, injection molding
Concentrations used		0.5-1.5%, 0.55-3%/cellulose/ polyolefins composite

Fusabond A560

PARAMETER	UNIT	VALUE
GENERAL INFORMATION		
Name		Fusabond A560
Composition		modified ethylene acrylate copolymer
Acronym		EA
Chemical class		maleic anhydride modified polymer
PHYSICAL PROPERTIES		
State	-	solid/pellets
Color	-	translucent
Odor	-	ester-like
Odor threshold	ppm	0.31
Melting point	°C	94
Maximum processing temperature	°C	260
Decomposition temperature	°C	>310
Density at 20°C	kg/m^3	930
Melt flow rate at 190°C/2.16 kg	g/10 min	5.6
Specific gravity at 25°C	-	0.93
HEALTH & SAFETY		
Exposure, personal protection		Safety glasses, protective clothing based on chemical resistance data, chemical-resistant gloves, general and local exhaust ventilation.
First aid, eye		Rinse cautiously with water for several minutes. Remove contact lenses, if present and easy to do. Continue rinsing. If eye irritation persists: Get medical advice/attention.
First aid, skin		Immediately wipe away excess material. Use a waterless hand cleaner to remove as much of the remaining material as possible. Wash with soap and water.
ACGIH, TLV	mg/m^3	3 (inhalable) 10 (respirable)
Carcinogenicity		None of the components present in this material at concentrations equal to or greater than 0.1% are listed by IARC as a carcinogen.
USE & PERFORMANCE		
Manufacturer		Dow

Fusabond A560

PARAMETER	UNIT	VALUE
Outstanding properties		improved adhesion between glass filler and the polyamide matrix. At low concentrations of 3.5%-7%, increased the Notched Izod impact strength of the glass-reinforced polyamide by as much as 30% to 50%, also reduced compound melt viscosity for improved mold filling and better surface appearance of the glass-filled molded parts. Filled compounds can retain more of their mechanical properties such as tensile strength, modulus, and heat deformation temperature (HDT).
Recommended for polymers		PA
Recommended for products		cable, wire
Recommended applications		intermediate performance, all-purpose PA modifier; adhesion promoter, filled PA compounds.
Food approval (FDA)		21CFR 175.105/may be used as a component of articles intended for use in packaging, transporting, or holding food, subject to the limitations and requirements therein.

Fusabond E100

PARAMETER	UNIT	VALUE
GENERAL INFORMATION		
Name		Fusabond E100
Composition		anhydride modified high density polyethylene
Acronym		HDPE
Chemical class		maleic anhydride modified polymer
PHYSICAL PROPERTIES		
State	-	solid/pellets
Color	-	clear to translucent
Odor	-	irritating
Melting point	°C	134
Freezing point	°C	115
Vicat softening point	°C	127
Maximum processing temperature	°C	300
Density at 20°C	kg/m^3	954
Melt flow rate at 190°C/2.16 kg	g/10 min	2
Specific gravity at 25°C	-	0.954
HEALTH & SAFETY		
Exposure, personal protection		Safety glasses, protective clothing based on chemical resistance data, chemical-resistant gloves, general and local exhaust ventilation.
First aid, eye		Rinse cautiously with water for several minutes. Remove contact lenses, if present and easy to do. Continue rinsing. If eye irritation persists: Get medical advice/attention.
First aid, skin		Immediately wipe away excess material. Use a waterless hand cleaner to remove as much of the remaining material as possible. Wash with soap and water.
Animal testing, acute toxicity, Rat oral LD50	mg/kg	>5000
Animal testing, acute toxicity, Rabbit dermal LD50	mg/kg	>2000
USE & PERFORMANCE		
Manufacturer		Dow
Recommended for polymers		PA, PE, PP
Recommended for products		wire, cable, packaging, pipe coating

Fusabond E100

PARAMETER	UNIT	VALUE
Recommended applications		primary use is for compounding (for example, in wire & cable) and as an adhesion promoter in a variety of other applications.
Processing methods		extrusions or coextrusions
Food approval (FDA)		21CFR 175.105/may be used as components of articles intended for use in packaging, transporting, or holding food, subject to the limitations and requirements therein.

Fusabond™ E158

PARAMETER	UNIT	VALUE
GENERAL INFORMATION		
Name		Fusabond™ E158
General description		anhydride modified polyethylene
PHYSICAL PROPERTIES		
State	-	pellets
Odor	-	clear to translucent
Color	-	irritating
Melting point	°C	128
Freezing point	°C	113
Vicat softening point	°C	102
Maximum processing temperature	°C	290
Decomposition temperature	°C	>310
Density at 20°C	kg/m³	927
Melt flow rate at 190°C/2.16 kg	g/10 min	1.8
HEALTH & SAFETY		
TDG class		not regulated for transport
ICAO/IATA class		not regulated for transport
IMDG class		not regulated for transport
Animal testing, acute toxicity, Rat oral LD50	mg/kg	>5000
Animal testing, acute toxicity, Rabbit dermal LD50	mg/kg	>2000
First aid, eye		Flush eyes thoroughly with water for several minutes. Remove contact lenses after the initial 1-2 minutes and continue flushing for several additional minutes. If effects occur, consult a physician, preferably an ophthalmologist.
First aid, inhalation		Move person to fresh air and keep comfortable for breathing; consult a physician.
First aid, skin		Wash off with plenty of water. Seek first aid or medical attention as needed. If molten material comes in contact with the skin, do not apply ice but cool under ice water or running stream of water
ECOLOGICAL PROPERTIES		
Bioaccumulative potential		No bioconcentration is expected be-cause of the relatively high molecular weight (M_w greater than 1000).

Fusabond™ E158

PARAMETER	UNIT	VALUE
Biodegradation probability		This water-insoluble polymeric solid is expected to be inert in the environment. Surface photodegradation is expected with exposure to sunlight. No appreciable biodegradation is expected.
USE & PERFORMANCE		
Manufacturer		Dow
Recommended for products		adhesives, pipe coatings, tie layer

Fusabond E204

PARAMETER	UNIT	VALUE
GENERAL INFORMATION		
Name		Fusabond E204
General description		very high MAH graft level
Composition		maleic anhydride grafted HDPE
Chemical class		maleic anhydride modified polymer
Active matter	wt%	>99
PHYSICAL PROPERTIES		
State	-	pellets
Odor	-	acidic
Color	-	white
Melting point	°C	127
Density at 20°C	kg/m^3	954
Melt flow rate at 190°C/2.16 kg	g/10 min	12
Solubility in water at 25°C	g/l	negligible
HEALTH & SAFETY		
ICAO/IATA class		not regulated
IMDG class		not regulated
Animal testing, acute toxicity, Rat oral LD50	mg/kg	>5000
Animal testing, acute toxicity, Rabbit dermal LD50	mg/kg	>2000
First aid, eye		Flush eyes thoroughly with water for several minutes. Remove contact lenses after the initial 1-2 minutes and continue flushing for several additional minutes. If effects occur, consult a physician, preferably an ophthalmologist
First aid, inhalation		Move person to fresh air and keep comfortable for breathing; consult a physician
First aid, skin		Wash off with plenty of water. Seek first aid or medical attention as needed.
ECOLOGICAL PROPERTIES		
Bioaccumulative potential		No bioconcentration is expected because of the relatively high molecular weight (M$_W$ greater than 1000).
Biodegradation probability		This water-insoluble polymeric solid is expected to be inert in the environment.
USE & PERFORMANCE		
Manufacturer		Dow
Recommended for polymers		PE to nylon, EVOH, metal, cellulose

Fusabond E204

PARAMETER	UNIT	VALUE
Recommended for products		food packaging, nonwoven, staple fiber, binder fiber

Fusabond™ E205

PARAMETER	UNIT	VALUE
GENERAL INFORMATION		
Name		Fusabond™ E205
General description		anhydride-modified high-density polyethylene.
Composition		compounded thermoplastic resin >= 87.0%, compounded thermoplastic copolymer <= 12%
Common synonym		ethylene copolymer
Graft content	wt%	>1.0
PHYSICAL PROPERTIES		
State	-	pellets
Odor	-	acidic
Color	-	white to off-white
Melting point	°C	130
Vicat softening point	°C	129
Maximum processing temperature	°C	285
Density at 25°C	kg/m³	960
Melt flow rate at 190°C/2.16 kg	g/10 min	2
HEALTH & SAFETY		
DOT class		not regulated for transport
TDG class		not regulated for transport
ICAO/IATA class		not regulated for transport
Animal testing, acute toxicity, Rat oral LD50	mg/kg	>5000
Animal testing, acute toxicity, Rabbit dermal LD50	mg/kg	>2000
First aid, eye		Flush eyes thoroughly with water for several minutes. Remove contact lenses.
First aid, inhalation		Move person to fresh air; if effects occur, consult a physician.
First aid, skin		Wash off with plenty of water. Seek first aid or medical attention as needed.
ECOLOGICAL PROPERTIES		
Bioaccumulative potential		No bioconcentration is expected because of the relatively high molecular weight (M_W greater than 1000).
Biodegradation probability		This water-insoluble polymeric solid is expected to be inert in the environment. Surface photodegradation is expected with exposure to sunlight. No appreciable biodegradation is expected.

Fusabond™ E205

PARAMETER	UNIT	VALUE
USE & PERFORMANCE		
Manufacturer		Dow
Outstanding properties		adhesion promoter between polyolefins and metal, cellulose and glass, compatibilizer
Recommended for products		sealant layer
Food approval (FDA)		21CFR177.1520 Olefin polymers

Fusabond™ N216

PARAMETER	UNIT	VALUE
GENERAL INFORMATION		
Name		Fusabond™ N216
General description		maleic anhydride grafted polymer.
Composition		compounded thermoplastic resin >= 87.0%, compounded thermoplastic copolymer <= 12%
Common synonym		ethylene copolymer
Graft content	wt%	0.5-1.0
PHYSICAL PROPERTIES		
State	-	pellets
Odor	-	odorless
Color	-	white
Melting point	°C	62.8
Decomposition temperature	°C	>310
Density at 25°C	kg/m³	875
Melt flow rate at 190°C/2.16 kg	g/10 min	1.3
HEALTH & SAFETY		
DOT class		not regulated for transport
TDG class		not regulated for transport
ICAO/IATA class		not regulated for transport
Animal testing, acute toxicity, Rat oral LD50	mg/kg	>5000
Animal testing, acute toxicity, Rabbit dermal LD50	mg/kg	>2000
First aid, eye		Flush eyes thoroughly with water for several minutes. Remove contact lenses.
First aid, inhalation		Move person to fresh air; if effects occur, consult a physician.
First aid, skin		Wash off with plenty of water. Seek first aid or medical attention as needed.
ECOLOGICAL PROPERTIES		
Bioaccumulative potential		No bioconcentration is expected because of the relatively high molecular weight (M_w greater than 1000).
Biodegradation probability		This water-insoluble polymeric solid is expected to be inert in the environment. Surface photodegradation is expected with exposure to sunlight. No appreciable biodegradation is expected.

Fusabond™ N216

PARAMETER	UNIT	VALUE
USE & PERFORMANCE		
Manufacturer		Dow
Outstanding properties		adhesion promoter between polyolefins and metal, cellulose and glass, compatibilizer
Recommended for products		sealant layer
Food approval (FDA)		21CFR177.1520 Olefin polymers

Fusabond E226

PARAMETER	UNIT	VALUE
GENERAL INFORMATION		
Name		Fusabond E226
Composition		anhydride modified high density polyethylene
Chemical class		maleic anhydride modified polymer
PHYSICAL PROPERTIES		
State	-	solid/pellets
Melting point	°C	120
Freezing point	°C	102
Vicat softening point	°C	95
Maximum processing temperature	°C	290
Density at 20°C	kg/m³	930
Melt flow rate at 190°C/2.16 kg	g/10 min	1.75
Specific gravity at 25°C	-	0.93
HEALTH & SAFETY		
Exposure, personal protection		Safety glasses, protective clothing based on chemical resistance data, chemical-resistant gloves, general and local exhaust ventilation.
First aid, eye		Rinse cautiously with water for several minutes. Remove contact lenses, if present and easy to do. Continue rinsing. If eye irritation persists: Get medical advice/attention.
First aid, skin		Immediately wipe away excess material. Use a waterless hand cleaner to remove as much of the remaining material as possible. Wash with soap and water.
USE & PERFORMANCE		
Manufacturer		Dow
Recommended for polymers		PE, PP
Recommended for products		wire, cable, packaging, pipe coating
Recommended applications		used as a compatibilizer: for non-halogen, flame-retarded wire & cable compounds containing fillers such as aluminum trihydrate (ATH) or magnesium hydroxide ($Mg(OH)_2$) or for natural fiber (wood) plastic compounds.

Fusabond E226

PARAMETER	UNIT	VALUE
Food approval (FDA)	21CFR 175.105/may be used as components of articles intended for use in packaging, transporting, or holding food, subject to the limitations and requirements therein.	

Fusabond E265

PARAMETER	UNIT	VALUE
GENERAL INFORMATION		
Name		Fusabond E265
Composition		anhydride modified high density polyethylene
Acronym		HDPE
Chemical class		maleic anhydride modified polymer
PHYSICAL PROPERTIES		
State	-	solid/pellets
Melting point	°C	131
Freezing point	°C	111
Density at 20°C	kg/m³	950
Melt flow rate at 190°C/2.16 kg	g/10 min	12
Specific gravity at 25°C	-	0.95
HEALTH & SAFETY		
Exposure, personal protection		Safety glasses, protective clothing based on chemical resistance data, chemical-resistant gloves, general and local exhaust ventilation.
First aid, eye		Rinse cautiously with water for several minutes. Remove contact lenses, if present and easy to do. Continue rinsing. If eye irritation persists: Get medical advice/attention.
First aid, skin		Immediately wipe away excess material. Use a waterless hand cleaner to remove as much of the remaining material as possible. Wash with soap and water.
USE & PERFORMANCE		
Manufacturer		Dow
Recommended for polymers		PE, PP
Recommended for products		wire, cable
Recommended applications		used as a compatibilizer for non-halogen, flame-retarded wire & cable compounds containing fillers such as aluminum trihydrate (ATH) or magnesium hydroxide ($Mg(OH)_2$) or for natural fiber (wood) plastic compounds.

Fusabond E528

PARAMETER	UNIT	VALUE
GENERAL INFORMATION		
Name		Fusabond E528
Composition		anhydride modified polyethylene
Chemical class		maleic anhydride modified polymer
PHYSICAL PROPERTIES		
State	-	solid/pellets
Color	-	clear to translucent
Odor	-	irritating
Melting point	°C	114
Freezing point	°C	97
Maximum processing temperature	°C	300
Decomposition temperature	°C	>310
Density at 20°C	kg/m^3	922
Melt flow rate at 190°C/2.16 kg	g/10 min	6.7
Specific gravity at 25°C	-	0.922
HEALTH & SAFETY		
Exposure, personal protection		Safety glasses, protective clothing based on chemical resistance data, chemical-resistant gloves, general and local exhaust ventilation.
First aid, eye		Rinse cautiously with water for several minutes. Remove contact lenses, if present and easy to do. Continue rinsing. If eye irritation persists: Get medical advice/attention.
First aid, skin		Immediately wipe away excess material. Use a waterless hand cleaner to remove as much of the remaining material as possible. Wash with soap and water.
Animal testing, acute toxicity, Rat oral LD50	mg/kg	>5000
Animal testing, acute toxicity, Rabbit dermal LD50	mg/kg	>2000
USE & PERFORMANCE		
Manufacturer		Dow
Recommended for polymers		PE, PP
Recommended for products		pipe coating
Recommended applications		adhesion promoter

Fusabond E564

PARAMETER	UNIT	VALUE
GENERAL INFORMATION		
Name		Fusabond E564
Composition		random ethylene copolymer incorporating a monomer classified as maleic anhydride
Chemical class		maleic anhydride modified polymer
PHYSICAL PROPERTIES		
State	-	solid/pellets
Melting point	°C	114
Freezing point	°C	99
Density at 20°C	kg/m^3	927
Melt flow rate at 190°C/2.16 kg	g/10 min	4
Specific gravity at 25°C	-	0.927
HEALTH & SAFETY		
Exposure, personal protection		Safety glasses, protective clothing based on chemical resistance data, chemical-resistant gloves, general and local exhaust ventilation.
First aid, eye		Rinse cautiously with water for several minutes. Remove contact lenses, if present and easy to do. Continue rinsing. If eye irritation persists: Get medical advice/attention.
First aid, skin		Immediately wipe away excess material. Use a waterless hand cleaner to remove as much of the remaining material as possible. Wash with soap and water.
USE & PERFORMANCE		
Manufacturer		Dow
Recommended for polymers		PE, PP
Recommended applications		used in a variety of applications where adhesion improvements are needed for coatings, or compatibilization is need for compounds

Fusabond M603

PARAMETER	UNIT	VALUE
GENERAL INFORMATION		
Name		Fusabond M603
Composition		maleic anhydride grafted high density polyethylene (high graft level)
Acronym		HDPE
Chemical class		maleic anhydride modified polymer
PHYSICAL PROPERTIES		
State	-	solid/pellets
Melting point	°C	108
Density at 20°C	kg/m³	940
Melt flow rate at 190°C/2.16 kg	g/10 min	20
HEALTH & SAFETY		
Exposure, personal protection		Safety glasses, protective clothing based on chemical resistance data, chemical-resistant gloves, general and local exhaust ventilation.
First aid, eye		Rinse cautiously with water for several minutes. Remove contact lenses, if present and easy to do. Continue rinsing. If eye irritation persists: Get medical advice/attention.
First aid, skin		Immediately wipe away excess material. Use a waterless hand cleaner to remove as much of the remaining material as possible. Wash with soap and water.
USE & PERFORMANCE		
Manufacturer		Dow
Recommended for polymers		HDPE, LLDPE, LDPE or combination of these polymers, resin can come from virgin, post industrial or post consumer recycled sources, recycle stream: PE/ PA, PE/ EVOH, PA/EVOH/ PE
Recommended applications		coupling agent which interacts with a variety of biofibers like both hardwood fibers, in particular, oak fibers, as well as pine softwood fibers, have also been shown to work well with rice hulls, many other cellulosic and organic materials may also interact with Fusabond M603
Concentrations used	wt%	1-2

Ricobond® 1731 HS

PARAMETER	UNIT	VALUE
GENERAL INFORMATION		
Name		Ricobond® 1731 HS
Composition		low vinyl polybutadiene functionalized with maleic anhydride dispersed on amorphous silica
Molecular mass, M_n	g/mol	5400
Active matter	wt%	70
Vinyl content	wt%	28
Maleic anhydride group content	groups/chain	9
PHYSICAL PROPERTIES		
State	-	liquid
Color	-	brown
Glass transition temperature	°C	-72
Viscosity at 20°C	mPas	50,000
USE & PERFORMANCE		
Manufacturer		Cray Valley
Outstanding properties		enhanced adhesion of rubber compounds to metal and textiles. The anhydride functionality can react with an epoxy, amine, and hydroxyl groups, enabling the creation of unique adhesives, sealants, encapsulants, and coatings. Compatibility with polar and non-polar materials
Recommended for polymers		epoxy, rubber
Recommended for products		adhesives, sealants, encapsulants, coatings, electronics

Xibond 120

PARAMETER	UNIT	VALUE
GENERAL INFORMATION		
Name	Xibond 120	
Composition	styrene and maleic anhydride random copolymer	
Molecular mass	daltons	245,000
Chemical class	maleic anhydride modified polymer	
PHYSICAL PROPERTIES		
State	-	solid/granules
Color	-	clear, white
Glass transition temperature	°C	120
Acid number	mg KOH/g	90
Thermal stability, TGA	°C	340
USE & PERFORMANCE		
Manufacturer	Polyscope	
Outstanding properties	optimized interfacial tension, stabilized morphology against high stresses, and enhanced adhesion between the phases in the solid-state. Compatibilization improved the mechanical properties of the polymer blend.	
Recommended for polymers	compatible with HIPS resins	
Recommended applications	adhesion promoter, compatibilizer, coupling agent, viscosity modifier, and coupling agent. It is designed to improve the blend morphology of specific polymer blends.	
Processing methods	twin-screw extruders with a mild screw configuration and vacuum degassing facility for good dispersion in styrenic polymers	
Concentrations used	wt%	1-5

Xibond 140

PARAMETER	UNIT	VALUE
GENERAL INFORMATION		
Name		Xibond 140
Composition		styrene and maleic anhydride copolymer
Acronym		SMA
Molecular mass	daltons	180,000
Chemical class		maleic anhydride modified polymer
PHYSICAL PROPERTIES		
State	-	solid/granules
Color	-	white
Glass transition temperature	°C	134
Acid number	mg KOH/g	170
Thermal stability, TGA	°C	340
USE & PERFORMANCE		
Manufacturer		Polyscope
Outstanding properties		optimized interfacial tension, stabilized morphology against high stresses, and enhanced adhesion between the phases in the solid-state. Compatibilization improved the mechanical properties of the polymer blend.
Recommended for polymers		compatible with HIPS resins
Recommended applications		adhesion promoter, compatibilizer, coupling agent, viscosity modifier, and coupling agent. It is designed to improve the blend morphology of specific polymer blends.
Processing methods		twin-screw extruders with a mild screw configuration and vacuum degassing facility for good dispersion in styrenic polymers
Concentrations used	wt%	1-5

Xibond 160

PARAMETER	UNIT	VALUE
GENERAL INFORMATION		
Name	Xibond 160	
Composition	styrene and maleic anhydride random copolymer	
Molecular mass	daltons	115,000
Chemical class	maleic anhydride modified polymer	
PHYSICAL PROPERTIES		
State	-	solid/granules
Color	-	clear, white
Glass transition temperature	°C	150
Acid number	mg KOH/g	250
Thermal stability, TGA	°C	320
USE & PERFORMANCE		
Manufacturer	Polyscope	
Outstanding properties	optimized interfacial tension, stabilized morphology against high stresses, and enhanced adhesion between the phases in the solid-state. Compatibilization improved the mechanical properties of the polymer blend.	
Recommended for polymers	compatible with ABS resins	
Recommended applications	adhesion promoter, compatibilizer, coupling agent, viscosity modifier, and coupling agent. It is designed to improve the blend morphology of specific polymer blends.	
Processing methods	twin-crew extruders with a mild screw configuration and vacuum degassing facility for good dispersion in styrenic polymers	
Concentrations used	wt%	1-5

Xibond 180

PARAMETER	UNIT	VALUE
GENERAL INFORMATION		
Name	Xibond 180	
Composition	styrene and maleic anhydride random copolymer	
Molecular mass	daltons	125,000
Chemical class	maleic anhydride modified polymer	
PHYSICAL PROPERTIES		
State	-	solid/granules
Color	-	clear, white
Glass transition temperature	°C	165
Acid number	mg KOH/g	320
Thermal stability, TGA	°C	320
USE & PERFORMANCE		
Manufacturer	Polyscope	
Outstanding properties	optimized interfacial tension, stabilized morphology against high stresses, and enhanced adhesion between the phases in the solid-state. Compatibilization improved the mechanical properties of the polymer blend.	
Recommended for polymers	compatible with ABS resins	
Recommended applications	adhesion promoter, compatibilizer, coupling agent, viscosity modifier, and coupling agent. It is designed to improve the blend morphology of specific polymer blends.	
Processing methods	twin-screw extruders with a mild screw configuration and vacuum degassing facility for good dispersion in styrenic polymers	
Concentrations used	wt%	1-5

Xibond 220

PARAMETER	UNIT	VALUE
GENERAL INFORMATION		
Name	Xibond 220	
Composition	styrene and maleic anhydride random copolymer	
Molecular mass	daltons	15,000
Chemical class	maleic anhydride modified polymer	
PHYSICAL PROPERTIES		
State	-	solid/granules
Color	-	clear, white
Glass transition temperature	°C	104
Acid number	mg KOH/g	120
Thermal stability, TGA	°C	270
USE & PERFORMANCE		
Manufacturer	Polyscope	
Outstanding properties	optimized interfacial tension, stabilized morphology against high stresses, and enhanced adhesion between the phases in the solid-state. Compatibilization improved the mechanical properties of the polymer blend.	
Recommended for polymers	compatible with ABS resins	
Recommended applications	adhesion promoter, compatibilizer, coupling agent, viscosity modifier, and coupling agent. It is designed to improve the blend morphology of specific polymer blends.	
Processing methods	twin-screw extruders with a mild screw configuration and vacuum degassing facility for good dispersion in styrenic polymers	
Concentrations used	wt%	1-5

Xibond 230

PARAMETER	UNIT	VALUE
GENERAL INFORMATION		
Name	Xibond 230	
Composition	styrene and maleic anhydride random copolymer	
Molecular mass	daltons	10,000
Chemical class	maleic anhydride modified polymer	
PHYSICAL PROPERTIES		
State	-	solid/granules
Color	-	clear, white
Glass transition temperature	°C	106
Acid number	mg KOH/g	155
Thermal stability, TGA	°C	270
USE & PERFORMANCE		
Manufacturer	Polyscope	
Outstanding properties	optimized interfacial tension, stabilized morphology against high stresses and enhanced adhesion between the phases in the solid-state. Compatibilization improved the mechanical properties of the polymer blend.	
Recommended for polymers	compatible with ABS resins	
Recommended applications	adhesion promoter, compatibilizer, coupling agent, viscosity modifier, and coupling agent. It is designed to improve the blend morphology of specific polymer blends.	
Processing methods	twin-screw extruders with a mild screw configuration and vacuum degassing facility for good dispersion in styrenic polymers	
Concentrations used	wt%	1-5

Xibond 240

PARAMETER	UNIT	VALUE
GENERAL INFORMATION		
Name	Xibond 240	
Composition	styrene and maleic anhydride random copolymer	
Molecular mass	daltons	10,000
Chemical class	maleic anhydride modified polymer	
PHYSICAL PROPERTIES		
State	-	solid/granules
Color	-	clear, white
Glass transition temperature	°C	115
Acid number	mg KOH/g	215
Thermal stability, TGA	°C	270
USE & PERFORMANCE		
Manufacturer	Polyscope	
Outstanding properties	optimized interfacial tension, stabilized morphology against high stresses, and enhanced adhesion between the phases in the solid-state. Compatibilization improved the mechanical properties of the polymer blend.	
Recommended for polymers	compatible with ABS resins	
Recommended applications	adhesion promoter, compatibilizer, coupling agent, viscosity modifier, and coupling agent. It is designed to improve the blend morphology of specific polymer blends.	
Processing methods	twin-screw extruders with a mild screw configuration and vacuum degassing facility for good dispersion in styrenic polymers	
Concentrations used	wt%	1-5

Xibond 250

PARAMETER	UNIT	VALUE
GENERAL INFORMATION		
Name		Xibond 250
Composition		styrene and maleic anhydride random copolymer
Molecular mass	daltons	10000
Chemical class		maleic anhydride modified polymer
PHYSICAL PROPERTIES		
State	-	solid/granules
Color	-	clear, white
Glass transition temperature	°C	130
Acid number	mg KOH/g	285
Thermal stability, TGA	°C	270
USE & PERFORMANCE		
Manufacturer		Polyscope
Outstanding properties		optimized interfacial tension, stabilized morphology against high stresses, and enhanced adhesion between the phases in the solid-state. Compatibilization improved the mechanical properties of the polymer blend.
Recommended for polymers		compatible with ABS resins
Recommended applications		adhesion promoter, compatibilizer, coupling agent, viscosity modifier, and coupling agent. It is designed to improve the blend morphology of specific polymer blends.
Processing methods		twin-screw extruders with a mild screw configuration and vacuum degassing facility for good dispersion in styrenic polymers
Concentrations used	wt%	1-5

Xibond 255

PARAMETER	UNIT	VALUE
GENERAL INFORMATION		
Name		Xibond 255
Composition		styrene and maleic anhydride random copolymer
Acronym		SMA
Molecular mass	daltons	10000
Chemical class		maleic anhydride modified polymer
PHYSICAL PROPERTIES		
State	-	solid/powder
Color	-	white
Glass transition temperature	°C	130
Acid number	mg KOH/g	285
Thermal stability, TGA	°C	400
HEALTH & SAFETY		
Hazardous combustion products		Carbon monoxide, Carbon dioxide, Nitrogen oxides
Exposure, personal protection		Safety glasses, protective clothing based on chemical resistance data, chemical-resistant gloves, general and local exhaust ventilation.
First aid, eye		Rinse cautiously with water for several minutes. Remove contact lenses, if present and easy to do. Continue rinsing. If molten material contacts the eye, immediately flush with plenty of water for at least 15 minutes. Get medical attention immediately.
First aid, inhalation		Remove to fresh air. If not breathing, give artificial respiration. If breathing is difficult, give oxygen. If irritation persists, obtain medical advice.
First aid, skin		Immediately flush with plenty of water for at least 15 minutes while removing contaminated clothing and shoes. Get medical attention. Wash contaminated clothing before reuse. Destroy or thoroughly clean contaminated shoes.
USE & PERFORMANCE		
Manufacturer		Polyscope

Xibond 255

PARAMETER	UNIT	VALUE
Outstanding properties		optimized interfacial tension, stabilized morphology against high stresses, and enhanced adhesion between the phases in the solid-state. Compatibilization improved the mechanical properties of the polymer blend.
Recommended for polymers		compatible with ABS resins
Recommended applications		adhesion promoter, compatibilizer, coupling agent, viscosity modifier, and coupling agent. It is designed to improve the blend morphology of specific polymer blends.
Processing methods		twin-screw extruders with a mild screw configuration and vacuum degassing facility for good dispersion in styrenic polymer
Concentrations used	wt%	1-5

Xibond 260

PARAMETER	UNIT	VALUE
GENERAL INFORMATION		
Name	Xibond 260	
Composition	styrene and maleic anhydride random copolymer	
Molecular mass	daltons	7,000
Chemical class	maleic anhydride modified polymer	
PHYSICAL PROPERTIES		
State	-	solid/granules
Color	-	white
Glass transition temperature	°C	130
Acid number	mg KOH/g	355
Thermal stability, TGA	°C	260
USE & PERFORMANCE		
Manufacturer	Polyscope	
Outstanding properties	optimized interfacial tension, stabilized morphology against high stresses, and enhanced adhesion between the phases in the solid-state. Compatibilization improved the mechanical properties of the polymer blend.	
Recommended for polymers	compatible with ABS resins	
Recommended applications	adhesion promoter, compatibilizer, coupling agent, viscosity modifier, and coupling agent. It is designed to improve the blend morphology of specific polymer blends.	
Processing methods	twin-screw extruders with a mild screw configuration and vacuum degassing facility for good dispersion in styrenic polymer	
Concentrations used	wt%	1-5

Xibond 280

PARAMETER	UNIT	VALUE
GENERAL INFORMATION		
Name	Xibond 280	
Composition	styrene and maleic anhydride random copolymer	
Molecular mass	daltons	5,000
Chemical class	maleic anhydride modified polymer	
PHYSICAL PROPERTIES		
State	-	solid/granules
Color	-	pale yellow
Glass transition temperature	°C	130
Acid number	mg KOH/g	480
Thermal stability, TGA	°C	260
USE & PERFORMANCE		
Manufacturer	Polyscope	
Outstanding properties	optimized interfacial tension, stabilized morphology against high stresses, and enhanced adhesion between the phases in the solid-state. Compatibilization improved the mechanical properties of the polymer blend.	
Recommended for polymers	compatible with ABS resins	
Recommended applications	adhesion promoter, compatibilizer, coupling agent, viscosity modifier, and coupling agent. It is designed to improve the blend morphology of specific polymer blends.	
Processing methods	twin-screw extruders with a mild screw configuration and vacuum degassing facility for good dispersion in styrenic polymer	
Concentrations used	wt%	1-5

Xibond 285

PARAMETER	UNIT	VALUE
GENERAL INFORMATION		
Name		Xibond 285
Composition		maleic anhydride polymer
Acronym		MAP
Molecular mass	daltons	5000
Chemical class		maleic anhydride modified polymer
PHYSICAL PROPERTIES		
State	-	solid/powder
Color	-	white
Glass transition temperature	°C	130
Acid number	mg KOH/g	480
Thermal stability, TGA	°C	260
USE & PERFORMANCE		
Manufacturer		Polyscope
Outstanding properties		optimized interfacial tension, stabilized morphology against high stresses, and enhanced adhesion between the phases in the solid-state. Compatibilization improved the mechanical properties of the polymer blend.
Recommended for polymers		compatible with ABS resins
Recommended applications		adhesion promoter, compatibilizer, coupling agent, viscosity modifier, and coupling agent. It is designed to improve the blend morphology of specific polymer blends.
Processing methods		twin-screw extruders with a mild screw configuration and vacuum degassing facility for good dispersion in styrenic polymer
Concentrations used	wt%	1-5

Xibond 315

PARAMETER	UNIT	VALUE
GENERAL INFORMATION		
Name		Xibond 315
Composition		styrene, maleic anhydride and N-phenylmaleimide random terpolymer
Molecular mass	daltons	155000
Chemical class		maleic anhydride modified polymer
PHYSICAL PROPERTIES		
State	-	solid/powder
Color	-	white
Glass transition temperature	°C	180
Acid number	mg KOH/g	23
Thermal stability, TGA	°C	380
USE & PERFORMANCE		
Manufacturer		Polyscope
Outstanding properties		optimized interfacial tension, stabilized morphology against high stresses, and enhanced adhesion between the phases in the solid-state. Compatibilization improved the mechanical properties of the polymer blend.
Recommended for polymers		compatible with HIPS resins
Recommended applications		adhesion promoter, compatibilizer, coupling agent, viscosity modifier, and coupling agent. It is designed to improve the blend morphology of specific polymer blends.
Processing methods		twin-screw extruders with a mild screw configuration and vacuum degassing facility for good dispersion in styrenic polymer
Concentrations used	wt%	1-5

Xibond 330

PARAMETER	UNIT	VALUE
GENERAL INFORMATION		
Name	Xibond 330	
Composition	styrene, maleic anhydride and N-phenylmaleimide terpolymer	
Acronym		SMANPMI
Molecular mass	daltons	155,000
Chemical class	maleic anhydride modified polymer	
PHYSICAL PROPERTIES		
State	-	solid/granules
Color	-	pale yellow
Glass transition temperature	°C	180
Acid number	mg KOH/g	66
Thermal stability, TGA	°C	360
USE & PERFORMANCE		
Manufacturer	Polyscope	
Outstanding properties	optimized interfacial tension, stabilized morphology against high stresses, and enhanced adhesion between the phases in the solid-state. Compatibilization improved the mechanical properties of the polymer blend.	
Recommended for polymers	compatible with HIPS resins	
Recommended applications	adhesion promoter, compatibilizer, coupling agent, viscosity modifier, and coupling agent. It is designed to improve the blend morphology of specific polymer blends.	
Processing methods	twin-screw extruders with a mild screw configuration and vacuum degassing facility for good dispersion in styrenic polymer	
Concentrations used	wt%	1-5

Xibond 370

PARAMETER	UNIT	VALUE
GENERAL INFORMATION		
Name	Xibond 330	
Composition	styrene, maleic anhydride and N-phenylmaleimide terpolymer	
Acronym		SMANPMI
Molecular mass	daltons	155,000
Chemical class	maleic anhydride modified polymer	
PHYSICAL PROPERTIES		
State	-	solid/granules
Color	-	pale yellow
Glass transition temperature	°C	175
Acid number	mg KOH/g	90
Thermal stability, TGA	°C	390
USE & PERFORMANCE		
Manufacturer	Polyscope	
Outstanding properties	optimized interfacial tension, stabilized morphology against high stresses, and enhanced adhesion between the phases in the solid-state. Compatibilization improved the mechanical properties of the polymer blend.	
Recommended for polymers	compatible with HIPS resins	
Recommended applications	adhesion promoter, compatibilizer, coupling agent, viscosity modifier, and coupling agent. It is designed to improve the blend morphology of specific polymer blends.	
Processing methods	twin-screw extruders with a mild screw configuration and vacuum degassing facility for good dispersion in styrenic polymer	
Concentrations used	wt%	1-5

Xibond 375

PARAMETER	UNIT	VALUE
GENERAL INFORMATION		
Name		Xibond 375
Composition		styrene, maleic anhydride and N-phenylmaleimide terpolymer
Acronym		SMANPMI
Molecular mass	daltons	150,000
Chemical class		maleic anhydride modified polymer
PHYSICAL PROPERTIES		
State	-	solid/powder
Color	-	pale yellow
Glass transition temperature	°C	175
Acid number	mg KOH/g	90
Thermal stability, TGA	°C	390
USE & PERFORMANCE		
Manufacturer		Polyscope
Outstanding properties		optimized interfacial tension, stabilized morphology against high stresses, and enhanced adhesion between the phases in the solid-state. Compatibilization improved the mechanical properties of the polymer blend.
Recommended for polymers		compatible with HIPS resins
Recommended applications		adhesion promoter, compatibilizer, coupling agent, viscosity modifier, and coupling agent. It is designed to improve the blend morphology of specific polymer blends.
Processing methods		twin-screw extruders with a mild screw configuration and vacuum degassing facility for good dispersion in styrenic polymer
Concentrations used	wt%	1-5

Xibond 830

PARAMETER	UNIT	VALUE
GENERAL INFORMATION		
Name	Xibond 830	
Composition	ethylene-octene copolymer grafted with styrene and glycidyl methacrylate copolymer	
Molecular mass	daltons	200,000
Chemical class	maleic anhydride modified polymer	
PHYSICAL PROPERTIES		
State	-	solid/pellets
Color	-	white
Glass transition temperature	°C	-53
USE & PERFORMANCE		
Manufacturer	Polyscope	
Outstanding properties	improved mechanical properties of the polymer blend.	
Recommended for polymers	polyolefins, thermoplastic polyester resins	
Recommended applications	adhesion promoter, compatibilizer, coupling agent, viscosity modifier, and coupling agent. It is designed to improve the blend morphology of specific polymer blends.	
Processing methods	twin-screw extruders with a mild screw configuration and vacuum degassing facility for good dispersion in styrenic polymer	
Concentrations used	wt%	1-20

3.16 Melamine
Actmix HMMM-50GE F140

PARAMETER	UNIT	VALUE
GENERAL INFORMATION		
Name		Actmix HMMM-50GE F140
CAS #	-	3089-11-0
EC number	-	221-422-3
Composition		50 wt% hexamethoxymethyl melamine
Acronym	-	HMMM
Empirical formula	-	C15H30N6O6
Formula		

PARAMETER	UNIT	VALUE
Molecular mass	daltons	390.51
Chemical class	melamine	
Active matter	wt%	50
Purity	wt%	>98.5
PHYSICAL PROPERTIES		
State	-	solid
Color	-	white, translucent
Melting point	°C	49
Ash content	wt%	<0.4
Density at 25°C	kg/m³	1,230
Particle size	µm	140
Mooney viscosity at 50°C	-	50
Volatility	-	0.3
HEALTH & SAFETY		
Flash point	°C	248.30
USE & PERFORMANCE		
Manufacturer		Ningbo Actmix Polymer
Recommended for polymers		NBR, SBR, AR, ECO
Recommended applications		adhesion promoter for rubber and framework materials due to reactions with methylene acceptor at vulcanizing temperature. It can react to thermosetting resin together with other adhesives such as GLR-18 resin, RE, RS, etc., which will lose effect if the reaction starts before vulcanization.

3.16 Melamine
Actmix HMMM-50GE F140

PARAMETER	UNIT	VALUE
Concentrations used		3-5 phr HMMM-50 with 2-5 phr methylene acceptor

Cohedur A 200

PARAMETER	UNIT	VALUE
GENERAL INFORMATION		
Name		Cohedur A 200
Composition		hexamethoxymethylmelamine ether
Formula		

R=CH$_3$, H, CH$_2$

PARAMETER	UNIT	VALUE
Chemical class	-	melamine
Nitrogen content	%	23.3
PHYSICAL PROPERTIES		
State	-	liquid
Color	clear to cloudy, colorless to yellow	
Density at 25°C	kg/m^3	1,200
pH	8 (50% aq. solution)	
Viscosity at 25°C	mPas	<= 8,000
USE & PERFORMANCE		
Manufacturer	RheinChemie Additives/Lanxess	
Outstanding properties	Cohedur bonding agents enhance adhesion by physical and chemical interaction. These direct bonding agents are mixed into the rubber compound, which can be applied directly to the substrate. Without bonding agents, adhesion is based only on mechanical interaction due to the penetration of the rubber between the fibers.	
Recommended for polymers	natural rubber, BR, CR, EPDM, NBR, SBR	
Concentrations used	phr	2.3
Recommended for products	tires, conveyor belting, V-belts, reinforcing hose, air springs, flexible containers and fabric proofings	

Cohedur A 200

PARAMETER	UNIT	VALUE
Recommended applications		promotes adhesion between elastic rubber and inelastic-but-durable material such as reinforcing carcasses or steel cords, which are combined to make long-lasting tires, conveyor belts, hoses, rubberized fabrics, containers, etc. Enhances adhesion by physical and chemical interaction.

Cohedur A 250

PARAMETER	UNIT	VALUE
GENERAL INFORMATION		
Name		Cohedur A 250
Composition		50% hexamethoxymethylmelamine ether, 50% filler
Acronym	-	HMMM
Formula		
		R=CH$_3$, H, CH$_2$
Chemical class	-	melamine
PHYSICAL PROPERTIES		
State	-	solid/powder
Color	-	white
Ash content	wt%	44.4
Density at 25°C	kg/m^3	1,600
Nitrogen content	wt%	11.6
USE & PERFORMANCE		
Manufacturer		RheinChemie Additives/Lanxess
Outstanding properties		Cohedur A 250 is a component of the direct bonding system, also known as Cohedur or RFS system. RFS bonding systems are multi-component systems. They are created by providing the rubber compound with a resorcinol component, a methylene component, and reinforcing silica, e.g., Vulkasil S
Recommended for polymers		rubber
Recommended applications		used as a bonding agent for rubber to fabric and rubber to steel cord bonding
Concentrations used	phr	4.6

Cohedur RDL

PARAMETER	UNIT	VALUE
GENERAL INFORMATION		
Name		Cohedur RDL
Composition		1/3 resorcinol, 1/3 hexamethoxymethyl-melamine ether, 1/3 silica
Acronym	-	HMMM
Empirical formula	-	C6H12N4
Chemical class	-	melamine
Mixture	-	yes
PHYSICAL PROPERTIES		
State		solid/powder, may contain soft lamps
Color	-	white to red brown
Ash content	wt%	30.7
Density at 25°C	kg/m³	1,500
Nitrogen content	wt%	7.7
USE & PERFORMANCE		
Manufacturer		RheinChemie Additives/Lanxess
Recommended for polymers		NBR, HNBR, CR, NR, IR, SBR, BR
Recommended for products		tires, conveyor belting, V-belts, reinforcing hose, air springs, flexible containers and fabric proofing's.
Recommended applications		used as a direct bonding agent for rubber to reinforcing materials, e.g. fabrics, steel cord, and glass fibers. It is employed as an additive to the rubber compound. It is used with most types of rubber. Best adhesion is obtained either with polar and nonpolar elastomers, such as NBR, HNBR, CR, NR, IR, SBR and BR. Oil-extended elastomers can yield in inferior results as also highly saturated polymers, e.g. IIR and EPDM, develop a considerably lower adhesion level. Fibers, such as cotton, rayon, polyamide, polyester (with special spin finish) and glass can be firmly adhered to rubber by using Cohedur® RDL in the rubber mix. For steel cord the bond strength increases in the order raw, zinc plated and brass-plated steel.
Concentrations used	phr	6-8

Cohedur RK

PARAMETER	UNIT	VALUE
GENERAL INFORMATION		
Name		Cohedur RK
Composition		resorcinol derivative on filler in the ratio 1:1
Chemical class	-	melamine
Mixture	-	yes
PHYSICAL PROPERTIES		
State	-	powder
Color	-	white
Ash content	wt%	44
MgO content	%	9
Density at 25°C	kg/m³	1,700
Nitrogen content	wt%	7-8.4
USE & PERFORMANCE		
Manufacturer		Lanxess
Recommended for polymers		chloroprene rubber
Recommended applications		used as the resorcinol component of the three-component system, but direct bonding compounds containing it have better scorch resistance than those made with resorcinol itself. Cohedur RK must be used in conjunction with Cohedur A grades and reinforcing silica, e.g., Vulkasil S. It is intended mainly for use with chloroprene rubber
Concentrations used	phr	7-9

Cyrez 963

PARAMETER	UNIT	VALUE
GENERAL INFORMATION		
Name		Cyrez 963
Composition		>98% hexamethoxymethyl melamine resin, no carrier
Acronym	-	HMMM
Chemical class	-	melamine
Active matter	wt%	98
PHYSICAL PROPERTIES		
State	-	liquid
Color	-	clear
Density at 25°C	kg/m^3	1195
Loss on heating (2 h at 125°C)	%	<=4
Refractive index	-	1.52175
Viscosity at 23°C	mPas	3000-6000
USE & PERFORMANCE		
Manufacturer		Allnex
Outstanding properties		low content of free formaldehyde. Cyrez resins are relatively nontoxic and present little hazard of dermatitis
Recommended applications		adhesion promoter and crosslinker. It is used for bonding rubber to organic cord and wire reinforcement materials. Curing agent for resorcinol and novolac resins.

Cyrez 964

PARAMETER	UNIT	VALUE
GENERAL INFORMATION		
Name		Cyrez 964
Composition		hexamethoxymethyl melamine resin, carrier precipitated amorphous silica, (65% HMMM)
Acronym	-	HMMM
Chemical class	-	melamine
PHYSICAL PROPERTIES		
State	-	solid/micro pearls
Color	-	white
Ash content	wt%	31-35
Water content	%	<=4
Particle size (going through 80 mesh)	%	>99.7
USE & PERFORMANCE		
Manufacturer		Allnex
Outstanding properties		good flow ability, easily dispersible. Cyrez resins are relatively nontoxic and present little hazard of dermatitis
Recommended applications		adhesion promoters and reinforcing systems in rubber and tires applications. Curing agent for resorcinol and novolac resins.
Recommended for polymers		rubber to organic cord and wire reinforcement
Concentrations used	phr	1.5-4

Cyrez CRA-100

PARAMETER	UNIT	VALUE
GENERAL INFORMATION		
Name		Cyrez CRA-100
Composition		hexamethoxymethyl melamine resin, 72% HMMM on precipitated amorphous silica as a carrier
Acronym	-	HMMM
Chemical class	-	melamine
PHYSICAL PROPERTIES		
State	-	solid
Color	-	white
Ash content	wt%	24-28
Water content	%	<=4
Particle size (through 80 mesh)	%	99.7
USE & PERFORMANCE		
Manufacturer		Allnex
Outstanding properties		high loading, easily dispersible. Cyrez resins are relatively nontoxic and present little hazard of dermatitis.
Recommended applications		adhesion promoters and reinforcing systems in rubber applications. Curing agent for resorcinol and novolac resins.
Recommended for polymers		rubber
Concentrations used	phr	1.5-4

Markoba HMMM

PARAMETER	UNIT	VALUE
GENERAL INFORMATION		
Name	Markoba HMMM	
CAS #	-	3089-11-0
EC number	-	221-422-3
Composition	hexamethoxymethyl melamine	
Common synonym	hexa(methoxymethyl)melamine	
Acronym		HMMM
Empirical formula	C15H30N6O6	
Formula		
Molecular mass	daltons	390.435
Chemical class	melamine	
PHYSICAL PROPERTIES		
State	clear liquid or waxy solid	
Color	-	white
HEALTH & SAFETY		
Effect of exposure, eye (human)	Causes serious eye irritation.	
USE & PERFORMANCE		
Manufacturer	Wholemark Fine Chemical	
Recommended for polymers	rubber	
Recommended applications	adhesion promoters	

3.17 Metal-organic complexes (non-silicon)
Chartwell B-505.1

PARAMETER	UNIT	VALUE
GENERAL INFORMATION		
Name		Chartwell B-505.1
General description		mercapto functional organic adhesion promoter synthesized using a stabilized bimetal precursor, carrier ethylene glycol (for rapid dispersion)
Active matter	wt%	25.5
Complexed organics	wt%	9.1 - 9.3
Metal content	wt%	5.2 - 5.9
Number of metals	-	2
Organofunctionality	-	mercapto
PHYSICAL PROPERTIES		
State	-	liquid
Odor	-	strong
Color	-	clear, pale yellow
Boiling point	°C	>198
Decomposition temperature	°C	>250
Density at 20°C	kg/m³	1,230
pH	-	4.0 (2% solution)
Specific gravity at 20°C	-	1.24
Viscosity at 20°C	mPas	<100
HEALTH & SAFETY		
Autoignition temperature	°C	>200
Flash point	°C	>100
Animal testing, acute toxicity, Rat oral LD50	mg/kg	1714
First aid, eye		In case of contact with eyes flush immediately with plenty of flowing water for 10 to 15 minutes holding eyelids apart and consult an ophthalmologist
First aid, inhalation		Provide fresh air.
First aid, skin		In case of skin irritation, consult a physician. Immediate medical treatment required because corrosive injuries that are not treated are hard to cure.
UN #	-	3265
ECOLOGICAL PROPERTIES		
Aquatic toxicity, *Daphnia magna*, 48-h LC50	mg/l	>100

3.17 Metal-organic complexes (non-silicon)
Chartwell B-505.1

PARAMETER	UNIT	VALUE
USE & PERFORMANCE		
Manufacturer		Protex International/Chartwell International, Inc.
Outstanding properties		increased T-peel strength. Improved resistance to moisture, heat and corrosive environments. Enhanced adhesion to metals plastics and elastomers. Improved vulcanization of all sulfur cured elastomers
Recommended for polymers		epoxy, PU, PVAc, neoprene
Recommended for products		adhesives, electronic encapsulants, metal primers
Recommended applications		epoxy, urethane, and rubber adhesives
Processing methods		high shear mixing in all solvent-borne system
Concentrations used		1.0-2.0 phr, post addition recommended under agitation

Chartwell B-515.1/2H

PARAMETER	UNIT	VALUE
GENERAL INFORMATION		
Name		Chartwell B-515.1/2H
General description		amino functional metal organic adhesion promoter synthesized using a stabilized bimetal precursor, carrier ethylene glycol
Chemical class		silane
Active matter	wt%	50.5
Functional organic group	-	primary amine
Metal content	wt%	10.4-11.8
Complexed organics	wt%	18.2-18.6
Number of metals	-	2
Organoreactive group		- (CH2)xNH2, x < 4
PHYSICAL PROPERTIES		
State	-	liquid
Odor	-	slight
Color	-	clear, pale yellow
Boiling point	°C	>198
Decomposition temperature	°C	>250
Density at 20°C	kg/m^3	1,380
pH	-	4.15 (2% solution)
Specific gravity at 20°C	-	1.39
Viscosity at 20°C	mPas	<100
HEALTH & SAFETY		
Autoignition temperature	°C	>200
Flash point	°C	>100
Animal testing, acute toxicity, Rat oral LD50	mg/kg	>10000
Animal testing, acute toxicity, Rabbit dermal LD50	mg/kg	>10000
First aid, eye		Rinse immediately carefully and thoroughly with eye-bath or water.
First aid, inhalation		Provide fresh air.
First aid, skin		Wash immediately with water.
ECOLOGICAL PROPERTIES		
Aquatic toxicity, *Daphnia magna*, 48-h LC50	mg/l	>100
USE & PERFORMANCE		
Manufacturer		Protex International/Chartwell International, Inc.

Chartwell B-515.1/2H

PARAMETER	UNIT	VALUE
Outstanding properties		enhanced adhesion to metals plastics and elastomers
Recommended for polymers		acrylics, alkyds, epoxy, polyester, PU, rubber
Recommended for products		molded rubber articles, roof membranes, automotive, electronic encapsulants, metal primers
Recommended applications		adhesives to metal, coatings, inks, plastics (filler pretreatment)
Concentrations used		0.5-1.0 phr, post add recommended under agitation

Chartwell B-515.1W

PARAMETER	UNIT	VALUE
GENERAL INFORMATION		
Name		Chartwell B-515.1W
General description		amino functional metal organic adhesion promoter synthesized using a stabilized bimetal precursor, carrier water
Active matter	wt%	22.0
Complexed organics	wt%	9.1 - 9.3
Functional organic group	-	primary amine
Metal content	wt%	5.2 - 5.9
Number of metals	-	2
Organoreactive group		-(CH2)xNH2, x < 4
PHYSICAL PROPERTIES		
State	-	liquid
Odor	-	slight
Color	-	clear, pale yellow
Boiling point	°C	100
Melting point	°C	<0
Decomposition temperature	°C	>250
Density at 20°C	kg/m^3	1,140
pH	-	4.35 (2% solution)
Specific gravity at 20°C	-	1.140
Vapor pressure at 20°C	kPa	2.5
HEALTH & SAFETY		
Autoignition temperature	°C	>200
Flash point	°C	>100
Animal testing, acute toxicity, Rat oral LD50	mg/kg	>4000
First aid, eye		Rinse immediately carefully and thoroughly with eye-bath or water.
First aid, inhalation		No special measures are necessary.
First aid, skin		Subsequently wash off with water.
USE & PERFORMANCE		
Manufacturer		Protex International/Chartwell International, Inc.
Outstanding properties		improved resistance to moisture, heat and corrosive environments. Enhanced adhesion to metals, plastics, concrete, elastomers, and ceramics
Recommended for polymers		PUD, PVAc, 2K epoxy WB, neoprene WB

Chartwell B-515.1W

PARAMETER	UNIT	VALUE
Recommended applications	adhesion to metal; metal coatings	
Concentrations used	1-3 phr, post add recommended under agitation	
Conditions to avoid	not recommended for untreated PP/PE, fiberglass or glass substrates	

Chartwell B-515.71W

PARAMETER	UNIT	VALUE
GENERAL INFORMATION		
Name		Chartwell B-515.71W
General description		amino functional metal organic adhesion promoter synthesized with stable neutralized metal complex, carrier water
Active matter	wt%	33
Complexed organics	wt%	19.9-20.4
Functional organic group	primary amine	
Metal content	wt%	7.3-7.9
Number of metals	-	1
Organoreactive group	- $(CH_2)_xNH_2$, x < 4	
PHYSICAL PROPERTIES		
State	-	liquid
Odor	-	slight
Color	-	clear, pale yellow
Boiling point	°C	100
Decomposition temperature	°C	>250
Density at 20°C	kg/m^3	1,200
Neutralizing agent	-	caustic
pH	-	7.25 (2% solution)
Specific gravity at 20°C	-	1.20
Volatility, VOC	-	0
Viscosity at 20°C	mPas	<100
HEALTH & SAFETY		
Autoignition temperature	°C	>200
Flash point	°C	>100
Animal testing, acute toxicity, Rat oral LD50	mg/kg	>4000
Animal testing, acute toxicity, Rabbit dermal LD50	mg/kg	>4000
First aid, eye		Rinse immediately carefully and thoroughly with eye-bath or water.
First aid, inhalation		No special measures are necessary.
First aid, skin		Subsequently wash off with water.
ECOLOGICAL PROPERTIES		
Aquatic toxicity, *Daphnia magna*, 48-h LC50	mg/l	>100

Chartwell B-515.71W

PARAMETER	UNIT	VALUE
USE & PERFORMANCE		
Manufacturer		Protex International/Chartwell International, Inc.
Outstanding properties		enhanced adhesion to metals, plastics, concrete, elastomers, and ceramics. Zero VOC
Recommended for polymers		acrylics, acrylic/styrenated acrylic latex, epoxy, urethane
Recommended for products		adhesives, packaging, inks, insulation, roof membrane
Recommended applications		coatings on: metal, plastic, wood; adhesives to metal, wood, plastic (treated PP/PE, Mylar), ceramic, concrete glass, rubber
Concentrations used		0.35 - 1.4 wt% based upon polymer solids + organic pigment weight + anti-corrosive pigment weight.
Guidelines for use		fully compatible with WB coatings/ inks/ adhesives having a pH of 5-11. it may be added directly to latex or polymer dispersion or post added in many cases. No special mixing or dilution is required. For solvent-borne coating, optimum performance is achieved when added directly to the grind stage resin and high shear mixed for 15 mins before adding other components
Conditions to avoid		not recommended for untreated PP/PE, fiberglass or glass substrates

Chartwell B-525.1

PARAMETER	UNIT	VALUE
GENERAL INFORMATION		
Name		Chartwell B-525.1
General description		carboxy functional metal organic adhesion promoter synthesized using a stabilized bimetal precursor, carrier ethylene glycol
Active matter	wt%	27.5
Complexed organics	wt%	9.1-9.3
Functional organic group	-	carboxyl
Metal content	wt%	5.2-5.9
Number of metals	-	2
Organoreactive group		-(CH2)xCOOH, x=4-6
PHYSICAL PROPERTIES		
State	-	liquid
Odor	-	slight
Color		opaque; colorless to light yellow
Boiling point	°C	>198
Decomposition temperature	°C	>2500
Density at 20°C	kg/m³	1230
pH	-	3.9 (2% solution)
Solubility in water at 25°C	g/l	miscible
Specific gravity at 20°C	-	1.23
Viscosity at 20°C	mPas	<100
HEALTH & SAFETY		
Autoignition temperature	°C	>200
Flash point	°C	>100
Animal testing, acute toxicity, Rat oral LD50	mg/kg	>7000
Animal testing, acute toxicity, Rabbit dermal LD50	mg/kg	>10000
First aid, eye		Rinse immediately carefully and thoroughly with eye-bath or water.
First aid, inhalation		Provide fresh air.
First aid, skin		Wash immediately with water.
ECOLOGICAL PROPERTIES		
Aquatic toxicity, *Daphnia magna*, 48-h LC50	mg/l	>100
Biodegradation probability		moderately/partially biodegradable.

Chartwell B-525.1

PARAMETER	UNIT	VALUE
USE & PERFORMANCE		
Manufacturer		Protex International/Chartwell International, Inc.
Outstanding properties		rapid dispersion and solubilization
Recommended for polymers		2K epoxy, 2K PU, acrylic, rubber
Recommended applications		adhesives, coatings
Concentrations used		1.0 - 2.0 phr, post add recommended under agitation; 1-3 phr in rubber
Guidelines for use		high shear mixing in all solvent-borne system

Chartwell C-515.71/1.5H

PARAMETER	UNIT	VALUE
GENERAL INFORMATION		
Name		Chartwell C-515.71/1.5H
General description		amino functional metal organic adhesion promoter synthesized with a stable neutralized metal complex, carrier propylene glycol
Active matter	wt%	51
Complexed organics	wt%	19.6-20.4
Functionality		primary amine
Metal content	wt%	7.3-7.9
Number of metals	-	1
Organoreactive group	- (CH2)xNH2, x < 4	
PHYSICAL PROPERTIES		
State	-	liquid
Odor	-	slight
Color	-	clear, pale yellow
Boiling point	°C	>100
Melting point	°C	0
Decomposition temperature	°C	>250
Density at 20°C	kg/m³	1310
pH	-	7.7 (1% solution)
Specific gravity at 20°C	-	1.31
Viscosity at 20°C	mPas	<100
HEALTH & SAFETY		
Autoignition temperature	°C	>200
Flash point	°C	>100
Animal testing, acute toxicity, Rat oral LD50	mg/kg	>10000
Animal testing, acute toxicity, Rabbit dermal LD50	mg/kg	>10000
First aid, eye		Rinse immediately carefully and thoroughly with eye-bath or water.
First aid, inhalation		Provide fresh air.
First aid, skin		Subsequently wash off with water.
ECOLOGICAL PROPERTIES		
Aquatic toxicity, *Daphnia magna*, 48-h LC50	mg/l	>100
Biodegradation probability		moderately/partially biodegradable

Chartwell C-515.71/1.5H

PARAMETER	UNIT	VALUE
USE & PERFORMANCE		
Manufacturer		Protex International/Chartwell International, Inc.
Outstanding properties		improved adhesion to all metals with accompanying improved salt fog resistance, reduction of creep at the scribe, and reduced blistering; and improved adhesion to plastics ceramics, concrete, and wood. In WB, exterior primers and deck stains will eliminate peeling on wood substrate. In foundry cores, will improve bonding between sand and phenolic, urethane and similar binders.
Recommended for polymers		alkyd WB, acrylic/styrene-acrylic emulsion, 2K epoxy SB, 2K PU
Recommended for products		coatings, adhesives, inks
Recommended applications		adhesives to metal, wood, plastic (treated PP/PE, Mylar), ceramic, concrete; coatings on: metal, plastic wood
Concentrations used		0.35 - 1.4 wt% on polymer solids.

Chartwell C-515.71.HR

PARAMETER	UNIT	VALUE
GENERAL INFORMATION		
Name		Chartwell C-515.71.HR
General description		amino functional metal organic adhesion promoter synthesized with a stable neutralized metal complex, carrier propylene glycol
Active matter	wt%	41.5
Complexed organics	wt%	25.6-26.4
Functionality	-	primary amine
Metal content	wt%	7.3-7.9
Number of metals	-	1
Organoreactive group		- (CH2)xNH2, x < 4
PHYSICAL PROPERTIES		
State	-	liquid
Odor	-	slight
Color	-	moderately yellow.
Boiling point	°C	>100
Melting point	°C	0
Decomposition temperature	°C	>250
Density at 20°C	kg/m^3	1230
Neutralizing agent	-	caustic
pH	-	7.5 (1% solution)
Specific gravity at 20°C	-	1.23
Viscosity at 20°C	mPas	<100
HEALTH & SAFETY		
Autoignition temperature	°C	>200
Flash point	°C	>100
Animal testing, acute toxicity, Rat oral LD50	mg/kg	>10000
Animal testing, acute toxicity, Rabbit dermal LD50	mg/kg	>10000
First aid, eye		Rinse immediately carefully and thoroughly with eye-bath or water.
First aid, inhalation		Provide fresh air.
First aid, skin		Subsequently wash off with water
ECOLOGICAL PROPERTIES		
Aquatic toxicity, *Daphnia magna*, 48-h LC50	mg/l	>100
Biodegradation probability		moderately/partially biodegradable.

Chartwell C-515.71.HR

PARAMETER	UNIT	VALUE
USE & PERFORMANCE		
Manufacturer		Chartwell International, Inc.
Outstanding properties		improved adhesion to all metals with accompanying improved salt fog resistance, reduction of creep at the scribe, and reduced blistering; and improved adhesion to plastics ceramics, concrete, and wood. In WB, exterior primers and deck stains will eliminate peeling on wood substrate. In foundry cores, will improve bonding between sand and phenolic, urethane and similar binders.
Recommended for polymers		alkyd WB, acrylic SB, acrylic/styrene-acrylic emulsion, 2K epoxy SB, polyester/melamine, acrylic/melamine, unsaturated polyester, 2K PU, SBR latex, urethane modified nitrocellulose
Recommended for products		coatings, adhesives, inks
Recommended applications		adhesives to metal, plastic, wood, ceramic, concrete, glass; coatings on: metal, plastic, wood, epoxy flooring
Processing methods		waterborne: mix with conventional paddle type mix; 2K Epoxy and 2K: high shear mixed with a Cowles type mix
Concentrations used		0.35-1.4 wt% on polymer solids.

Chartwell C-515.72HRW

PARAMETER	UNIT	VALUE
GENERAL INFORMATION		
Name		Chartwell C-515.72HRW
General description		amino functional metal organic adhesion promoter synthesized with a stabilized neutralized metal complex. The product is supplied as a solution in water.
Active matter	wt%	33.12
Complexed organics	wt%	33.8-34.6
Metal content	wt%	7.3-7.9
PHYSICAL PROPERTIES		
State	-	liquid
Color	-	clear, pale yellow
Density at 20°C	kg/m^3	1180
Neutralizing agent	-	MEA
pH	-	8.3 (1% solution)
Specific gravity at 20°C	-	1.21
USE & PERFORMANCE		
Manufacturer		Protex International/Chartwell International, Inc.
Outstanding properties		improved salt fog resistance, reduction of creep at the scribe, and reduced blistering; and improved adhesion to ceramics, concrete, wood and plastics
Recommended for polymers		acrylic, epoxy, styrenated acrylic, phenolic, alkyd, urethane
Recommended for products		adhesives, coatings, inks, WB exterior primers and deck stains
Concentrations used		0.35-1.4 wt % based upon polymer solids + organic pigment weight + anti-corrosive pigment weight
Guidelines for use		fully compatible with coatings/inks having a pH of 7-11. May be added directly to latex or polymer dispersion or post added in many cases

Chartwell C-515.72.HRX

PARAMETER	UNIT	VALUE
GENERAL INFORMATION		
Name		Chartwell C-515.72.HRX
General description		amino functional metal organic adhesion promoter synthesized with a stabilized MEA neutralized metal complex, carrier propylene glycol
Active matter	wt%	34
Neutralizing agent	-	MEA
Complexed organics	wt%	21.3-21.9
Functionality	-	primary amine
Metal content	wt%	7.3-7.9
Number of metals	-	1
PHYSICAL PROPERTIES		
State	-	liquid
Odor	-	slight
Color	-	clear, moderate yellow
Boiling point	°C	>185
Melting point	°C	<0
Decomposition temperature	°C	>250
Density at 20°C	kg/m^3	1180
pH	-	6 (1% solution)
Specific gravity at 20°C	-	1.18
Viscosity at 20°C	mPas	<100
HEALTH & SAFETY		
Autoignition temperature	°C	>200
Flash point	°C	>100
Animal testing, acute toxicity, Rat oral LD50	mg/kg	>2000
Animal testing, acute toxicity, Rabbit dermal LD50	mg/kg	>2000
First aid, eye		Rinse immediately carefully and thoroughly with eye-bath or water.
First aid, inhalation		Provide fresh air.
First aid, skin		Subsequently wash off with water.
ECOLOGICAL PROPERTIES		
Aquatic toxicity, *Daphnia magna*, 48-h LC50	mg/l	>100
USE & PERFORMANCE		
Manufacturer		Protex International/Chartwell International, Inc.

Chartwell C-515.72.HRX

PARAMETER	UNIT	VALUE
Outstanding properties		low cost. Improved adhesion to all metals with accompanying improved salt fog resistance, reduction of creep at the scribe, and reduced blistering; and improved adhesion to plastics, ceramics, concrete, and wood. In WB, exterior primers and deck stains will eliminate peeling on wood substrate.
Recommended for polymers		2K epoxy, 2K PU, acrylic, styrenated acrylic, and alkyds
Recommended applications		adhesives, coatings, inks
Processing methods		waterborne: mix with conventional paddle type mix; 2K epoxy and 2K: high shear mixed with a Cowles type mix
Concentrations used		0.35-1.4 wt% on polymer solids.

Chartwell C-545.1

PARAMETER	UNIT	VALUE
GENERAL INFORMATION		
Name		Chartwell C-545.1
General description		methacrylate-functional organic adhesion promoter synthesized using a stabilized bimetal precursor, carrier propylene glycol
Active matter	wt%	23
Complexed organics	wt%	9.1-9.3
Functional group	-	methacrylatic
Metal content	wt%	5.2-5.9
Number of metals	-	2
PHYSICAL PROPERTIES		
State	-	liquid
Odor	-	slight
Color	-	pale yellow
Boiling point	°C	>188
Melting point	°C	<59
Decomposition temperature	°C	>250
Density at 20°C	kg/m^3	1150
pH	-	3.9 (1% solution)
Specific gravity at 20°C	-	1.15
HEALTH & SAFETY		
Autoignition temperature	°C	>200
Flash point	°C	>103
Animal testing, acute toxicity, Rat oral LD50	mg/kg	>2000
First aid, eye		Rinse immediately carefully and thoroughly with eye-bath or water.
First aid, inhalation		Provide fresh air.
First aid, skin		Subsequently wash off with water.
USE & PERFORMANCE		
Manufacturer		Protex International/Chartwell International, Inc.
Recommended for polymers		acrylate/methacrylate, EPDM, polyesters, PU, silicone, rubber
Recommended applications		adhesives, coatings, sealants
Concentrations used		1-2 phr, post addition recommended under agitation; 1-3 phr rubber

Chartwell C-600

PARAMETER	UNIT	VALUE
GENERAL INFORMATION		
Name		Chartwell C-600
General description		sulfido-functional metal organic adhesion promoter synthesized using a stabilized bimetal precursor; carrier dipropylene glycol
Active matter	wt%	24.2
Complexed organics	wt%	6.9-7.1
Functionality	-	sulfide
Metal content	wt%	4.3-4.5
Number of metals	-	2
PHYSICAL PROPERTIES		
State	-	liquid
Odor	-	slight
Color	-	pale yellow
Boiling point	°C	185-189
Decomposition temperature	°C	>250
Density at 20°C	kg/m^3	1160
pH	-	4.1 (1% solution)
Solubility in water at 25°C	wt%	soluble
Viscosity at 20°C	mPas	130
HEALTH & SAFETY		
Autoignition temperature	°C	>200
Flash point	°C	>121
Animal testing, acute toxicity, Rat oral LD50	mg/kg	880
Animal testing, acute toxicity, Rabbit dermal LD50	mg/kg	1060
First aid, eye		Rinse thoroughly with plenty of water for at least 15 minutes and consult a physician. If easy to do, remove contact lenses.
First aid, inhalation		If breathed in, move person into fresh air. If not breathing, give artificial respiration. Consult a physician.
First aid, skin		Wash off with soap and plenty of water. Remove contaminated clothing and shoes. Consult a physician.
ECOLOGICAL PROPERTIES		
Aquatic toxicity, *Daphnia magna*, 48-h EC50	mg/l	1550
Aquatic toxicity, *Rainbow trout*, 96-h LC50	mg/l	1474

Chartwell C-600

PARAMETER	UNIT	VALUE
Biodegradation probability		90.4%/28d, readily biodegradable
USE & PERFORMANCE		
Manufacturer		Protex International/Chartwell International, Inc.
Recommended for polymers		2K epoxy, 2K PU, elastomers
Recommended applications		adhesives, dispersion of difficult to disperse pigments
Concentrations used		1-3 phr elastomers

Chartwell D-535.1

PARAMETER	UNIT	VALUE
GENERAL INFORMATION		
Name		Chartwell D-535.1
General description		hydrocarbon functional metal organic adhesion promoter synthesized using a stabilized bimetal precursor; carrier dipropylene glycol
Active matter	wt%	24
Complexed organics	wt%	9.9-10.1
Functionality	-	C-14 carboxyl
Metal content	wt%	4.3-4.5
Number of metals	-	2
PHYSICAL PROPERTIES		
State	-	liquid
Odor	-	slight
Color	-	clear, pale yellow
Boiling point	°C	>230
Melting point	°C	<-20
Decomposition temperature	°C	>250
Density at 20°C	kg/m³	1130
pH	-	3.5-4 (2% solution)
Solubility in water at 25°C	wt%	10
Specific gravity at 20°C	-	1.13
Viscosity at 20°C	mPas	<100
HEALTH & SAFETY		
Autoignition temperature	°C	>200
Flash point	°C	>121
Animal testing acute toxicity, Rat oral LD50	mg/kg	>10000
Animal testing, acute toxicity, Rabbit dermal LD50	mg/kg	>10000
First aid, eye		Rinse immediately carefully and thoroughly with eye-bath or water.
First aid, inhalation		Provide fresh air.
First aid, skin		Subsequently wash off with water.
ECOLOGICAL PROPERTIES		
Aquatic toxicity, *Daphnia magna*, 48-h LC50	mg/l	>100
USE & PERFORMANCE		
Manufacturer		Protex International/Chartwell International, Inc.

Chartwell D-535.1

PARAMETER	UNIT	VALUE
Outstanding properties		use of D-535.1 to treat mineral fillers, such as calcium carbonate, mica, silica, clay, etc., results in improved physical properties of the compounded polyolefin.
Recommended for polymers		2K epoxy, 2K PU, polyolefins
Recommended applications		adhesives, coatings, dispersion of difficult to disperse pigments, inks
Processing methods		high shear mixed mixing in all solventborne system
Concentrations used		1-2 phr, post add recommended under agitation; 3-5 phf (parts per hundred fillers) for high surface area pigments/fillers; 1-3 phr in rubbers

3.18 Metal-organic complexes (non-silicon)+silica
Chartsil B-515.1/2H

PARAMETER	UNIT	VALUE
GENERAL INFORMATION		
Name		Chartsil B-515.1/2H
CAS #		107-21-1, 112926-00-8/7631-86-9
General description		amino functional metal organic adhesion promoter absorbed upon a high surface area precipitated silica carrier
Composition		72% Chartwell B-515.1/2H (liquid) + 28% HiSil ABS precipitated silica; ethylene glycol (absorbed solvent)
Chemical class	-	silane+silica
Active matter	wt%	36.8
Complexed organics	wt%	13.0-13.4
Functional organic group	-	primary amine
Metal content	wt%	7.5-8.5
Organoreactive group		-(CH2)xNH2, x < 4
PHYSICAL PROPERTIES		
State	-	free flowing powder
Color	-	white
pH	-	4.15 (2% aq.)
Solubility, water	-	partial
Decomposition temperature	°C	>250
Bulk density	kg/m^3	570
HEALTH & SAFETY		
Autoignition temperature	°C	>200
Animal testing, acute toxicity, Rat oral LD50	mg/kg	>10000
Animal testing, acute toxicity, Rabbit dermal LD50	mg/kg	>10000
ECOLOGICAL PROPERTIES		
Aquatic toxicity, *Daphnia magna*, 48-h LC50	mg/l	>100
Biodegradation		Moderately/partially biodegradable
USE & PERFORMANCE		
Manufacturer		Protex International/Chartwell International, Inc.
Outstanding properties		improved corrosion resistance (salt fog), reduced creep at the scribe; rubber: improved adhesion, abrasion resistance, tensile and tear strengths
Recommended for polymers		alkyd, epoxy, hybrid, EPDM, polyester, PA, PU, rubber
Recommended applications		adhesives, epoxy powder coatings

3.18 Metal-organic complexes (non-silicon)+silica
Chartsil B-515.1/2H

PARAMETER	UNIT	VALUE
Concentrations used		powder coatings: 0.5-1 phr. High shear mixing (Henschel, etc.) is strongly recommended (do not exceed recommended use level); adhesives: 0.5-1 phr, add to resin and mix; plastics: 0.5-1 phf (parts per hundred filler), may be added directly to the extruder with resin, filler, and other additives (for high surface area pigments/fillers, i.e., fumed silica, carbon black, phthalo, and similar use 1-2 phf; rubber: 0.7-1.4 phr, add directly and compound in a Banbury mixer.

Chartsil C-505.1/2H

PARAMETER	UNIT	VALUE
GENERAL INFORMATION		
Name		Chartsil C-505.1/2H
CAS #		57-55-6, 112926-00-8/7631-86-9
General description		mercapto functional metal organic adhesion promoter absorbed on a high surface area precipitated silica carrier.
Composition		72% Chartwell C-523.2H (liquid) + 28% HiSil ABS precipitated silica (absorbed solvent propylene glycol)
Chemical class	-	silane+silica
Active matter	wt%	33.10
Complexed organics	wt%	11.4 - 11.7
Functional organic group	-	mercapto
Metal content	wt%	6.6 - 7.4
Organoreactive group	-(CH2)x SH, x<6	
PHYSICAL PROPERTIES		
State	-	free flowing solid
Color	-	white
Odor	-	mercaptan
pH	-	3.5 (2% in aq.)
Decomposition temperature	°C	>250
Bulk density	kg/m^3	570
HEALTH & SAFETY		
Autoignition temperature	°C	>200
Animal testing, acute toxicity, Rat oral LD50	mg/kg	>2000
USE & PERFORMANCE		
Manufacturer		Protex International/Chartwell International, Inc.
Outstanding properties		product is a dry, free-flowing solid, which physically breaks down on compounding (Banbury, Henschel, etc.) to release the active mercapto functional adhesion promoter. Particularly useful for enhancing adhesion of epoxy and urethane powder coatings to many metal substrates (CRS, aluminum, brass, copper, etc.), where liquid additives cannot easily be handled. Will improve salt fog and blistering resistance and reduce creep at the scribe.

Chartsil C-505.1/2H

PARAMETER	UNIT	VALUE
Recommended for polymers		all sulfur cured rubbers (SBR etc.), epoxy urethane, EPR, EPDM
Recommended applications		adhesives, powder coatings
Processing methods		set extruder for maximum shear and maximum torque
Concentrations used		adhesives: 0.6 - 1.2 phr, add to resin and mix; powder coatings: 0.6 - 1.2 phr (parts per hundred resin); plastics: 0.6 - 1.2 phf (parts per hundred filler/ pigment); rubber: 0.6 - 1.2 phr; add directly and compound in a Banbury mixer

Chartsil C-523.2H

PARAMETER	UNIT	VALUE
GENERAL INFORMATION		
Name		Chartsil C-523.2H
CAS #		57-55-6, 112926-00-8/7631-86-9
General description		hybrid carboxy/hydroxy functional metal organic adhesion promoter absorbed upon a high surface area precipitated silica carrier
Composition		72% Chartwell C-523.2H (liquid) + 28% HiSil ABS precipitated silica (absorbed solvent propylene glycol), propylene glycol (absorbed solvent)
Chemical class	-	silane+silica
Active matter	wt%	25.4
Complexed organics	wt%	9.2-9.4
Functional organic group	-	carboxyl/hydroxyl
Metal content	wt%	5.2-5.9
Organoreactive group		-CH(OH)-CH(OH)-COOH
PHYSICAL PROPERTIES		
State	-	free flowing solid
Color	-	white
Odor	-	slight
pH	-	3.5 (2% in aq.)
Decomposition temperature	°C	>250
Bulk density	kg/m^3	570
HEALTH & SAFETY		
Autoignition temperature	°C	>200
Animal testing, acute toxicity, Rat oral LD50	mg/kg	>2000
USE & PERFORMANCE		
Manufacturer		Protex International/Chartwell International, Inc.
Outstanding properties		improved corrosion resistance (salt fog), reduced creep at the scribe. Enhanced adhesion to metals, plastics, and elastomers. Enhanced adhesion of polyester powder coatings to many metal substrates (CRS, aluminum, brass, copper, etc.).
Recommended for polymers		polyester melamine, polyester TGIC, polyester Primid

Chartsil C-523.2H

PARAMETER	UNIT	VALUE
Recommended applications		adhesives and sealants, powder coatings, plastics (improved dispersion of pigments and fillers), inks
Processing methods		set extruder for maximum shear and maximum torque
Concentrations used		1.0 - 1.4 phr (parts per hundred resin), for each specific application, the optimum level of additive should be determined by testing.
Guidelines for use		The product is a dry, free-flowing solid, which physically breaks down upon compounding (Banbury, Henschel, etc.) to release active carboxy/hydroxy functional adhesion promoter.

3.19 Monomers
Bis(2-methacryloxyethyl) phosphate

PARAMETER	UNIT	VALUE
GENERAL INFORMATION		
Name		Bis(2-methacryloxyethyl) phosphate
CAS #	-	32435-46-4
EC number	-	251-040-2
IUPAC name		2-[hydroxy-[2-(2-methylprop-2-enoyloxy) ethoxy]phosphoryl]oxyethyl 2-methyl-prop-2-enoate
Empirical formula		C12H19O8P
Formula		
Molecular mass	daltons	322.25
PHYSICAL PROPERTIES		
State	-	liquid
Color	-	colorless
Initial boiling point	°C	221
Density at 25°C	kg/m³	1280
HEALTH & SAFETY		
NFPA classification	Flammability	0
	Health	2
	Instability	0
HMIS classification	Flammability	0
	Health	2
	Physical hazard	0
Carcinogenicity		No component of this product present at levels greater than or equal to 0.1% is identified as probable, possible or confirmed human carcinogen by IARC.
Flash point	°C	218.8
First aid, eye		Rinse thoroughly with plenty of water for at least 15 minutes and consult a physician.
First aid, inhalation		If breathed in, move person into fresh air. If not breathing, give artificial respiration. Consult a physician.
First aid, skin		Wash off with soap and plenty of water. Consult a physician.

3.19 Monomers
Bis(2-methacryloxyethyl) phosphate

PARAMETER	UNIT	VALUE
USE & PERFORMANCE		
Manufacturer		generic
Outstanding properties		crosslinking monomer. Adhesion promoter through free phosphoric acid group.
Recommended for products		adhesives, paints (automotive, furniture)

Fancryl FA-512AS

PARAMETER	UNIT	VALUE
GENERAL INFORMATION		
Name		Fancryl FA-512AS
CAS #	-	65983-31-5
EC number	-	251-678-1
Common synonym		Dicyclopentenyloxyethyl acrylate
Empirical formula		C15H20O3
Formula		
Molecular mass	daltons	248.32
Chemical class	monomer	
Purity	wt%	90
PHYSICAL PROPERTIES		
State	-	liquid
Color	-	clear, pale yellow
Color, Platinum-cobalt scale	-	<100
Boiling point	°C	113 (0.13 kPa)
Freezing point	°C	<-40
Glass transition temperature	°C	10-15
Refractive index at 25°C	-	1.499-1.501
Specific gravity at 25°C	-	1.085
Viscosity at 25°C	mPas	15-25
HEALTH & SAFETY		
Flash point	°C	166
Animal testing, acute toxicity, Rat oral LD50	mg/kg	<5000
Animal testing, acute toxicity, Rabbit dermal LD50	mg/kg	2500
USE & PERFORMANCE		
Manufacturer		Showa Denko
Outstanding properties		low volatility and low viscosity
Recommended for polymers		acrylics
Recommended for products		UV adhesives
Recommended applications		provides flexibility, good adhesion to metal and good compatibility with other materials

4-Methacryloxyethyl trimellitic anhydride

PARAMETER	UNIT	VALUE
GENERAL INFORMATION		
Name		4-Methacryloxyethyl trimellitic anhydride
CAS #	-	70293-55-9
EC number	-	274-547-0
Common synonym		2-[(2-methyl-1-oxoallyl)oxy]ethyl 1,3-dihydro-1,3-dioxoisobenzofuran-5-carboxylate
Acronym	-	4META
Empirical formula	-	C15H12O7
Formula		
Molecular mass	daltons	304.25
PHYSICAL PROPERTIES		
State	-	crystalline solid
Color	-	white, off-white
Melting point	°C	95
HEALTH & SAFETY		
Carcinogenicity		IARC: No ingredient of this product present at levels greater than or equal to 0.1% is identified as probable, possible or confirmed human carcinogen by IARC.
Animal testing, acute toxicity, Rat oral LD50	mg/kg	>2000
Animal testing, acute toxicity, Mouse oral LD50	mg/kg	>2000
Animal testing, acute toxicity, Rat dermal LD50	mg/kg	>2000
First aid, eye		Flush eyes with water as a precaution.
First aid, inhalation		If breathed in, move person into fresh air. If not breathing, give artificial respiration. Consult a physician.
First aid, skin		Wash off with soap and plenty of water. Consult a physician.
USE & PERFORMANCE		
Manufacturer		generic
Outstanding properties		Reactive monomer, especially used in dental applications as adhesion promoter.

3.20 Oligomers
Sarbox SB400

PARAMETER	UNIT	VALUE
GENERAL INFORMATION		
Name		Sarbox SB400
Composition		highly functional, carboxylic acid and anhydride containing methacrylate oligomer, blended in 2-methoxy propanol
Chemical class	-	oligomer
Functionality, average	-	8
PHYSICAL PROPERTIES		
Color, Platinum-cobalt scale	-	140
Acid number	mg KOH/g	138
Density at 20°C	kg/m³	1,120
Refractive index at 25°C	-	1.492
Viscosity at 25°C	mPas	10,000
USE & PERFORMANCE		
Manufacturer		Sartomer/Arkema Group
Outstanding properties		cure speed and excellent copper adhesion, alkali strippability, high acid number for alkali strippability, good solvent and acid resistance, and excellent high gloss aqueous development
Recommended for products		coatings, electronic, inks

UA-1605N

PARAMETER	UNIT	VALUE
GENERAL INFORMATION		
Name		UA-1605N
Composition		aliphatic urethane acrylate oligomer
Molecular mass	daltons	800-900
Chemical class	oligomer	
Functionality, average	-	6
PHYSICAL PROPERTIES		
Acid number	mg KOH/g	<1
Refractive index at 25°C	-	1.498
Viscosity at 25°C	mPas	80,000-90,000
USE & PERFORMANCE		
Manufacturer		Royal Gent
Outstanding properties		excellent adhesion, alcohol resistance and good hardness

3.21 Phenol novolac resins
Alnovol PN 760

PARAMETER	UNIT	VALUE
GENERAL INFORMATION		
Name		Alnovol PN 760
Composition		functionalized phenol resin. Phenol content <=0.5.
Chemical class		phenol novolac resin
PHYSICAL PROPERTIES		
State	-	solid
Color	-	yellow-brown
Melting point	°C	95-115/softening point
Solvent solubility		readily soluble in commonly used alcohols, ketones, and ethers. It is insoluble in aliphatic and aromatic hydrocarbons.
Viscosity at 25°C	mPas	800-1800/50% in methoxy propanol
USE & PERFORMANCE		
Manufacturer		Allnex
Outstanding properties		very good aging, improved adhesion, environmentally friendly.
Recommended applications		used in rubber compounds as adhesion promoter for steel and textile cord
Processing		Alnovol® PN 760 is incorporated into the mix in the first stage of compounding. Due to the low content of free phenol, processing does not generate any vapors, and it is not accompanied by odorous annoyance.
Concentrations used	phr	1.5-5

3.22 Phosphoric acid esters
Sipomer PAM 100

PARAMETER	UNIT	VALUE
GENERAL INFORMATION		
Name		Sipomer PAM 100
Common synonym		phosphate ester of polyethylene glycol monomethacrylate
Chemical class		phosphoric acid ester
Active matter	wt%	96-97
Functional organic group	-	methacrylate
PHYSICAL PROPERTIES		
State	-	liquid
Odor	-	mild acrylic
Color	-	dark brown
Glass transition temperature	°C	20
Acid number	mg KOH/g	131.5
Density at 25°C	kg/m³	1,230
pH	-	1.8-1.9
Specific gravity at 25°C	-	1
Viscosity at 25°C	mPas	1,700
HEALTH & SAFETY		
NFPA classification	Flammability	1
	Health	3
	Reactivity	1
DOT class	-	8 III
Flash point	°C	>93
Flash point method	-	CC
Effect of exposure, eye (human)		Corrosive. Can cause redness, irritation, pain, burns, irreversible eye damage
Effect of exposure, inhalation (human)		Mist may cause upper respiratory tract irritation
Effect of exposure, skin (human)		Can cause redness, inflammation, burns
UN/NA class	-	3265
USE & PERFORMANCE		
Manufacturer		Solvay
Outstanding properties		incorporated into acrylic latex systems improves adhesion to metallic substrates and glass. Sipomer PAM-100 also enhances properties such as freeze/thaw stability, mechanical stability, and gloss.

3.22 Phosphoric acid esters
Sipomer PAM 100

PARAMETER	UNIT	VALUE
Recommended for products		specialty purpose coatings, including industrial maintenance, OEM coil, light industrial, as well as high performance structural adhesives
Food contact approval		175.105

Sipomer PAM 200

PARAMETER	UNIT	VALUE
GENERAL INFORMATION		
Name		Sipomer PAM 200
Common synonym		phosphate esters of polyethylene glycol monomethacrylate
Chemical class		phosphoric acid ester
Active matter	wt%	96-97
Functional organic group		methacrylate
PHYSICAL PROPERTIES		
State	-	liquid
Odor	-	mild acrylic
Color	-	dark brown
Glass transition temperature	°C	0
Acid number	mg KOH/g	119
Acidity (as HCl)	wt%	1.5-3.5 (as H_3PO_4)
pH	-	1.7-2
Specific gravity	-	1.1
Viscosity at 25°C	mPas	4,700
USE & PERFORMANCE		
Manufacturer		Solvay
Outstanding properties		promotes adhesion of resins to metal and glass
Recommended for products		industrial maintenance, OEM coil, light industrial, and high performance structural adhesives

Sipomer PAM 4000

PARAMETER	UNIT	VALUE
GENERAL INFORMATION		
Name		Sipomer PAM 4000
CAS #		52628-03-2, 7664-38-2, 868-77-9
Common synonym		ethyl methacrylate phosphate
Chemical class		phosphoric acid ester
Composition		2-propenoic acid, 2-methyl-, 2-hydroxy-ethyl ester, phosphate - 90-95%, phosphoric acid - 3-5%, 2-propenoic acid, 2-methyl-, 2-hydroxyethyl ester - 1-5%
Functional organic group		methacrylate
PHYSICAL PROPERTIES		
State	-	liquid
Odor	-	mild acrylic
Color	-	clear, colorless to pale yellow
Boiling point	°C	155-760
Density	kg/m^3	1240-1280
Acid number	mg KOH/g	171-187
pH	-	1
HEALTH & SAFETY		
Flash point	°C	123.5
Animal testing, acute toxicity, Rat oral LD50	mg/kg	>2000
USE & PERFORMANCE		
Manufacturer		Solvay
Outstanding properties		incorporated into acrylic or styrene-acrylic systems improves adhesion to metallic substrates. It also enhances properties such as polymer stability and gloss. This monomer has been used mainly in special purpose coatings, including industrial maintenance, OEM, as well as high-performance structural adhesives.
Recommended for products		structural adhesives, industrial maintenance coatings
Food contact approval		175.105

Sipomer PAM 5000

PARAMETER	UNIT	VALUE
GENERAL INFORMATION		
Name	Sipomer PAM 5000	
CAS #	-	60497-09-08
Chemical class	phosphoric acid ester	
Functional organic group	allyl ether	
PHYSICAL PROPERTIES		
State	-	liquid
HEALTH & SAFETY		
DOT class	8 III	
USE & PERFORMANCE		
Manufacturer	Solvay	
Outstanding features	efficient specialty monomers added during emulsion polymerization to improve performance in adhesives and architectural and industrial coatings applications. Excellent adhesion to a variety of metals, glass, concrete and inorganic substrates	
Recommended for polymers	acrylics, styrenic, VEOVA/acrylic, vinyl VEOVA, polyurethane	
Recommended for products	adhesives, architectural coatings	
Concentrations used	wt%	1-4

3.23 Polymers and copolymers
Bomar BR-3741AJ

PARAMETER	UNIT	VALUE
GENERAL INFORMATION		
Name		Bomar BR-3741AJ
Common synonym		1.3 functional aliphatic polyether urethane acrylate
Chemical class		polymer
PHYSICAL PROPERTIES		
Color, Platinum-cobalt scale	-	15
Density at 25°C	kg/m³	1,010
Refractive index at 25°C	-	1.460
Viscosity at 60°C	mPas	23,000
USE & PERFORMANCE		
Manufacturer		Dymax Corporation
Outstanding properties		optical clarity, non-yellowing, enhances flexibility
Recommended for polymers		ABS, HDPE, PC, PET, PMMA, PP, PVC
Recommended for products		pressure-sensitive adhesives, reactive tackifier
Recommended applications		aluminum, cold rolled steel, glass, stainless steel

Cecabase RT 2N1

PARAMETER	UNIT	VALUE
GENERAL INFORMATION		
Name		Cecabase RT 2N1
Chemical class		(co)polymer
PHYSICAL PROPERTIES		
State	-	liquid
Density at 25°C	kg/m^3	995
Viscosity	cP	800
HEALTH & SAFETY		
Exposure, personal protection		Safety glasses, protective clothing based on chemical resistance data, chemical-resistant gloves, general and local exhaust ventilation.
USE & PERFORMANCE		
Manufacturer		Arkema
Outstanding properties		good moisture resistance even with difficult aggregates, improved TSR results (mechanical performances), removal of standard antistrip.
Recommended applications		adhesion promoter, additives for warm mixed asphalts. Very good workability of the asphalt mix.
Concentrations used		3-5 kg/t of bitumen
Guidelines for use		can be produced and paved at temperatures 40°C cooler than standard hot mixes

Cecabase RT 945

PARAMETER	UNIT	VALUE
GENERAL INFORMATION		
Name		Cecabase RT 945
CAS #	-	68991-84-4
Composition		contains fatty acids tetraethylenepentamine polyamides
Chemical class		(co)polymer
PHYSICAL PROPERTIES		
State	-	liquid
Melting point	°C	-10
Density at 25°C	kg/m^3	995
Viscosity at 25°C	mPas	600
HEALTH & SAFETY		
ICAO/IATA class		Amines, liquid, corrosive, n.o.s.,(contains fatty acids/ tetraethyl-enepentamine polyamides) 8, II
IMDG class		AMINES, LIQUID, CORROSIVE, N.O.S, (contains fatty acids/ tetraethylene-pentamine polyamides) 8, II
Flash point	°C	178
Animal testing, acute toxicity, Rat dermal LD50	mg/kg	>6800
Effect of exposure, eye (human)		Causes serious eye damage.
Effect of exposure, inhalation (human)		May cause allergy or asthma symptoms or breathing difficulties if inhaled.
Effect of exposure, skin (human)		Causes severe skin burns. May cause an allergic skin reaction. No skin sensitizer.
Effect of exposure, swallowing (human)		Harmful if swallowed.
Exposure, personal protection		Safety glasses, protective clothing based on chemical resistance data, chemical-resistant gloves, general and local exhaust ventilation.
First aid, eye		Immediately flush eyes with plenty of water, occasionally lifting the upper and lower eyelids. Check for and remove any contact lenses. Rinse opened eye for several minutes under running water. Obtain medical attention if irritation develops.
First aid, inhalation		Move person into the fresh air and keep him warm, let him rest; if there is difficulty in breathing, medical advice is required.

Cecabase RT 945

PARAMETER	UNIT	VALUE
First aid, skin		Immediately flush skin with plenty of water for at least 15 minutes while removing contaminated clothing. If symptoms persist, call a physician.
UN risk phrases, R	-	R38,R41,R34
USE & PERFORMANCE		
Manufacturer		Arkema
Outstanding properties		decreased application temperature, increased mix workability. Good for all kinds of traffic, all kinds of weather, many different mix type (dense grade, SMA, OGFC and more) and all kind of weather.
Recommended applications		additives for warm mixed asphalts. High workability of the asphalt mix, excellent antistripping effect, flexible to all techniques and asphalt mixes, effective at very low dosage.
Concentrations used		3-5 kg/t of bitumen
Guidelines for use		can be produced and paved at temperatures 40°C cooler than standard hot mixes

Cecabase RT BIO 10

PARAMETER	UNIT	VALUE
GENERAL INFORMATION		
Name		Cecabase RT BIO 10
CAS #		39464-692-2, 9004-98-2, 7664-38-2
Composition		>50% fatty alcohol derivatives, <20% poly(oxy-1,2-ethanediyl), .alpha.-(9Z)-9-octadecenyl.omega.-hydroxy- (<2.5 OE), <5% orthophosphoric acid
Chemical class		(co)polymer
PHYSICAL PROPERTIES		
State	-	liquid
Color	-	yellow to brown
Boiling point	°C	>205 (Initial boiling point)
Melting point	°C	-10
Density at 25°C	kg/m³	1000
HEALTH & SAFETY		
Carcinogenicity		IARC, OSHA, NTP: no ingredient of this product present at levels greater than or equal to 0.1% is identified as probable, possible or confirmed human carcinogen
Mutagenicity		no evidence
ICAO/IATA class		Corrosive liquid, n.o.s. 8, III, Marine Pollutant
Flash point	°C	174
Effect of exposure, eye (human)		Causes serious eye damage.
Effect of exposure, inhalation (human)		Not respiratory sensitizer.
Effect of exposure, skin (human)		Causes severe skin burns. May cause an allergic skin reaction. No skin sensitizer.
Effect of exposure, swallowing (human)		Harmful if swallowed.
Exposure, personal protection		Safety glasses, protective clothing based on chemical resistance data, chemical-resistant gloves, general and local exhaust ventilation.
First aid, eye		Immediately flush eyes with plenty of water, occasionally lifting the upper and lower eyelids. Check for and remove any contact lenses. Rinse opened eye for several minutes under running water. Obtain medical attention if irritation develops.

Cecabase RT BIO 10

PARAMETER	UNIT	VALUE
First aid, inhalation	Move person into the fresh air and keep him warm, let him rest; if there is difficulty in breathing, medical advice is required.	
First aid, skin	Immediately flush skin with plenty of water for at least 15 minutes while removing contaminated clothing. If symptoms persist, call a physician.	
ACGIH, TLV	mg/m^3	1/orthophosphoric acid, STEL3/ortho-phosphoric acid
UN risk phrases, R	-	R38,R41,R34
UN/NA class	-	1760
USE & PERFORMANCE		
Manufacturer	Arkema	
Outstanding properties	odor-free, biodegradable, and mostly biosourced, not dangerous to the environment and human beings, long storage stability in asphalt	
Recommended applications	additives for warm mixed asphalts. High workability of the asphalt mix.	
Concentrations used	3-5 kg/t of bitumen	
Guidelines for use	can be produced and paved at temperatures 40°C cooler than standard hot mixes	

Cleartack® W130

PARAMETER	UNIT	VALUE
GENERAL INFORMATION		
Name		Cleartack® W130
General description		hydrocarbon resin
Composition		pure styrenic monomer resin produced by cationic polymerization
PHYSICAL PROPERTIES		
State	-	pellets
Odor	-	low
Color	-	water white
Color, Platinum-cobalt scale	-	<40
Softening point	°C	130
Acid number	mg KOH/g	<0.1
Density at 20°C	kg/m³	1060
USE & PERFORMANCE		
Manufacturer		Cray Valley
Outstanding properties		tackifying properties, thermal stability, compatibility with many polymers
Recommended for products		hotmelts, pressure-sensitive adhesives, coatings & paints, polymer modification
Food approval (FDA)		Cleartack® W130 may be permitted for use by one or more FDA regulations governing substances used in food-contact articles

Escorex 2173

PARAMETER	UNIT	VALUE
GENERAL INFORMATION		
Name		Escorex 2173
Common synonym		aromatic-modified aliphatic hydrocarbon resin
Chemical class		polymer
PHYSICAL PROPERTIES		
State	-	solid
Color	-	yellow
Cloud point	°C	-14
Softening point	°C	89.7
Viscosity at 160°C	mPas	450
USE & PERFORMANCE		
Manufacturer		ExxonMobil
Recommended for polymers		SB(S) elastomers, SBR, natural rubber EVA
Recommended for products		hot melt adhesives

Poly DNB

PARAMETER	UNIT	VALUE
GENERAL INFORMATION		
Name		Poly DNB
Composition		poly-p-dinitrosobenzene in wax
Acronym	-	DNB
Chemical class		benzene derivatives
Active matter	wt%	21-26
Purity	wt%	97
Solids content	wt%	98
PHYSICAL PROPERTIES		
State	-	waxy chips
Color	-	brown
Boiling point	°C	259.7
Ash content	wt%	1
Density at 25°C	kg/m^3	960
HEALTH & SAFETY		
Flash point	°C	103.5
USE & PERFORMANCE		
Manufacturer		Lord Corporation
Outstanding properties		broad temperature range, durable
Recommended for polymers		butyl, EPDM
Recommended for products		inner tubes, curing bladders, cements, molded goods, mechanical goods, wire/cable covers, electrical blankets, membranes, hose, tank liners, gaskets, tapes and load bearing pads

Polytex E-100

PARAMETER	UNIT	VALUE
GENERAL INFORMATION		
Name		Polytex E-100
Common synonym		tosylamide-epoxy resin
PHYSICAL PROPERTIES		
State	-	solid
Odor	-	very mild, esteric
Color	-	clear
Boiling point	°C	>200
Softening point	°C	70-76
Decomposition temperature	°C	>250
Density at 25°C	kg/m³	1270
Solubility (diluents)		ketones, esters, and a variety of acrylic monomers and diluents for UV applications
HEALTH & SAFETY		
Carcinogenicity		not listed as a carcinogen by ACGIH, IARC, NTP, or CA Prop 65
Flash point	°C	>93.3
First aid, eye		Mildly irritating to the eyes. Wash with plenty of water for 15 minutes. Consult a physician.
First aid, inhalation		Not applicable, unless heated past the decomposition point or dust is allowed to accumulate in the air while handling. If so, remove to fresh air.
First aid, skin		Mildly irritating to the skin. Wash thoroughly with soap and water. If skin irritation persists consult a doctor.
USE & PERFORMANCE		
Manufacturer		Estron Chemical
Outstanding properties		Polytex E-100 enhances gloss, DOI (Distinctness of Image), adhesion to numerous substrates, compatibility, and overall coating durability.
Recommended for products		lacquers, inks, adhesives, and UV curable coatings
Concentrations used	wt%	5-15; up to 30 in adhesives
Guidelines for use		bond strengths are maintained at temperatures of up to 60°C (140°F), dependent upon the specific formulation.

Sulfonex M-80

PARAMETER	UNIT	VALUE
GENERAL INFORMATION		
Name	-	Sulfonex M-80
CAS #	-	1338-51-8
Composition		78-82% tosylamide/formaldehyde resin, 18-22 n-butyl acetate, <600 ppm free formaldehyde
Common synonym		tosylamide/formaldehyde resin
PHYSICAL PROPERTIES		
State	-	liquid
Odor	-	strong esteric
Color	-	clear
Boiling point	°C	124
Decomposition temperature	°C	>250
Evaporation rate (butyl acetate=1)	-	1
Vapor density	-	4
Volatility	%	18-22
HEALTH & SAFETY		
Autoignition temperature	°C	407
Flash point	°C	26
Flash point method	-	TCC
Explosive LEL	wt%	1.7
Explosive UEL	wt%	7.6
First aid, eye		Causes serious eye irritation. Rinse cautiously with water for several minutes. Remove contact lenses if present and easy to do – continue rinsing. Wash with plenty of water for 15 minutes. If eye irritation persists, get medical advice/attention.
First aid, inhalation		Harmful if inhaled. May cause drowsiness or dizziness. If inhaled, Remove victim to fresh air and keep at rest in a position comfortable for breathing. Call a poison center or physician if you feel unwell. If breathing is labored or with coughing, give 100% supplemental oxygen. If not breathing begin artificial respiration and get medical aid.

Sulfonex M-80

PARAMETER	UNIT	VALUE
First aid, skin		Causes skin irritation. Remove/take off immediately all contaminated clothing and wash before re-use. Wash with plenty of soap and water. Wash thoroughly with soap and water. If skin irritation persists, consult a doctor
UN/NA #	-	1866
ECOLOGICAL PROPERTIES		
Partition coefficient, log P_{ow}	-	1.82
USE & PERFORMANCE		
Manufacturer	Estron Chemical	
Outstanding properties		promotes gloss, adhesion, and overall durability. Sulfonex M-80 has excellent compatibility with most commercial polymeric binders, pigments, dyes, and plasticizers. Solutions of Sulfonex M-80 have a low viscosity and offer the advantage of formulating high solids lacquers. Sulfonex M-80 improves the compatibility of resins with plasticizers and pigments, which may otherwise remain incompatible. In adhesives formulations, Sulfonex M-80 helps in bonding many substrates such as cellophane and aluminum foils to themselves or to paper.
Recommended for products		lacquers, inks, adhesives, nail enamel
Guidelines for use		Bond strengths are maintained at temperatures of up to 135°F (57°C).

3.24 Polyols
Hypomer FX-2460A

PARAMETER	UNIT	VALUE
GENERAL INFORMATION		
Name		Hypomer FX-2460AF
Composition		acrylic polyol in solvent xylene/n-butyl acetate
Active matter	wt%	59-62
PHYSICAL PROPERTIES		
State	-	liquid, viscous
Color	-	clear
Color, Platinum-cobalt scale	-	<50
Glass transition temperature	°C	48.9
Acid number	mg KOH/g	11.5-14
Density at 25°C	kg/m^3	1,030
Melt flow rate at 190°C/2.16 kg	g/10 min	1.03
Solubility (diluents)		toluene, xylene, ethyl acetate, n-butyl acetate
HEALTH & SAFETY		
Exposure, personal protection		Safety glasses, protective clothing based on chemical resistance data, chemical-resistant gloves, general and local exhaust ventilation.
USE & PERFORMANCE		
Manufacturer		Elementis
Outstanding properties		provides hardness and solvent resistance. Possesses high film build.
Recommended for products		coatings
Recommended applications		used in metal coating, wood finish, automobile, motorcycle base coat and pigment coatings, transportations and industrial applications.

Hypomer FX-2860A

PARAMETER	UNIT	VALUE
GENERAL INFORMATION		
Name	Hypomer FX-2860A	
Composition	acrylic polyol resin in solvent xylene/n-butyl acetate	
Chemical class	polyol	
Active matter	wt%	58-62
PHYSICAL PROPERTIES		
State	-	liquid
Odor	-	acidic
Color	-	clear
Color, Platinum-cobalt scale	-	<50
Glass transition temperature	°C	57
Acid number	mg KOH/g	10-15
Density at 25°C	kg/m³	1,020
Melt flow rate at 190°C/2.16 kg	g/10 min	1.02
Solubility (diluents)	toluene, xylene, ethyl acetate, n-butyl acetate, MEK, acetate, propylene glycol monomethyl ether	
Viscosity at 25°C	mPas	2,500-4,000
HEALTH & SAFETY		
Exposure, personal protection	Safety glasses, protective clothing based on chemical resistance data, chemical-resistant gloves, general and local exhaust ventilation.	
USE & PERFORMANCE		
Manufacturer	Elementies/Deuchem	
Outstanding properties	fast curing speed hardness. Offers good gloss and levelling. Provides good pigment dispersability.	
Recommended for products	coatings	
Recommended applications	used in plastic coatings, automobile refinish, transportation, and industrial application.	

Hypomer FX-4365

PARAMETER	UNIT	VALUE
GENERAL INFORMATION		
Name		Hypomer FX-4365
Composition		acrylic polyol resin in solvent xylene/n-butyl acetate
Chemical class		polyol
Active matter	wt%	62.5-65.5
PHYSICAL PROPERTIES		
State	-	liquid, viscous
Odor	-	acidic
Color	-	clear
Color, Platinum-cobalt scale	-	<80
Glass transition temperature	°C	25
Acid number	mg KOH/g	8-15
Density at 25°C	kg/m^3	1,030
Melt flow rate at 190°C/2.16 kg	g/10 min	1.03
Solubility (diluents)		xylene, n-butyl acetate, ethylene glycol monoethyl ether acetate
Viscosity at 25°C	mPas	2,600-4,300
HEALTH & SAFETY		
Exposure, personal protection		Safety glasses, protective clothing based on chemical resistance data, chemical-resistant gloves, general and local exhaust ventilation.
USE & PERFORMANCE		
Manufacturer		Elementis
Outstanding properties		provides hardness and solvent resistance. Possesses high film build.
Recommended for products		coatings, automotive refinish
Recommended applications		used in automobile refinish, transportation, and industrial application

Polypol 610

PARAMETER	UNIT	VALUE
GENERAL INFORMATION		
Name		Polypol 610
Composition		acrylic polyol resin in xylene
Chemical class		polyol
Active matter	wt%	58-62
PHYSICAL PROPERTIES		
State	-	liquid, viscous
Acid number	mg KOH/g	2-6
OH content	wt%	1.3-1.9
Viscosity at 25°C	mPas	3,000-5,000
HEALTH & SAFETY		
Exposure, personal protection		Safety glasses, protective clothing based on chemical resistance data, chemical-resistant gloves, general and local exhaust ventilation.
USE & PERFORMANCE		
Manufacturer		Polychem Resins
Outstanding properties		excellent durability, gloss, drying, adhesion, and chemical & stain resistance
Recommended for products		industrial paints & coatings
Recommended applications		used in polyurethane coatings for metals and plastics.

Polypol 615

PARAMETER	UNIT	VALUE
GENERAL INFORMATION		
Name		Polypol 615
Composition		acrylic polyol resin in butyl acetate and xylene
Chemical class		polyol
Active matter	wt%	48-52
PHYSICAL PROPERTIES		
State	-	liquid, viscous
Acid number	mg KOH/g	2-6
OH content	wt%	0.8-1.2
Viscosity at 25°C	mPas	3,500-5,500
HEALTH & SAFETY		
Exposure, personal protection		Safety glasses, protective clothing based on chemical resistance data, chemical-resistant gloves, general and local exhaust ventilation.
USE & PERFORMANCE		
Manufacturer		Polychem Resins
Outstanding properties		excellent hardness, good weather, chemical and solvent resistance, and quick drying.
Recommended for products		coatings (aerospace, automotive, rail, vehicle refinish) and many other coatings
Recommended applications		used in two-component polyurethane coatings for metals and plastics.

Polypol 653

PARAMETER	UNIT	VALUE
GENERAL INFORMATION		
Name		Polypol 653
Composition		acrylic polyol resin in butyl acetate and xylene
Chemical class		polyol
Active matter	wt%	64-66
PHYSICAL PROPERTIES		
State	-	liquid, viscous
Acid number	mg KOH/g	5-10
OH content	wt%	2.9
Viscosity at 25°C	mPas	2,500-3,500
HEALTH & SAFETY		
Exposure, personal protection		Safety glasses, protective clothing based on chemical resistance data, chemical-resistant gloves, general and local exhaust ventilation.
USE & PERFORMANCE		
Manufacturer		Polychem Resins
Outstanding properties		excellent hardness and flexibility, excellent gloss & gloss retention, high chalking resistance and light stability, excellent weather and chemical resistance
Recommended for products		paints & coatings (2 component)
Recommended applications		used in two-component polyurethane coatings, vehicle repair paints

Polypol 663

PARAMETER	UNIT	VALUE
GENERAL INFORMATION		
Name		Polypol 663
Composition		acrylic polyol resin in xylene
Chemical class		polyol
Active matter	wt%	62-64
PHYSICAL PROPERTIES		
State	-	liquid, viscous
Acid number	mg KOH/g	<10
OH content	wt%	2.2-2.6
Viscosity at 25°C	mPas	7,000-9,000
HEALTH & SAFETY		
Exposure, personal protection		Safety glasses, protective clothing based on chemical resistance data, chemical-resistant gloves, general and local exhaust ventilation.
USE & PERFORMANCE		
Manufacturer		Polychem Resins
Outstanding properties		excellent adhesion and levelling properties and chemical and stain resistance.
Recommended for products		paints & coatings
Recommended applications		used in polyurethane coatings for metals and plastics, excellent adhesion and levelling properties.

Polypol 676

PARAMETER	UNIT	VALUE
GENERAL INFORMATION		
Name	Polypol 676	
Composition	acrylic polyol resin in butyl acetate and xylene	
Chemical class	polyol	
Active matter	wt%	59-61
PHYSICAL PROPERTIES		
State	-	liquid, viscous
Acid number	mg KOH/g	6.5
OH content	wt%	2.4
Viscosity at 25°C	mPas	1,700-2,300
HEALTH & SAFETY		
Exposure, personal protection	Safety glasses, protective clothing based on chemical resistance data, chemical-resistant gloves, general and local exhaust ventilation.	
USE & PERFORMANCE		
Manufacturer	Polychem Resins	
Outstanding properties	good light stability, chalking resistance and good gloss retention.	
Recommended for products	coatings	
Recommended applications	used in polyurethane coatings, automotive coatings	

Polypol 693

PARAMETER	UNIT	VALUE
GENERAL INFORMATION		
Name		Polypol 693
Composition		acrylic polyol resin in xylene
Chemical class		polyol
Active matter	wt%	63-67
PHYSICAL PROPERTIES		
State	-	liquid, viscous
Acid number	mg KOH/g	<12
OH content	wt%	4.2-4.8
Viscosity at 25°C	mPas	3,500-5,000
HEALTH & SAFETY		
Exposure, personal protection		Safety glasses, protective clothing based on chemical resistance data, chemical-resistant gloves, general and local exhaust ventilation.
USE & PERFORMANCE		
Manufacturer		Polychem Resins
Outstanding properties		good light stability, chalking resistance, and gloss retention.
Recommended for products		coatings, automotive refinishing
Recommended applications		used in polyurethane coatings

Priplast 1837

PARAMETER	UNIT	VALUE
GENERAL INFORMATION		
Name	Priplast 1837	
Common synonym	polyester polyol	
Biobased	92% of renewable carbon content	
Molecular mass, M_w	daltons	1000
Chemical class	polyol	
PHYSICAL PROPERTIES		
State	-	solid, wax-like
Odor	-	mild
Color	-	light yellow
Color, Gardner scale	-	6
Boiling point	°C	>200
Melting point	°C	42
Acid number	mg KOH/g	1
Hydroxyl number	mg KOH/g	110
Solubility in water at 25°C	g/l	insoluble
Vapor pressure at 20°C	kPa	<0.01
Viscosity at 25°C	mPas	340
HEALTH & SAFETY		
NFPA classification	Flammability	1
	Health	1
	Reactivity	0
HMIS classification	Flammability	1
	Health	1
	Reactivity	0
DOT class	not regulated	
TDG class	not regulated	
ICAO/IATA class	not regulated	
IMDG class	not regulated	
Flash point	°C	>200
Flash point method	-	OC
Effect of exposure, eye (human)	May irritate eyes	
Effect of exposure, inhalation (human)	May cause irritation of respiratory tract	
Effect of exposure, skin (human)	May irritate skin	
Effect of exposure, swallowing (human)	Ingestion may cause irritation to mucous membranes	

Priplast 1837

PARAMETER	UNIT VALUE
First aid, eye	In case of contact, immediately flush eyes with plenty of water for at least 15 minutes.
First aid, inhalation	If breathed in, move person into fresh air. If symptoms persist, call a physician.
First aid, skin	Take off contaminated clothing and shoes immediately. Wash off with soap and plenty of water.
USE & PERFORMANCE	
Manufacturer	Croda
Outstanding properties	excellent wetting of rigid and fibrous substrates providing hardness, flexibility, hydrophobicity, thermooxidative stability, and good adhesion
Recommended for polymers	PU
Recommended for products	adhesives & sealants, protective coatings
Food contact approval	EU 10/2011, FDA 175.105 and 175.300

Terrin 168

PARAMETER	UNIT	VALUE
GENERAL INFORMATION		
Name		Terrin 168
Composition		100% aliphatic polyester polyol containing a minimum of 50% recycled content
Chemical class		polyol
Functionality, average	-	1.8
Hydroxyl type	-	mainly primary
Moisture content	wt%	0.1
PHYSICAL PROPERTIES		
State	-	liquid
Color	-	brown
Glass transition temperature	°C	-75
Acid number	mg KOH/g	<1.5
Density at 25°C	kg/m^3	1,100
Equivalent weight	g	312-351
Hydroxyl number	mg KOH/g	160-180
Viscosity at 23°C	mPas	350
HEALTH & SAFETY		
Exposure, personal protection		Safety glasses, protective clothing based on chemical resistance data, chemical-resistant gloves, general and local exhaust ventilation.
USE & PERFORMANCE		
Manufacturer		Invista
Outstanding properties		cost-competitive in comparison to conventional polyols. Have similar hydroxyl values to castor oil and can be substituted on a nearly equal weight basis
Recommended applications		used in lieu of or in combination with polyether or polyester polyols to formulate a variety of polyurethane products. The resulting polyurethanes can be designed to be soft and flexible or hard and stiff.

Terrin 168G

PARAMETER	UNIT	VALUE
GENERAL INFORMATION		
Name		Terrin 168G
Composition		100% aliphatic polyester polyol containing a minimum of 60% recycled content
Chemical class		polyol
Functionality, average	-	2
Hydroxyl type	-	primary/secondary
Moisture content	wt%	0.1
PHYSICAL PROPERTIES		
State	-	liquid
Color	-	brown
Glass transition temperature	°C	-70
Acid number	mg KOH/g	<1.5
Density at 25°C	kg/m^3	1,100
Hydroxyl number	mg KOH/g	1312-35160-180
Viscosity at 23°C	mPas	830
HEALTH & SAFETY		
Exposure, personal protection		Safety glasses, protective clothing based on chemical resistance data, chemical-resistant gloves, general and local exhaust ventilation.
First aid, inhalation		Move person to fresh air. If symptoms persist, call a physician.
USE & PERFORMANCE		
Manufacturer		Invista
Outstanding properties		cost-competitive in comparison to conventional polyols. Have similar hydroxyl values to castor oil and can be substituted on a nearly equal weight basis
Recommended applications		used in lieu of or in combination with polyether or polyester polyols to formulate a variety of polyurethane products. The resulting polyurethanes can be designed to be soft and flexible or hard and stiff.

Terrin 170

PARAMETER	UNIT	VALUE
GENERAL INFORMATION		
Name		Terrin 170
Composition		100% aliphatic polyester polyol made from raw materials that are all either recycled or renewable
Chemical class		polyol
Functionality, average	-	2.2
Hydroxyl type	-	primary/secondary
Moisture content	wt%	0.1
PHYSICAL PROPERTIES		
State	-	liquid
Color	-	brown
Glass transition temperature	°C	-60
Acid number	mg KOH/g	<1.5
Density at 25°C	kg/m³	1,100
Equivalent weight	g	312-351
Hydroxyl number	mg KOH/g	160-180
Viscosity at 23°C	mPas	5,500
HEALTH & SAFETY		
Exposure, personal protection		Safety glasses, protective clothing based on chemical resistance data, chemical-resistant gloves, general and local exhaust ventilation.
First aid, inhalation		Move person to fresh air. If symptoms persist, call a physician.
USE & PERFORMANCE		
Manufacturer		Invista
Outstanding properties		cost-competitive in comparison to conventional polyols. Have similar hydroxyl values to castor oil and can be substituted on a nearly equal weight basis
Recommended applications		used in lieu of or in combination with polyether or polyester polyols to formulate a variety of polyurethane products.

3.25 Resorcinol
Cofill 11 GR

PARAMETER	UNIT	VALUE
GENERAL INFORMATION		
Name		Cofill 11 GR
CAS #		108-46-3/112926-00-8
Composition		resorcinol and silica in the ratio of 1:1. Resorcinol content is 50%
Formula		
Chemical class	-	resorcinol
PHYSICAL PROPERTIES		
State	-	powder
Ash content	wt%	45
Density at 20°C	kg/m³	1650
pH	-	6.5
USE & PERFORMANCE		
Manufacturer		Evonik
Recommenced for polymers		rubber
Recommended for products		steel-belted radial tires, all-steel truck tires, conveyor belts, transmission belts, hoses, rubberized fabrics
Recommended applications		combination with a formaldehyde donor (Hexa K) and fine particle size silica whenever high static and dynamic bonding strength of rubber compounds to textiles or steel cord is necessary

Cohedur RS

PARAMETER	UNIT	VALUE
GENERAL INFORMATION		
Name		Cohedur RS
Composition		homogeneous solidified melt of resorcinol and stearic acid in the ratio 2:1
Empirical formula		CH3(CH2)16COOH
Formula		
Chemical class	-	resorcinol
Mixture	-	yes
Resorcinol content	%	67
PHYSICAL PROPERTIES		
State		solid/lentil-shaped granules
Color		beige to slightly reddish brown
Melting point	°C	109
Density at 25°C	kg/m³	1,200
USE & PERFORMANCE		
Manufacturer		RheinChemie Additives/Lanxess
Outstanding properties		Cohedur RS is a component of the direct bonding system, also known as RFS system. RFS bonding systems are multi-component systems. They are created by providing the rubber component with a resorcinol component, a methylene component, and reinforcing silica, e.g., Vulkasil® S.
Recommended for polymers		natural rubber, BR, CR, EPDM, NBR, SBR
Recommended applications		used as a direct bonding agent for rubber to fabric and rubber to steel cord bonding.
Concentrations used	phr	3.4

Cohedur VP KA 9197

PARAMETER	UNIT	VALUE
GENERAL INFORMATION		
Name		Cohedur VP KA 9197
Composition		homogeneous solidified melt of resorcinol and stearic acid in the ratio 2:1
Chemical class	-	resorcinol
Mixture	-	yes
PHYSICAL PROPERTIES		
State	-	liquid
Color		light yellow to brown
Density at 20°C	kg/m^3	1,180
Refractive index at 20°C	-	1.505
USE & PERFORMANCE		
Manufacturer		RheinChemie Additives/Lanxess
Outstanding properties		Cohedur® VP KA 9197, in conjunction with Cohedur® A grades and reinforcing silica, e.g., Vulkasil S, gives vulcanizates based on chloroprene rubber (e. g., Baypren®) good adhesion to all the normal reinforcing materials (rayon, polyamide, polyester with the special spin finish, glass fibers, and bare, galvanized and brass-coated steel cord) without their first having to be treated with a bonding agent. The bonds are highly resistant to dynamic and thermal stresses.
Recommended for polymers		chloroprene rubber, NR, SBR, BR
Recommended applications		used as a direct bonding agent for rubber to fabric and rubber to steel cord bonding.
Concentrations used	phr	3.5-4.5
Guidelines for use		The bonding is further improved to a small extent by preliminary dipping of the

Markoba RSC

PARAMETER	UNIT	VALUE
GENERAL INFORMATION		
Name	Markoba RSC	
CAS #	-	108-46-3
EC number	-	203-585-2
Composition	m-dihydrobenzene	
Common synonym	resorcinol	
Empirical formula	C6H4(OH)2	
Formula		
Molecular mass	daltons	110.1
RTECS number	-	VG9625000
Chemical class	resorcinol	
PHYSICAL PROPERTIES		
State	solid/needle shaped crystals	
Color	-	white
Melting point	°C	108-112
Density at 25°C	kg/m³	1280
HEALTH & SAFETY		
Flash point	°C	127
Animal testing, acute toxicity, Rat oral LD50	mg/kg	331
Animal testing, acute toxicity, Rabbit dermal LD50	mg/kg	>2000
Effect of exposure, eye (human)	Causes serious eye irritation.	
Effect of exposure, skin (human)	Causes skin irritation.	
Effect of exposure, swallowing (human)	Harmful if swallowed.	
Exposure, personal protection	Safety glasses, protective clothing based on chemical resistance data, chemical-resistant gloves, general and local exhaust ventilation.	
First aid, eye	Immediately flush eyes with plenty of water for at least 15 minutes. If eye irritation persists, consult a specialist.	
First aid, inhalation	Move person to fresh air. If symptoms persist, call a physician.	
First aid, skin	Immediately flush skin with plenty of water for at least 15 minutes while removing contaminated clothing. If symptoms persist, call a physician.	

Markoba RSC

PARAMETER	UNIT	VALUE
UN risk phrases, R	-	R22,R36/38,R50
US safety phrases, S	-	S26,S61
UN/NA class	-	2876
USE & PERFORMANCE		
Manufacturer	Wholemark Fine Chemical	
Recommended for polymers	rubber	
Recommended applications	adhesion promoter	

3.26 Rosin
Eastman ester gum 8D resin

PARAMETER	UNIT	VALUE
GENERAL INFORMATION		
Name		Eastman ester gum 8D resin
CAS #	-	8050-31-5
EC number	-	232-482-5
Composition	100% glycerol ester of rosin	
Chemical class	-	rosin
Mixture	-	yes
PHYSICAL PROPERTIES		
State	-	solid
Color	-	amber
Color, Gardner scale	7 (50 % resin solids in toluene)	
Odor	-	low odor
Freezing point	°C	93-94/ softening point
Ring and ball softening point	°C	89
Softening point (Hercules drop method)	°C	95
Acid number	mg KOH/g	7
Density at 20°C	kg/m³	1,080
Solubility (diluents)	aromatic and aliphatic hydrocarbons, esters, ketones and carbon tetrachloride, insoluble in methanol and ethanol.	
HEALTH & SAFETY		
NFPA classification	Flammability	1
	Health	1
	Reactivity	0
HMIS classification	Flammability	1
	Health	1
	Reactivity	0
Carcinogenicity	no data available	
Mutagenicity	no data available	
Teratogenicity	no data available	
DOT class	not regulated	
TDG class	not regulated	
ICAO/IATA class	not regulated	
IMDG class	not regulated	
Flash point	°C	>190
Flash point method	-	OC

3.26 Rosin
Eastman ester gum 8D resin

PARAMETER	UNIT	VALUE
Hazardous combustion products	Carbon monoxide, carbon dioxide, aldehydes	
Agency rating, listed	TSCA USA, DSL Canada, AICS Australia, MITI Japan, ECL Korea, IECSC China	
Animal testing, acute toxicity, Rat oral LD50	mg/kg	>2000/glycerol ester of rosin
Animal testing, acute toxicity, Rabbit dermal LD50	mg/kg	slight irritant
Animal testing, acute toxicity, Rat dermal LD50	mg/kg	>5000/glycerol ester of rosin
Effect of exposure, eye (human)	Avoid contact with dust. May cause eye irritation by mechanical abrasion.	
Effect of exposure, inhalation (human)	May cause respiratory irritation by mechanical abrasion.	
Effect of exposure, skin (human)	No irritant effect. Avoid contact with dust. May cause skin irritation by mechanical abrasion.	
Exposure, personal protection	Safety glasses, protective clothing based on chemical resistance data, chemical-resistant gloves, general and local exhaust ventilation.	
First aid, eye	Rinse cautiously with water for several minutes. Remove contact lenses, if present and easy to do. Continue rinsing. If molten material contacts the eye, immediately flush with plenty of water for at least 15 minutes. Get medical attention immediately	
First aid, inhalation	Remove to fresh air. If not breathing, give artificial respiration. If breathing is difficult, give oxygen. If irritation persists, obtain medical advice.	
First aid, skin	Immediately flush with plenty of water for at least 15 minutes while removing contaminated clothing and shoes. Get medical attention.	
ECOLOGICAL PROPERTIES		
Aquatic toxicity *Green algae*, 96-h EC50	mg/l	>1000/72H/glycerol ester of rosin
Aquatic toxicity, *Daphnia magna*, 48-h LC50	mg/l	>100/glycerol ester of rosin
Biodegradation probability	0%/28d/glycerol ester of rosin	

3.26 Rosin
Eastman ester gum 8D resin

PARAMETER	UNIT	VALUE
USE & PERFORMANCE		
Manufacturer		Eastman Chemical Company
Outstanding properties		good adhesion and tack-producing properties, pale color, low odor, low acid number, broad solubility and compatibility ranges.
Recommended for polymers		compatible with natural rubber, synthetic rubbers, ethyl cellulose, PS, many natural and synthetic resins, waxes and most vinyl resins.
Recommended for products		chewing gum, cosmetics (antiperspirants and deodorants), depilatories
Recommended applications		typical application: graphic arts, coatings, cosmetics, as a masticatory ingredient in chewing gum compositions.

Foral 85-E CG hydrogenated rosinate

PARAMETER	UNIT	VALUE
GENERAL INFORMATION		
Name		Foral 85-E CG hydrogenated rosinate
CAS #	-	65997-13-9
EC number	-	266-042-9
Composition		100% glyceryl hydrogenated rosinate
Common synonym		ester of hydrogenated rosin
Chemical class		rosin
PHYSICAL PROPERTIES		
State	-	solid
Odor	-	slight
Color	-	light amber
Color, Gardner scale	-	3
Acid number	mg KOH/g	7
Softening point	°C	74
Solubility in water at 25°C	g/l	0.00015
Specific gravity at 25°C	-	1.06
HEALTH & SAFETY		
HMIS classification	Flammability	1
	Health	1
	Reactivity	0
Carcinogenicity	no data available	
Mutagenicity	negative/glyceryl hydrogenated rosinate	
Teratogenicity	no data available	
DOT class	not regulated	
TDG class	not regulated	
ICAO/IATA class	not regulated	
IMDG class	not regulated	
Autoignition temperature	°C	402
Hazardous combustion products	carbon monoxide, carbon dioxide	
Agency rating, listed	TSCA USA, DSL Canada, AICS Australia, MITI Japan, ECL Korea, IECSC China	
Animal testing, acute toxicity, Rat oral LD50	mg/kg	>2000/glyceryl hydrogenated rosinate
Animal testing, acute toxicity, Rat dermal LD50	mg/kg	>2000/ester of hydrogenated rosin

Foral 85-E CG hydrogenated rosinate

PARAMETER	UNIT	VALUE
Animal testing, acute toxicity, Rat inhalation, LC50	mg/m^3	10000 ppm/90d/ glyceryl hydrogenated rosinate
Effect of exposure, eye (human)		Avoid contact with molten material. Contact with molten substance/product may cause severe eyes damage.
Effect of exposure, inhalation (human)		Avoid contact with dust.
Effect of exposure, skin (human)		Non irritant. Avoid contact with molten material. Contact with molten substance/ product may cause severe burns to skin.
Exposure, personal protection		Safety glasses, protective clothing based on chemical resistance data, chemical-resistant gloves, general and local exhaust ventilation.
First aid, eye		Rinse cautiously with water for several minutes. Remove contact lenses, if present and easy to do. Continue rinsing. If molten material contacts the eye, immediately flush with plenty of water for at least 15 minutes. Get medical attention immediately.
First aid, inhalation		Remove to fresh air. If not breathing, give artificial respiration. If breathing is difficult, give oxygen. If irritation persists, obtain medical advice.
First aid, skin		Immediately flush with plenty of water for at least 15 minutes while removing contaminated clothing and shoes. Get medical attention. Wash contaminated clothing before reuse. Destroy or thoroughly clean contaminated shoes.
ECOLOGICAL PROPERTIES		
Aquatic toxicity, *Green algae*, 96-h EC50	mg/l	>1000/72H/glyceryl hydrogenated rosinate
Biodegradation probability	not readily biodegradable	
Partition coefficient, log K$_{oc}$	-	1.851/glyceryl hydrogenated rosinate
Partition coefficient, log K$_{ow}$	-	4.7-5.8/glyceryl hydrogenated rosinate
USE & PERFORMANCE		
Manufacturer	Eastman Chemical Company	

Foral 85-E CG hydrogenated rosinate

PARAMETER	UNIT	VALUE
Outstanding properties		delivered from a natural, renewable source. Excellent oxidative and color stability, excellent gloss, light color, low odor, wide solubility Derived from a natural and compatibility range.
Recommended for products		cosmetics: depilatory wax, eye make-up, lipstick and gloss, mascara, nails care
Recommended applications		adhesion promoter in cosmetic and depilatory wax formulations.
Food approval (FDA)		FDA approved: 175.105; 175.300; 175.390; 176.200; 176.210; 177.1200; 177.1210(b); 177.1400; 177.2600; 178.3120; 178.3800; 178.3850; 178.3870

Foral 105-E CG hydrogenated rosinate

PARAMETER	UNIT	VALUE
GENERAL INFORMATION		
Name		Foral 105-E CG hydrogenated rosinate
CAS #	-	65997-13-9
EC number	-	264-848-5
Composition		100% pentaerythrityl hydrogenated rosinate
Common synonym		ester of hydrogenated rosin
Chemical class		rosin
PHYSICAL PROPERTIES		
State	-	solid/pellets
Odor	-	slight
Color	-	amber
Color, Gardner scale		6 (50% resin solids in toluene or xylene)
Softening point	°C	98
Acid number	mg KOH/g	14
Solubility in water at 25°C	g/l	0.00022
Specific gravity at 25°C	-	1.06
Vapor pressure at 20°C	kPa	0.10
HEALTH & SAFETY		
HMIS classification	Flammability	1
	Health	1
	Reactivity	0
Carcinogenicity		IARC, OSHA, NTP: no ingredient of this product present at levels greater than or equal to 0.1% is identified as probable, possible or confirmed human carcinogen
Mutagenicity		no evidence
Teratogenicity		no data available
DOT class		not regulated
TDG class		not regulated
ICAO/IATA class		not regulated
IMDG class		not regulated
Autoignition temperature	°C	396
Hazardous combustion products		Carbon monoxide, carbon dioxide
Agency rating, listed		TSCA USA, DSL Canada, AICS Australia, MITI Japan, ECL Korea, IECSC China
Animal testing, acute toxicity, Rat oral LD50	mg/kg	>2000/ester of hydrogenated rosin

Foral 105-E CG hydrogenated rosinate

PARAMETER	UNIT	VALUE
Animal testing, acute toxicity, Rat dermal LD50	mg/kg	>2000/ester of hydrogenated rosin
Animal testing, acute toxicity, Rat inhalation, LC50	mg/m^3	10000 ppm/90d/ glyceryl hydrogenated rosinate
Effect of exposure, eye (human)		Avoid contact with molten material. Contact with molten substance/product may cause severe eyes damage.
Effect of exposure, inhalation (human)		Avoid contact with dust.
Effect of exposure, skin (human)		Non irritant. Avoid contact with molten material. Contact with molten substance/ product may cause severe burns to skin.
Exposure, personal protection		Safety glasses, protective clothing based on chemical resistance data, chemical-resistant gloves, general and local exhaust ventilation.
First aid, eye		Rinse cautiously with water for several minutes. Remove contact lenses, if present and easy to do. Continue rinsing. If molten material contacts the eye, immediately flush with plenty of water for at least 15 minutes. Get medical attention immediately.
First aid, inhalation		Remove to fresh air. If not breathing, give artificial respiration. If breathing is difficult, give oxygen. If irritation persists, obtain medical advice.
First aid, skin		Wash with soap and water. Get medical attention if symptoms occur. If burned by contact with hot material, cool molten material adhering to skin as quickly as possible with water, and see a physician for removal of adhering material and treatment of burn. Get medical attention.
ECOLOGICAL PROPERTIES		
Aquatic toxicity, *Green algae*, 96-h EC50	mg/l	>1000/72H/ester of hydrogenated rosin
Aquatic toxicity, *Daphnia magna*, 48-h LC50	mg/l	>1000/ester of hydrogenated rosin
Partition coefficient, log K$_{oc}$	-	1.867/ester of hydrogenated rosin

Foral 105-E CG hydrogenated rosinate

PARAMETER	UNIT	VALUE
USE & PERFORMANCE		
Manufacturer		Eastman Chemical Company
Outstanding properties		delivered from a natural, renewable source. Excellent oxidative and color stability, excellent gloss, light color, low odor, wide solubility
Recommended for products		cosmetics: eye make-up, lipstick, and mascara
Recommended applications		adhesion promoter in color cosmetic formulations (Foral 105-E CG is a resin derived from the esterification of a highly stabilized gum rosin and pentaerythritol-grade with inherent adhesive properties)
Food approval (FDA)		FDA approved: 175.105; 175.125 (b); 175.300; 175.380; 175.390; 177.1210(b); 177.2600

Foralyn 5020-F CG hydrogenated rosinate

PARAMETER	UNIT	VALUE
GENERAL INFORMATION		
Name		Foralyn 5020-F CG hydrogenated rosinate
CAS #	-	8050-15-5
EC number	-	232-476-2
Composition		100% methyl hydrogenated rosinate
Chemical class		rosin
PHYSICAL PROPERTIES		
State	-	liquid
Odor	-	piney
Color	-	light amber
Color, Gardner scale	-	3 (neat melt)
Acid number	mg KOH/g	6
Density at 20°C	kg/m³	1,030
Refractive index at 20°C	-	1.519
Saponification value	mg KOH/g	160
Solubility in water at 25°C	g/l	negligible
Viscosity at 25°C	mPas	5,400
HEALTH & SAFETY		
HMIS classification	Flammability	1
	Health	1
	Reactivity	0
Carcinogenicity		no data available
Mutagenicity		negative/glyceryl hydrogenated rosinate
Teratogenicity		no data available
DOT class		not regulated
TDG class		not regulated
ICAO/IATA class		not regulated
IMDG class		not regulated
Autoignition temperature	°C	402
Flash point	°C	170
Flash point method	-	SFCC
Hazardous combustion products		Carbon monoxide, carbon dioxide
Agency rating, listed		TSCA USA, DSL Canada, AICS Australia, MITI Japan, ECL Korea, IECSC China

Foralyn 5020-F CG hydrogenated rosinate

PARAMETER	UNIT	VALUE
Animal testing, acute toxicity, Rat oral LD50	mg/kg	>2000/methyl hydrogenated rosinate
Animal testing, acute toxicity, Rat dermal LD50	mg/kg	>2000/ester of hydrogenated rosin
Effect of exposure, eye (human)		Avoid contact with molten material. Contact with molten substance/product may cause severe eyes damage.
Effect of exposure, inhalation (human)		
Effect of exposure, skin (human)		Non irritant. Avoid contact with molten material. Contact with molten substance/product may cause severe burns to skin.
Exposure, personal protection		Safety glasses, protective clothing based on chemical resistance data, chemical-resistant gloves, general and local exhaust ventilation.
First aid, eye		Rinse cautiously with water for several minutes. Remove contact lenses, if present and easy to do. Continue rinsing. If molten material contacts the eye, immediately flush with plenty of water for at least 15 minutes. Get medical attention immediately.
First aid, inhalation		Remove to fresh air. If not breathing, give artificial respiration. If breathing is difficult, give oxygen. If irritation persists, obtain medical advice.
First aid, skin		Immediately flush with plenty of water for at least 15 minutes while removing contaminated clothing and shoes. Get medical attention. Wash contaminated clothing before reuse. Destroy or thoroughly clean contaminated shoes.
ECOLOGICAL PROPERTIES		
Aquatic toxicity, *Green algae*, 96-h EC50	mg/l	>1000/72H/methyl hydrogenated rosinate
Biodegradation probability		54%28d//methyl hydrogenated rosinate
USE & PERFORMANCE		
Manufacturer		Eastman Chemical Company
Outstanding properties		delivered from a natural, renewable source. Good oxidative stability, low odor and color, low vapor pressure, high gloss (high refractive index), wide solubility, and compatibility range

Foralyn 5020-F CG hydrogenated rosinate

PARAMETER	UNIT	VALUE
Recommended for products		cosmetics: depilatory wax, fragrance, lipstick and gloss
Recommended applications		cosmetic grade resin is the methyl ester of hydrogenated gum rosin, has good oxidative stability, and is given a special steam-sparging treatment to assure minimum odor. Foralyn 5020-F CG is particularly useful as a fragrance fixative. It has excellent solubility and compatibility with non-polar and many polar ingredients in cosmetic applications, contributing to both adhesion and gloss.
Food approval (FDA)		FDA approved: 175.105; 175.125 (b); 175.300; 175.380; 175.390; 176.170(a); 176.180; 176.200; 176.210; 177.1200; 177.1210(b); 177.1400; 177.2600; 178.3120; 178.3800; 178.3850; 178.3870

Pexalyn 9085

PARAMETER	UNIT	VALUE
GENERAL INFORMATION		
Name	Pexalyn 9085	
CAS #	-	8050-31-5
EC number	-	232-482-5
General description	biobased	
Composition	resin acids and rosin acids, esters with glycerol, wood pulp-based	
Common synonym	resin acid pentaerythritol esters ester of rosin	
Empirical formula	C3H8O3	
Chemical class	rosin	
PHYSICAL PROPERTIES		
State	-	solid/pastilles
Odor	-	rosin, slight
Color	-	pale yellow
Color, Gardner	-	6
Softening point	°C	82-90
Acid number	mg KOH/g	3-9
Density at 25°C	kg/m³	1,070
Solubility in water at 25°C	g/l	negligible
HEALTH & SAFETY		
NFPA classification	Flammability	1
	Health	0
	Reactivity	0
HMIS classification	Flammability	1
	Health	0
	Reactivity	0
Carcinogenicity	IARC, OSHA, NTP: no ingredient of this product present at levels greater than or equal to 0.1% is identified as probable, possible or confirmed human carcinogen	
DOT class	not regulated	
ICAO/IATA class	not regulated	
IMDG class	not regulated	
Animal testing, acute toxicity, Rat oral LD50	mg/kg	>2000
Animal testing, acute toxicity, Rabbit dermal LD50	mg/kg	>2000
Effect of exposure, eye (human)	No irritant effect. May cause eye irritation by mechanical abrasion.	

Pexalyn 9085

PARAMETER	UNIT	VALUE
Effect of exposure, inhalation (human)		May cause respiratory irritation by mechanical abrasion.
Effect of exposure, skin (human)		No irritant effect. May cause skin by mechanical abrasion.
Effect of exposure, swallowing (human)		No hazard expected.
Effect of repeated or overexposure (human)		Rosin and some rosin derivatives have been reported to cause allergic skin reaction (sensitization) in susceptible individuals after repeated or prolonged contact.
Exposure, personal protection		Safety glasses, protective clothing based on chemical resistance data, chemical-resistant gloves, general and local exhaust ventilation.
First aid, eye		Immediately flush eyes with plenty of water, occasionally lifting the upper and lower eyelids. Check for and remove any contact lenses. Rinse opened eye for several minutes under running water. Obtain medical attention if irritation develops.
First aid, inhalation		Move to fresh air. Get medical attention if nasal, throat or lung irritation develops.
First aid, skin		HOT MOLTEN product: Immediately cool skin burns with water and cold packs for at least 15 minutes. Do NOT put ice directly on the skin. Do NOT attempt to remove solidified resin from the skin as severe tissue damage may result. Get immediate medical attention. SOLID product at ambient temperature: Wash thoroughly with soap and water. Remove contaminated clothing. Get medical attention if irritation develops or persists. Thoroughly wash clothing before reuse.

ECOLOGICAL PROPERTIES

Biodegradation probability		readily biodegradable

USE & PERFORMANCE

Manufacturer		Pinova
Outstanding properties		promotes adhesion. Good color stability, low acid number. Broad solubility and wide compatibility range.

Pexalyn 9085

PARAMETER	UNIT	VALUE
Recommended for products		cosmetics, pressure sensitive and solvent or water-based adhesives, resin modifier for coatings, depilatory formulations, elastomers, and waxes
Recommended applications		used as a tackifier for solvent or water-based adhesive systems, including acrylics, SBR, and neoprene. Tackifier for hot melt and pressure-sensitive adhesives. Resin modifier for coatings, depilatory formulations, elastomers, and waxes.

Pexalyn 9100

PARAMETER	UNIT	VALUE
GENERAL INFORMATION		
Name	Pexalyn 9100	
CAS #	-	8050-26-8
EC number	-	232-479-9
General description	biobased	
Composition	resin acids and rosin acids, esters with pentaerythritol, wood pulp-based	
Empirical formula	C25H34O2	
Molecular mass	daltons	366.54
Chemical class	rosin	
PHYSICAL PROPERTIES		
State	-	solid/pastilles
Odor	-	rosin, slight
Color	-	pale yellow
Color, Gardner scale	5 (50% solids in toluene)	
Melting point	°C	97-106/ softening point
Acid number	mg KOH/g	7-16
Density at 25°C	kg/m³	1,060
Solubility in water at 25°C	g/l	negligible
HEALTH & SAFETY		
NFPA classification	Flammability	1
	Health	0
	Reactivity	0
HMIS classification	Flammability	1
	Health	0
	Reactivity	0
Carcinogenicity	IARC, OSHA, NTP: no ingredient of this product present at levels greater than or equal to 0.1% is identified as probable, possible or confirmed human carcinogen	
DOT class	not regulated	
ICAO/IATA class	not regulated	
IMDG class	not regulated	
Flash point	°C	>200
Hazardous combustion products	Carbon monoxide (CO), Carbon dioxide (CO2), aldehydes, carboxylic acid, smoke	
Animal testing, acute toxicity, Rat oral LD50	mg/kg	>2000

Pexalyn 9100

PARAMETER	UNIT	VALUE
Animal testing, acute toxicity, Rabbit dermal LD50	mg/kg	>2000
Effect of exposure, eye (human)	No irritant effect. May cause eye irritation by mechanical abrasion.	
Effect of exposure, inhalation (human)	May cause respiratory irritation by mechanical abrasion.	
Effect of exposure, skin (human)	No irritant effect. May cause skin by mechanical abrasion.	
Effect of exposure, swallowing (human)	No hazard expected.	
Effect of repeated or overexposure (human)	Rosin and some rosin derivatives have been reported to cause allergic skin reaction (sensitization) in susceptible individuals after repeated or prolonged contact.	
Exposure, personal protection	Safety glasses, protective clothing based on chemical resistance data, chemical-resistant gloves, general and local exhaust ventilation.	
First aid, eye	Immediately flush eyes with plenty of water, occasionally lifting the upper and lower eyelids. Check for and remove any contact lenses. Rinse opened eye for several minutes under running water. Obtain medical attention if irritation develops.	
First aid, inhalation	Move to fresh air. Get medical attention if nasal, throat or lung irritation develops.	
First aid, skin	HOT MOLTEN product: Immediately cool skin burns with water and cold packs for at least 15 minutes. Do NOT put ice directly on the skin. Do NOT attempt to remove solidified resin from the skin as severe tissue damage may result. Get immediate medical attention. SOLID product at ambient temperature: Wash thoroughly with soap and water. Remove contaminated clothing. Get medical attention if irritation develops or persists. Thoroughly wash clothing before reuse.	

ECOLOGICAL PROPERTIES

PARAMETER	UNIT	VALUE
Aquatic toxicity, *Green algae*, 96-h EC50	mg/l	1000/72H
Aquatic toxicity *Fathead minnow*, 96-h LC50	mg/l	1000
Partition coefficient, log K_{oc}	-	>4

Pexalyn 9100

PARAMETER	UNIT	VALUE
USE & PERFORMANCE		
Manufacturer		Pinova
Outstanding properties		promotes adhesion to porous and polar substrates. Light initial color, good resistance to oxidation and good thermal stability/pot life, low volatility and odor, widely compatible, dispersible in aqueous media.
Recommended applications		used as a cost-effective tackifier resin for EVA hot melt adhesives and tackifier resin for pressure-sensitive adhesives formulated with polar polymers and elastomers. It is also useful as a resin modifier for elastomers and waxes.
Food approval (FDA)		US FDA 21 CFR 175.105, 21 CFR 175.300, 21 CFR 175.380, 21 CFR 175.390, 21 CFR 176.210, 21 CFR 177.1210, 21 CFR 177.2600

Pexalyn Ester 10

PARAMETER	UNIT	VALUE
GENERAL INFORMATION		
Name		Pexalyn Ester 10
CAS #	-	68475-37-6
General description		biobased
Composition		glycerol ester of partially dimerized rosin
Chemical class		rosin
PHYSICAL PROPERTIES		
State	-	solid/pastilles
Odor	-	rosin, slight
Color, Gardner scale		8 (50% solids in toluene)
Softening point	°C	100-110
Acid number	mg KOH/g	8
Ash content	wt%	< 0.04
Density at 25°C	kg/m³	1,080
Solubility (diluents)		C9-C11 isoparaffins
Solubility in water at 25°C	g/l	negligible
HEALTH & SAFETY		
HMIS classification	Flammability	1
	Health	0
	Reactivity	0
Carcinogenicity		IARC, OSHA, NTP: no ingredient of this product present at levels greater than or equal to 0.1% is identified as probable, possible or confirmed human carcinogen
DOT class		not regulated
ICAO/IATA class		not regulated
IMDG class		not regulated
Flash point	°C	225
Hazardous combustion products		Carbon monoxide, carbon dioxide, aldehydes, carboxylic acid, smoke
Animal testing, acute toxicity, Rat oral LD50	mg/kg	>2000
Animal testing, acute toxicity, Rabbit dermal LD50	mg/kg	>2000
Effect of exposure, eye (human)		No irritant effect. May cause eye irritation by mechanical abrasion.
Effect of exposure, inhalation (human)		May cause respiratory irritation by mechanical abrasion.
Effect of exposure, skin (human)		No irritant effect. May cause skin by mechanical abrasion.

Pexalyn Ester 10

PARAMETER	UNIT	VALUE
Effect of exposure, swallowing (human)		No hazard expected.
Effect of repeated or overexposure (human)		Rosin and some rosin derivatives have been reported to cause allergic skin reaction (sensitization) in susceptible individuals after repeated or prolonged contact.
Exposure, personal protection		Safety glasses, protective clothing based on chemical resistance data, chemical-resistant gloves, general and local exhaust ventilation.
First aid, eye		Immediately flush eyes with plenty of water, occasionally lifting the upper and lower eyelids. Check for and remove any contact lenses. Rinse opened eye for several minutes under running water. Obtain medical attention if irritation develops.
First aid, inhalation		Move to fresh air. Get medical attention if nasal, throat or lung irritation develops.
First aid, skin		HOT MOLTEN product: Immediately cool skin burns with water and cold packs for at least 15 minutes. Do NOT put ice directly on the skin. Do NOT attempt to remove solidified resin from the skin as severe tissue damage may result. Get immediate medical attention. SOLID product at ambient temperature: Wash thoroughly with soap and water. Remove contaminated clothing. Get medical attention if irritation develops or persists. Thoroughly wash clothing before reuse.
ECOLOGICAL PROPERTIES		
Biodegradation probability		readily biodegradable
USE & PERFORMANCE		
Manufacturer		Pinova
Outstanding properties		promotes adhesion, hardness, and water resistance. Good initial color and oxidation resistance, good solution properties and emulsion stability, broad solubility and compatibility ranges, good solubility in C9-C11 isoparaffins, low ash content.
Recommended for products		cosmetics (mascara), adhesives, hot-melts, lacquers, varnishes

Pexalyn Ester 10

PARAMETER	UNIT	VALUE
Recommended applications		used as a modifier for mascara and similar cosmetic applications, where its broad solubility and compatibility are particularly beneficial, and as a tackifier resin for various types of adhesives.
Food approval (FDA)		US FDA 21 CFR 175.105

Pexalyn SR

PARAMETER	UNIT	VALUE
GENERAL INFORMATION		
Name		Pexalyn SR
General description		biobased
Composition		stabilized aromatic/aliphatic hybrid resin
Chemical class		rosin
PHYSICAL PROPERTIES		
State	-	solid/pastilles
Odor	-	rosin, slight
Color	-	pale yellow
Color, Gardner scale	5 (50% solids in toluene)	
Color, Platinum-cobalt scale	-	<200 (in melt)
Softening point	°C	77-87
Acid number	mg KOH/g	90-100
Density at 25°C	kg/m^3	1,030
HEALTH & SAFETY		
NFPA classification	Flammability	1
	Health	0
	Reactivity	0
HMIS classification	Flammability	1
	Health	0
	Reactivity	0
Carcinogenicity	IARC, OSHA, NTP: no ingredient of this product present at levels greater than or equal to 0.1% is identified as probable, possible or confirmed human carcinogen	
DOT class	not regulated	
ICAO/IATA class	not regulated	
IMDG class	not regulated	
Animal testing, acute toxicity, Rat oral LD50	mg/kg	>4000
Animal testing, acute toxicity, Rabbit dermal LD50	mg/kg	>2500
Effect of exposure, eye (human)	No irritant effect. May cause eye irritation by mechanical abrasion.	
Effect of exposure, inhalation (human)	May cause respiratory irritation by mechanical abrasion.	
Effect of exposure, skin (human)	No irritant effect. May cause skin by mechanical abrasion.	
Effect of exposure, swallowing (human)	No hazard expected.	

Pexalyn SR

PARAMETER	UNIT	VALUE
Effect of repeated or overexposure (human)		Rosin and some rosin derivatives have been reported to cause allergic skin reaction (sensitization) in susceptible individuals after repeated or prolonged contact.
Exposure, personal protection		Safety glasses, protective clothing based on chemical resistance data, chemical-resistant gloves, general and local exhaust ventilation.
First aid, eye		Immediately flush eyes with plenty of water, occasionally lifting the upper and lower eyelids. Check for and remove any contact lenses. Rinse opened eye for several minutes under running water. Obtain medical attention if irritation develops.
First aid, inhalation		Move to fresh air. Get medical attention if nasal, throat or lung irritation develops.
First aid, skin		HOT MOLTEN product: Immediately cool skin burns with water and cold packs for at least 15 minutes. Do NOT put ice directly on the skin. Do NOT attempt to remove solidified resin from the skin as severe tissue damage may result. Get immediate medical attention. SOLID product at ambient temperature: Wash thoroughly with soap and water. Remove contaminated clothing. Get medical attention if irritation develops or persists. Thoroughly wash clothing before reuse.
USE & PERFORMANCE		
Manufacturer		Pinova
Outstanding properties		excellent specific adhesion. Compatible with natural and synthetic polymers, rosins, rosin esters, EVA resins, and thermoplastic block copolymers. Light color and good oxidative stability.
Recommended applications		used as a tackifier for hot melt adhesives where compatibility is required with copolymers exhibiting polar and nonpolar character and tackifier for solvent-based adhesives.
Food approval (FDA)		US FDA 21 CFR 175.105, 21 CFR 177.2600

Pexalyn T100

PARAMETER	UNIT	VALUE
GENERAL INFORMATION		
Name	Pexalyn T100	
CAS #	-	8050-26-8
EC number	-	232-479-9
General description	biobased	
Composition	resin acids and rosin acids, esters with pentaerythritol, wood pulp-based	
Common synonym	resin acid pentaerythritol esters, ester of rosin	
Empirical formula	C25H34O2	
Molecular mass	daltons	366.54
Chemical class	rosin	
PHYSICAL PROPERTIES		
State	-	solid/pastilles
Odor	-	rosin, slight
Color, Gardner scale	-	6
Melting point	°C	100/softening point
Acid number	mg KOH/g	12
Density at 25°C	kg/m³	1,060
Solubility in water at 25°C	g/l	negligible
HEALTH & SAFETY		
NFPA classification	Flammability	1
	Health	0
	Reactivity	0
HMIS classification	Flammability	1
	Health	0
	Reactivity	0
Carcinogenicity	IARC, OSHA, NTP: no ingredient of this product present at levels greater than or equal to 0.1% is identified as probable, possible or confirmed human carcinogen	
DOT class	not regulated	
ICAO/IATA class	not regulated	
IMDG class	not regulated	
Animal testing, acute toxicity, Rat oral LD50	mg/kg	>2000
Animal testing, acute toxicity, Rabbit dermal LD50	mg/kg	>2000
Effect of exposure, eye (human)	No irritant effect. May cause eye irritation by mechanical abrasion.	

Pexalyn T100

PARAMETER	UNIT	VALUE
Effect of exposure, inhalation (human)	May cause respiratory irritation by mechanical abrasion.	
Effect of exposure, skin (human)	No irritant effect. May cause skin by mechanical abrasion.	
Effect of exposure, swallowing (human)	No hazard expected.	
Effect of repeated or overexposure (human)	Rosin and some rosin derivatives have been reported to cause allergic skin reaction (sensitization) in susceptible individuals after repeated or prolonged contact.	
Exposure, personal protection	Safety glasses, protective clothing based on chemical resistance data, chemical-resistant gloves, general and local exhaust ventilation.	
First aid, eye	Immediately flush eyes with plenty of water, occasionally lifting the upper and lower eyelids. Check for and remove any contact lenses. Rinse opened eye for several minutes under running water. Obtain medical attention if irritation develops.	
First aid, inhalation	Move to fresh air. Get medical attention if nasal, throat or lung irritation develops.	
First aid, skin	HOT MOLTEN product: Immediately cool skin burns with water and cold packs for at least 15 minutes. Do NOT put ice directly on the skin. Do NOT attempt to remove solidified resin from the skin as severe tissue damage may result. Get immediate medical attention. SOLID product at ambient temperature: Wash thoroughly with soap and water. Remove contaminated clothing. Get medical attention if irritation develops or persists. Thoroughly wash clothing before reuse.	

ECOLOGICAL PROPERTIES

Aquatic toxicity, *Green algae*, 96-h EC50	mg/l	1000/72H
Biodegradation probability	readily biodegradable	

USE & PERFORMANCE

Manufacturer	Pinova	
Outstanding properties	promotes adhesion to porous and polar substrates. Resistant to oxidation, good thermal stability/pot life, low volatility, widely compatible.	

Pexalyn T100

PARAMETER	UNIT	VALUE
Recommended applications		used as a tackifier resin for hot melt adhesives and pressure-sensitive adhesives. Modifier for elastomers and waxes.
Food approval (FDA)		US FDA 21 CFR 175.105, 21 CFR 175.300, 21 CFR 175.380, 21 CFR 175.390, 21 CFR 176.210, 21 CFR 177.1210, 21 CFR 177.2600

Regalite R1090 hydrocarbon resin

PARAMETER	UNIT	VALUE
GENERAL INFORMATION		
Name		Regalite R1090 hydrocarbon resin
Composition		99.5% hydrocarbon resin, <0.5% additives
Chemical class		rosin
Mixture	-	yes
Molecular weight, M_n	-	540
Molecular weight, M_w	-	700
Molecular weight, M_z	-	920
Polydispersity, M_w/M_n	-	1.3
PHYSICAL PROPERTIES		
State	-	solid
Color	-	water-white
Color, Hunterlab b	-	0.5
Softening point	°C	87
Glass transition temperature	°C	36
Cloud point, MMAP	°C	76
Density at 25°C	kg/m³	980
Solubility in water at 25°C	g/l	negligible
Viscosity at 140°C	cP	800
HEALTH & SAFETY		
HMIS classification	Flammability	1
	Health	1
	Reactivity	0
Carcinogenicity	no data available	
Mutagenicity	no data available	
Teratogenicity	no data available	
DOT class	not regulated	
TDG class	not regulated	
ICAO/IATA class	not regulated	
IMDG class	not regulated	
Hazardous combustion products	Carbon monoxide, carbon dioxide	
Agency rating, listed	TSCA USA, DSL Canada, AICS Australia, MITI Japan, ECL Korea, IECSC China	
Effect of exposure, eye (human)	Avoid contact with molten material. Contact with molten substance/product may cause severe eyes damage.	

Regalite R1090 hydrocarbon resin

PARAMETER	UNIT	VALUE
Effect of exposure, skin (human)		Avoid contact with molten material. Contact with molten substance/product may cause severe burns to skin.
Exposure, personal protection		Safety glasses, protective clothing based on chemical resistance data, chemical-resistant gloves, general and local exhaust ventilation.
First aid, eye		Rinse cautiously with water for several minutes. Remove contact lenses, if present and easy to do. Continue rinsing. If molten material contacts the eye, immediately flush with plenty of water for at least 15 minutes. Get medical attention immediately.
First aid, inhalation		Remove to fresh air. If not breathing, give artificial respiration. If breathing is difficult, give oxygen. If irritation persists, obtain medical advice.
First aid, skin		Wash with soap and water. Get medical attention if symptoms occur. If burned by contact with hot material, cool molten material adhering to skin as quickly as possible with water, and see a physician for removal of adhering material and treatment of burn. Get medical attention.
USE & PERFORMANCE		
Manufacturer		Eastman Chemical Company
Outstanding properties		extremely light color, excellent adhesion, very good resistance to thermal and oxidative degradation, excellent compatibility.
Recommended for products		adhesives/sealants, bookbinding, carpet construction, case and carton closures, casting wax, commercial printing inks, depilatories ingredients, hygiene adhesives, labels non-food contact, polymer modification, protective coatings, specialty tape, tires

Regalite R1090 hydrocarbon resin

PARAMETER	UNIT	VALUE
Recommended applications		typical application: hot melt adhesives, sealants, coatings, bookbinding, rubber and plastic compounding, wax blends, food packaging adhesives, cosmetics (antiperspirants and deodorants), food packaging adhesives. Specially designed as tackifiers in hot melt adhesives based on EVA copolymers and SIS block copolymers requiring excellent color retention upon aging.

Regalite R1100 CG hydrocarbon resin

PARAMETER	UNIT	VALUE
GENERAL INFORMATION		
Name		Regalite R1100 CG hydrocarbon resin
Composition		>99% hydrogenated styrene/methyl styrene/indene copolymer, <1% antioxidant
Chemical class		rosin
Mixture	-	yes
Molecular weight, M_n	-	600
Molecular weight, M_w	-	830
Molecular weight, M_z	-	830
Polydispersity, M_w/M_n		1.4
PHYSICAL PROPERTIES		
State	-	solid/pellets
Odor	-	odorless
Color	-	water-white
Color, Gardner scale	-	<1
Softening point	°C	100
Cloud point, MMAP	°C	80
Density at 20°C	kg/m^3	990
Solubility in water at 25°C	g/l	negligible
Viscosity at 140°C	mPas	2,500
HEALTH & SAFETY		
HMIS classification	Flammability	1
	Health	1
	Reactivity	0
Carcinogenicity	no data available	
Mutagenicity	no data available	
Teratogenicity	no data available	
DOT class	not regulated	
TDG class	not regulated	
ICAO/IATA class	not regulated	
IMDG class	not regulated	
Flash point	°C	>200
Flash point method	-	SFCC
Hazardous combustion products	Carbon monoxide (CO), Carbon dioxide (CO2)	
Agency rating, listed	TSCA USA, DSL Canada, AICS Australia, MITI Japan, ECL Korea, IECSC China	

Regalite R1100 CG hydrocarbon resin

PARAMETER	UNIT	VALUE
Animal testing, acute toxicity, Rat oral LD50	mg/kg	>3200/antioxidant
Animal testing, acute toxicity, Guinea pig dermal LD50	mg/kg	>1000/antioxidant, non-sensitizing
Effect of exposure, eye (human)		Avoid contact with molten material. Contact with molten substance/product may cause severe eyes damage.
Effect of exposure, skin (human)		Non irritant. Avoid contact with molten material. Contact with molten substance/product may cause severe burns to skin.
Exposure, personal protection		Safety glasses, protective clothing based on chemical resistance data, chemical-resistant gloves, general and local exhaust ventilation.
First aid, eye		Rinse cautiously with water for several minutes. Remove contact lenses, if present and easy to do. Continue rinsing. If molten material contacts the eye, immediately flush with plenty of water for at least 15 minutes. Get medical attention immediately.
First aid, inhalation		Remove to fresh air. If not breathing, give artificial respiration. If breathing is difficult, give oxygen. If irritation persists, obtain medical advice.
First aid, skin		Wash with soap and water. Get medical attention if symptoms occur. If burned by contact with hot material, cool molten material adhering to skin as quickly as possible with water, and see a physician for removal of adhering material and treatment of burn. Get medical attention.
USE & PERFORMANCE		
Manufacturer		Eastman Chemical Company
Outstanding properties		resistant to thermal and oxidative degradation, contributes to adhesion and transfer resistance
Recommended for products		cosmetics: eye makeup, face makeup, lipstick and gloss, hair care, nails
Recommended applications		cosmetic grade resin derived from petrochemical feedstocks. This thermoplastic resin is fully hydrogenated, providing excellent stability. It has excellent solubility and compatibility with non-polar ingredients in cosmetic applications, contributing both adhesion and gloss, forms clear glossy films.

3.27 Silanes
Carbo NXT

PARAMETER	UNIT	VALUE
GENERAL INFORMATION		
Name		Carbo NXT
Composition		3-octanoylthio-1-propyltriethoxysilane
PHYSICAL PROPERTIES		
State	-	solid/powder
Color	-	black
Density at 25°C	kg/m^3	1,800
Specific gravity at 20°C	-	1.8
HEALTH & SAFETY		
Flash point	°C	176
Hazardous products of hydrolysis	-	ethanol
Effect of repeated or overexposure (human)		Repeated exposure to ethanol may aggravate liver injury produced from other causes.
Exposure, personal protection		Safety glasses, protective clothing based on chemical resistance data, chemical-resistant gloves, general and local exhaust ventilation.
First aid, eye		Immediately flush eyes with plenty of water, occasionally lifting the upper and lower eyelids. Check for and remove any contact lenses. Continue to rinse for at least 10 minutes. Get medical attention if irritation occurs.
First aid, inhalation		Remove to fresh air. If not breathing, give artificial respiration. If breathing is difficult, give oxygen. If irritation persists, obtain medical advice.
First aid, skin		Immediately flush with plenty of water for at least 15 minutes while removing contaminated clothing and shoes. Wash contaminated clothing before reuse. Get medical attention.
USE & PERFORMANCE		
Manufacturer		Momentive

3.27 Silanes
Carbo NXT

PARAMETER	UNIT	VALUE
Outstanding properties		NXT* silane is in powder form for use when liquid silanes are inconvenient. Improved silica dispersion, easier mixing, and faster, pliable processing. Improved Payne Effect. Improved rolling resistance without loss of wet traction in tires. Reduced tan and max. Excellent dynamic properties at low temperature (-20oC to +10oC). Green compounds made with Carbo NXT silane have demonstrated stability for a significantly longer time without reflocculation. Reduced ethanol (VOC) emissions during tire manufacturing and use.
Recommended for products		tires
Recommended applications		superior coupling agent for silica-reinforced tire tread compounds. It is used in tire tread compounds to enable high silica loading while typically managing compound viscosity, improving processibility, and increasing mixing temperatures, even in functionalized polymer and high surface area silica compounds. This NXT silane is in powder form for use when liquid silanes are inconvenient. The excellent dynamic properties and improved resilience make it an ideal candidate to consider for silica tire production when liquid systems are occupied or unavailable.

CoatOSil 1770 Silane

PARAMETER	UNIT	VALUE
GENERAL INFORMATION		
Name		CoatOSil 1770 Silane
CAS #	-	10217-34-2
Common synonym		β-(3,4-epoxycyclohexyl) ethyltriethoxysilane
Formula		$(CH_3CH_2O)_3SiCH_2CH_2-$
Molecular mass	daltons	288.46
PHYSICAL PROPERTIES		
State	-	liquid
Odor	-	faint
Color	-	clear, colorless
Boiling point	°C	>300
Melting point	°C	<0
Density at 20°C	kg/m³	1003.5
Vapor pressure at 20°C	kPa	<0.133
VOC	g/l	507
HEALTH & SAFETY		
HMIS classification	Flammability	1
	Health	2
	Physical hazard	1
Flash point	°C	129
Flash point method	-	PMCC
Animal testing, acute toxicity, Rat oral LD50	mg/kg	>5000
Animal testing, acute toxicity, Rabbit dermal LD50	mg/kg	>2000
Animal testing, acute toxicity, Rat dermal LD50	mg/kg	>2000
First aid, eye		Rinse immediately with plenty of water, also under the eyelids, for at least 15 minutes. Get medical attention.
First aid, inhalation		Move the exposed person to fresh air at once. If respiratory problems, artificial respiration/oxygen. Call a physician or poison control center immediately.
First aid, skin		Wash off promptly and flush contaminated skin with water. Promptly remove clothing if soaked through and flush skin with water. Wash contaminated clothing before reuse. Get medical attention.

CoatOSil 1770 Silane

PARAMETER	UNIT	VALUE
ECOLOGICAL PROPERTIES		
Aquatic toxicity, *Carp*, 96-h LC50	mg/l	42.3
Aquatic toxicity, *Daphnia magna*, 48-h LC50	mg/l	58
USE & PERFORMANCE		
Manufacturer	Momentive	
Outstanding properties	shelf stable, one-pack and two-pack systems are possible. When properly formulated, crosslinking can occur under both room temperature and elevated temperature conditions. Excellent gloss retention upon aging	
Recommended for products	automotive plastic coatings, metal coatings, waterborne coatings, wood coatings, coatings for leather, vinyl and plastics	
Concentrations used	wt%	0.8-5

CoatOSil 2287 Silane

PARAMETER	UNIT	VALUE
GENERAL INFORMATION		
Name		CoatOSil 2287 Silane
CAS #	-	2897-60-1
Common synonym		3-glycidoxypropylmethyldiethoxysilane
Formula		
PHYSICAL PROPERTIES		
State	-	liquid
Odor	-	ester-like
Color	-	clear, pale yellow
Initial boiling point	°C	260
Melting point	°C	-80
Density at 20°C	kg/m^3	979
Vapor pressure at 20°C	hPa	<1.33
HEALTH & SAFETY		
HMIS classification	Flammability	1
	Health	2
	Physical hazard	1
Autoignition temperature	°C	210
Flash point	°C	107
Flash point method	-	PMCC
Animal testing, acute toxicity, Rat oral LD50	mg/kg	>2000
Animal testing, acute toxicity, Rabbit dermal LD50	mg/kg	>2000
First aid, eye		Rinse immediately with plenty of water and seek medical advice. Get medical attention.
First aid, inhalation		Move the exposed person to fresh air at once. Get medical attention.
First aid, skin		Take off immediately all contaminated clothing. Wash with soap and water. Get medical attention.
ECOLOGICAL PROPERTIES		
Aquatic toxicity, *Carp*, 96-h LC50	mg/l	139
Partition coefficient, logP$_{ow}$	-	2.7
USE & PERFORMANCE		
Manufacturer		Momentive

CoatOSil 2287 Silane

PARAMETER	UNIT	VALUE
Outstanding properties		shelf-stable non-yellowing crosslinking performance to enhance the physical properties of waterborne dispersion of polymers such as polyurethane and acrylic latexes. When CoatOSil 2287 silane is incorporated as a crosslinker or adhesion promoter, it may provide improved water resistance and wet adhesion, with good shelf stability.
Recommended for polymers		acrylics, polyurethanes, SBR
Recommended for products		flooring finishes, metal coatings, waterborne dispersion of polymers, wood furniture
Concentrations used	wt%	1.2-2
Guidelines for use		carboxylated latexes exhibiting a pH between 6 and 8.5

CoatOSil DRI Waterborne Silicone Resin

PARAMETER	UNIT	VALUE
GENERAL INFORMATION		
Name		CoatOSil DRI Waterborne Silicone Resin
Active contents	%	45
PHYSICAL PROPERTIES		
State	-	liquid
Color	-	white
Density at 25°C	kg/m³	1100
Particle size	nm	120
pH		11
Viscosity at 25°C	mPas	20
USE & PERFORMANCE		
Manufacturer		Momentive
Outstanding properties		reduces water uptake and improves UV resistance in organic waterborne coating compositions. Its chemical structure enables CoatOSil DRI waterborne silicone resin to overcome the difficulties of combining silicone materials with organic waterborne resins.
Recommended for polymers		alkyds, acrylics, epoxy, polyurethanes, styrene acrylics
Recommended for products		architectural, concrete, roof and wood coatings and sealants
Concentrations used	wt%	5-30

CoatOSil MP 200 Silane

PARAMETER	UNIT	VALUE
GENERAL INFORMATION		
Name		CoatOSil MP 200 Silane
CAS #	-	2530-83-8
Common synonym		3-glycidyl-oxypropyltrimethoxysilane
Formula		
Molecular mass	daltons	236.34
PHYSICAL PROPERTIES		
State	-	liquid
Odor	-	ester-like
Color	-	yellow
Initial boiling point	°C	290
Melting point	°C	<-70
Density at 20°C	kg/m³	1003.5
pH	-	8.5-10.5
Vapor pressure at 20°C	kPa	<0.133
VOC	g/l	250
HEALTH & SAFETY		
HMIS classification	Flammability	1
	Health	3
	Physical hazard	1
Flash point	°C	>107
Flash point method	-	PMCC
Animal testing, acute toxicity, Rat inhalation LC50	mg/l	>5.3
First aid, eye		Rinse immediately with plenty of water, also under the eyelids, for at least 15 minutes. Get medical attention.
First aid, inhalation		Move the exposed person to fresh air at once. If respiratory problems, artificial respiration/oxygen. Call a physician or poison control center immediately.
First aid, skin		Wash off promptly and flush contaminated skin with water. Promptly remove clothing if soaked through and flush skin with water. Wash contaminated clothing before reuse. Get medical attention.
ECOLOGICAL PROPERTIES		
Aquatic toxicity, *Fish*, 96-h LC50	mg/l	55

CoatOSil MP 200 Silane

PARAMETER	UNIT	VALUE
Aquatic toxicity, *Daphnia magna*, 48-h LC50	mg/l	324
Partition coefficient $logK_{ow}$	-	0.5
USE & PERFORMANCE		
Manufacturer	Momentive	
Outstanding properties	polyfunctional structure bearing gamma-glycidoxy groups is an excellent candidate to consider to reduce emissions of methanol upon hydrolysis of the material as compared with monomeric epoxy silanes. It typically aids adhesion promotion and crosslinking of water borne or solvent based coatings as well as dispersion of metallic pigments in water borne systems.	
Recommended for polymers	polysulfide, urethane, epoxy and acrylics	
Recommended for products	caulks, sealants, adhesives and coatings	
Concentrations used	wt%	1-2

Dowsil™ AZ-720

PARAMETER	UNIT	VALUE
GENERAL INFORMATION		
Name		Dowsil™ AZ-720
CAS #	-	919-30-2
General description		3-aminopropyltriethoxysilane with organic surfactant to improve wettability onto substrates.
Common synonym		3-aminopropyltriethoxysilane
Mixture	-	yes
Active matter	wt%	42-58
PHYSICAL PROPERTIES		
State	-	liquid
Odor	-	amine-like
Color	clear to pale yellow liquid	
Boiling point	°C	>35
Kinematic viscosity at 25°C	cSt	24.5
Specific gravity at 25°C	-	0.97
HEALTH & SAFETY		
NFPA classification	Flammability	2
	Health	3
	Instability	0
HMIS classification	Flammability	2
	Health	1
	Physical hazard	0
Carcinogenicity	Contains component(s) which did not cause cancer in laboratory animals.	
Mutagenicity	Genetic toxicity studies in animals were negative for component(s) tested.	
Teratogenicity	Did not cause birth defects or other effects in the fetus even at doses which caused toxic effects in the mother.	
Flash point	°C	78
Flash point method	-	CC
Animal testing, acute toxicity, Rat oral LD50	mg/kg	>2000
Animal testing, acute toxicity, Rabbit dermal LD50	mg/kg	>5000

Dowsil™ AZ-720

PARAMETER	UNIT	VALUE
First aid, eye		Wash immediately and continuously with flowing water for at least 30 minutes. Remove contact lenses after the first 5 minutes and continue washing. Obtain prompt medical consultation, preferably from an ophthalmologist. Suitable emergency eye wash facility should be immediately available.
First aid, inhalation		Move person to fresh air and keep comfortable for breathing. If not breathing, give artificial respiration; if by mouth to mouth use rescuer protection (pocket mask, etc). If breathing is difficult, oxygen should be administered by qualified personnel. Call a physician or transport to a medical facility.
First aid, skin		Immediately flush skin with plenty of water for at least 15 minutes while removing contaminated clothing. Seek medical attention if symptoms occur or irritation persists.
Dow IHG, TWA	mg/m^3	0.5
UN/NA class	-	1760
ECOLOGICAL PROPERTIES		
Aquatic toxicity, *Green algae*, 72-h EC50	mg/l	>1000 3-aminopropyltriethoxysilane
Aquatic toxicity, *Daphnia magna*, 48-h LC50	mg/l	331 3-aminopropyltriethoxysilane
Aquatic toxicity, *Zebra fish*, 96-h LC50	mg/l	>934 3-aminopropyltriethoxysilane
Bioconcentration factor	BCF	<100 3-aminopropyltriethoxysilane
Biodegradation probability		Based on stringent OECD test guidelines, this material cannot be considered as readily biodegradable; however, these results do not necessarily mean that the material is not biodegradable under environmental conditions.
USE & PERFORMANCE		
Manufacturer		Dow
Outstanding properties		improved surface appearance
Recommended for products		glass fiber composites

Dowsil Primer C OS

PARAMETER	UNIT	VALUE
GENERAL INFORMATION		
Name		Dowsil Primer-C OS
CAS #	-	26936-30-1
Composition		3-4% methylmethacrylate, 3-(trimethoxysilyl)propyl methacrylate polymer, >=82% methyl acetate, 2-2.8% xylene, 1.8-2.4% ethylbenzene, 0.72-0.98% cyclohexanone
PHYSICAL PROPERTIES		
State	-	liquid
Odor	-	solvent-like
Color	-	colorless
Boiling point	°C	57
Kinematic viscosity at 25°C	cSt	1
Specific gravity at 23°C	-	0.9
Viscosity at 23°C	mPas	>10
HEALTH & SAFETY		
NFPA classification	Flammability	3
	Health	2
	Instability	0
HMIS classification	Flammability	3
	Health	2
	Physical hazard	0
Carcinogenicity		Ethylbenzene: IARC Group 2B: Possibly carcinogenic to humans; ACGIH A3: Confirmed animal carcinogen with unknown relevance to humans. Cyclohexanone: ACGIH A3: Confirmed animal carcinogen with unknown relevance to humans.
Mutagenicity		Contains a component(s) which were negative in *in vitro* genetic toxicity studies.
Teratogenicity		Contains component(s) which caused birth defects in laboratory animals only at doses toxic to the mother.
Flash point	°C	-9
Flash point method	-	TCC
Animal testing, acute toxicity, Rat oral LD50	mg/kg	>5000
Animal testing, acute toxicity, Rabbit dermal LD50	mg/kg	>2000

Dowsil Primer C OS

PARAMETER	UNIT	VALUE
First aid, eye		Immediately flush eyes with water; remove contact lenses, if present, after the first 5 minutes, then continue flushing eyes for at least 15 minutes. Obtain medical attention without delay, preferably from an ophthalmologist. Suitable emergency eye wash facility should be immediately available.
First aid, inhalation		Move person to fresh air and keep comfortable for breathing. If not breathing, give artificial respiration; if by mouth to mouth use rescuer protection (pocket mask, etc). If breathing is difficult, oxygen should be administered by qualified personnel. Call a physician or transport to a medical facility.
First aid, skin		Wash off with plenty of water.
UN/NA class	-	1993
ECOLOGICAL PROPERTIES		
Aquatic toxicity, *Green algae*, 72-h EC50	mg/l	>120 (methyl acetate)
Aquatic toxicity, *Daphnia magna*, 48-h LC50	mg/l	1027 (methyl acetate)
Aquatic toxicity *Fathead minnow*, 96-h LC50	mg/l	320-399 (methyl acetate)
Aquatic toxicity, *Zebra fish*, 96-h LC50	mg/l	250-350 (methyl acetate)
Bioaccumulative potential		Bioconcentration potential is low
Bioconcentration factor	BCF	<100
Biodegradation probability		Material is readily biodegradable. Passes OECD test
Partition coefficient, log K_{oc}	-	8-50 (methyl acetate)
USE & PERFORMANCE		
Manufacturer		Dow
Outstanding properties		low VOC (49 g/l) and unique fluorescing feature, allowing for a visual quality control check to ensure primer has been applied. Fluoresces under a 365 nm wavelength. Quick cure time
Recommended for polymers		silicone
Recommended for products		adhesives, sealants

Dowsil Primer C OS

PARAMETER	UNIT	VALUE
Recommended applications		accelerated adhesion of Dowsil™ 983 structural glazing sealant to coated aluminum substrates such as poly(vinylidene fluoride) or Kynar based paints
Processing methods		apply a single coat of Dowsil™ Primer-C OS to the substrate. Dowsil™ Primer-C OS should be applied with a lint-free cloth to maximize primer coverage rate and obtain a consistent film thickness. While a brush may be used to apply Dowsil™ Primer-C OS, the coverage rate will be lessened, and it will be more difficult to obtain consistent film thickness.
Guidelines for use		allow the primer to dry for 20 minutes at room temperature or 1 hour at 5°C before applying and tooling the sealant.

Dowsil™ SZ-6030 Silane

PARAMETER	UNIT	VALUE
GENERAL INFORMATION		
Name	Dowsil™ SZ-6030 Silane	
CAS #	-	2530-85-0
Common synonym	methacryloxypropyl trimethoxysilane	
Chemical name	3-trimethoxysilylpropyl methacrylate	
Active matter	wt%	98-100
PHYSICAL PROPERTIES		
State	-	liquid
Color	colorless to pale yellow	
Odor	-	aromatic
Boiling point	°C	250
Kinematic viscosity at 25°C	cSt	2.5
HEALTH & SAFETY		
NFPA classification	Flammability	1
	Health	0
	Instability	0
HMIS classification	Flammability	1
	Health	0
	Physical hazard	0
Mutagenicity	*In vitro* genetic toxicity studies were negative in some cases and positive in other cases. Animal genetic toxicity studies were negative.	
Teratogenicity	Has caused birth defects in laboratory animals only at doses toxic to the mother.	
Autoignition temperature	°C	360
Flash point	°C	>100
Flash point method	-	TCC
Animal testing, acute toxicity, Rat oral LD50	mg/kg	>2000
Animal testing, acute toxicity, Rat dermal LD50	mg/kg	>2000
Animal testing, acute toxicity, Rat inhalation, LC50	mg/l	>2.28/4H
First aid, eye	Flush eyes thoroughly with water for several minutes. Remove contact lenses after the initial 1-2 minutes and continue flushing for several additional minutes. If effects occur, consult a physician, preferably an ophthalmologist.	

Dowsil™ SZ-6030 Silane

PARAMETER	UNIT	VALUE
First aid, inhalation	Move person to fresh air and keep comfortable for breathing; consult a physician.	
First aid, skin	Wash off with plenty of water. Suitable emergency safety shower facility should be available in work area.	
Dow IHG, TWA	mg/m^3	0.1
ECOLOGICAL PROPERTIES		
Aquatic toxicity, *Green algae*, 72-h EC50	mg/l	>100
Aquatic toxicity, *Daphnia magna*, 48-h LC50	mg/l	>876
Aquatic toxicity, *Zebra fish*, 96-h LC50	mg/l	>1042
Bioaccumulative potential	Bioconcentration potential is low	
Bioconcentration factor	BCF	<100
Biodegradation probability	Material is expected to be readily biodegradable.	
Partition coefficient, log P$_{ow}$	-	2.1
USE & PERFORMANCE		
Manufacturer	Dow	
Outstanding properties	used in glass fiber reinforced composites to improve mechanical strength and surface smoothness	
Recommended for polymers	unsaturated resin systems	
Recommended for products	adhesives	

Dowsil™ Z-6026 Silane

PARAMETER	UNIT	VALUE
GENERAL INFORMATION		
Name		Dowsil™ Z-6026 Silane
CAS #	-	1760-24-3
Composition		15-23% N-(3-(trimethoxysilyl) propyl)-1,2-ethanediamine, 51-69% methanol, 3-5% oligomers of (ethylene-diaminepropyl)trimethoxysilane, 0.17-0.23% ethylenediamine
Chemical name		N-(3-(trimethoxysilyl) propyl)-1,2-ethanediamine
Active matter	wt%	40
Amine equivalent	-	500
PHYSICAL PROPERTIES		
State	-	liquid
Odor	-	fishy
Color	-	straw
Boiling point	°C	>65
Kinematic viscosity at 25°C	cSt	2.4
Refractive index at 25°C	-	1.375
Specific gravity at 25°C	-	0.88
HEALTH & SAFETY		
NFPA classification	Flammability	3
	Health	3
	Instability	0
HMIS classification	Flammability	3
	Health	4
	Physical hazard	0
Carcinogenicity		Contains component(s) which did not cause cancer in laboratory animals.
Mutagenicity		Contains a component(s) which were negative in *in vitro* genetic toxicity studies. Contains component(s) which were negative in some animal genetic toxicity studies and positive in others.
Teratogenicity		Contains component(s) which, in laboratory animals, have been toxic to the fetus only at doses toxic to the mother. Methanol has caused birth defects in mice at doses nontoxic to the mother as well as slight behavioral effects in offspring of rats.

Dowsil™ Z-6026 Silane

PARAMETER	UNIT	VALUE
Flash point	°C	15.5
Flash point method	-	PMCC
Animal testing, acute toxicity, Rat oral LD50	mg/kg	2295
Animal testing, acute toxicity, Rabbit dermal LD50	mg/kg	>2000
Animal testing, acute toxicity, Rat inhalation, LC50	mg/l	3/4H
First aid, eye		Wash immediately and continuously with flowing water for at least 30 minutes. Remove contact lenses after the first 5 minutes and continue washing. Obtain prompt medical consultation, preferably from an ophthalmologist. Suitable emergency eye wash facility should be immediately available.
First aid, inhalation		Move person to fresh air and keep comfortable for breathing. If not breathing, give artificial respiration; if by mouth to mouth use rescuer protection (pocket mask, etc). If breathing is difficult, oxygen should be administered by qualified personnel. Call a physician or transport to a medical facility.
First aid, skin		Immediately flush skin with plenty of water for at least 15 minutes while removing contaminated clothing and shoes. Obtain medical attention without delay. Wash clothing before reuse. Properly dispose of contaminated leather items, such as shoes, belts, and watchbands. Suitable emergency eye wash facility should be immediately available.
UN/NA class	-	1230
ECOLOGICAL PROPERTIES		
Aquatic toxicity, *Green algae*, 96-h EC50	mg/l	22,000
Aquatic toxicity, *Bluegill sunfish*, 96-h LC50	mg/l	15,400
Aquatic toxicity, *Daphnia magna*, 48-h LC50	mg/l	>10,000
Bioaccumulative potential	: Bioconcentration potential is low	
Bioconcentration factor	BCF	<100
Biodegradation probability	Material is readily biodegradable. Passes OECD test(s) for ready biodegradability.	
Chemical oxygen demand	g/g	1.49
Theoretical oxygen demand	g/g	1.5

Dowsil™ Z-6026 Silane

PARAMETER	UNIT	VALUE
Partition coefficient, log P_{ow}	-	<3
USE & PERFORMANCE		
Manufacturer	Dow	
Outstanding properties	superior wetting of siliceous surfaces	
Recommended for polymers	phenolic, melamine	
Recommended for products	fabrics	
Guidelines for use	Dowsil Z-6026 Silane will cure properly under the ordinary high-temperature cycles used to cure most thermosetting resins. Adequate cure of the silane will usually take place during the cure or molding of most resin/glass laminates.	

Dowsil™ Z-6062 Silane

PARAMETER	UNIT	VALUE
GENERAL INFORMATION		
Name		Dowsil™ Z-6062 Silane
CAS #	-	4420-74-0
Composition		95-100% 3-mercaptopropyltrimethoxysilane, <5% chloropropyltrimethoxysilane, <0.5% methanol
Common synonym		3-mercaptopropyltrimethoxysilane
Formula		
Functional organic group	-	mercaptopropyl
Purity	wt%	>96
PHYSICAL PROPERTIES		
State	-	liquid
Odor	-	characteristic
Color		colorless to pale yellow
Boiling point	°C	>35
Kinematic viscosity at 25°C	cSt	2
Refractive index at 25°C	-	1.441
Specific gravity at 25°C	-	1.05
HEALTH & SAFETY		
NFPA classification	Flammability	2
	Health	2
	Instability	0
HMIS classification	Flammability	2
	Health	2
	Physical hazard	0
Mutagenicity		*In vitro* genetic toxicity studies were predominantly negative.
Flash point	°C	75
Flash point method	-	TCC
Animal testing, acute toxicity, Rat oral LD50	mg/kg	758-914
Animal testing, acute toxicity, Rat dermal LD50	mg/kg	2348
First aid, eye		Flush eyes thoroughly with water for several minutes. Remove contact lenses after the initial 1-2 minutes and continue flushing for several additional minutes. If effects occur, consult a physician, preferably an ophthalmologist.

Dowsil™ Z-6062 Silane

PARAMETER	UNIT	VALUE
First aid, inhalation		Move person to fresh air and keep comfortable for breathing; consult a physician
First aid, skin		Remove material from skin immediately by washing with soap and plenty of water. Remove contaminated clothing and shoes while washing. Seek medical attention if irritation or rash occurs. Wash clothing before reuse. Discard items which cannot be decontaminated, including leather articles such as shoes, belts and watchbands.
Dow IHG, TWA	ppm	0.1
UN/NA class	-	1993, 3082
ECOLOGICAL PROPERTIES		
Aquatic toxicity, *Green algae*, 72-h EC50	mg/l	267
Aquatic toxicity, *Bluegill sunfish*, 96-h LC50	mg/l	12.3
Aquatic toxicity, *Daphnia magna*, 48-h LC50	mg/l	6.7
Bioaccumulative potential		bioconcentration potential is low
Bioconcentration factor	BCF	<100
Biodegradation probability		based on stringent OECD test guidelines, this material cannot be considered as readily biodegradable; however, these results do not necessarily mean that the material is not biodegradable under environmental conditions.
Partition coefficient, log P_{ow}	-	0.25
USE & PERFORMANCE		
Manufacturer		Dow
Recommended for polymers		sulfur cured elastomers, epoxy, EPDM, SBR, natural rubber

Dowsil™ Z-6094 Silane

PARAMETER	UNIT	VALUE
GENERAL INFORMATION		
Name		Dowsil™ Z-6094 Silane
CAS #	-	1760-24-3
Chemical name		3-(2-aminoethyl)aminopropyltrime-thoxysilane
Formula		
Chemical class	-	silane
Functional organic group	-	amine
Purity	wt%	>97
PHYSICAL PROPERTIES		
State	-	liquid
Odor	-	amine-like
Color	-	clear
Boiling point	°C	264
Kinematic viscosity at 25°C	cSt	4.2
Refractive index at 25°C	-	1.445
Specific gravity at 25°C	-	1.02
HEALTH & SAFETY		
NFPA classification	Flammability	1
	Health	3
	Instability	0
HMIS classification	Flammability	1
	Health	3
	Physical hazard	0
Mutagenicity	*In vitro* genetic toxicity studies were negative. Animal genetic toxicity studies were negative.	
Teratogenicity	Did not cause birth defects in laboratory animals	
Autoignition temperature	°C	320
Flash point	°C	94
Flash point method	-	SCC
Animal testing, acute toxicity, Rat oral LD50	mg/kg	2295
Animal testing, acute toxicity, Rabbit dermal LD50	mg/kg	>2000

Dowsil™ Z-6094 Silane

PARAMETER	UNIT	VALUE
Animal testing, acute toxicity, Rat inhalation, LC50	mg/l	1.49-2.44/4H
First aid, eye		Wash immediately and continuously with flowing water for at least 30 minutes. Remove contact lenses after the first 5 minutes and continue washing. Obtain prompt medical consultation, preferably from an ophthalmologist. Suitable emergency eye wash facility should be immediately available.
First aid, inhalation		Move person to fresh air. If not breathing, give artificial respiration; if by mouth to mouth use rescuer protection (pocket mask, etc). If breathing is difficult, oxygen should be administered by qualified personnel. Call a physician or transport to a medical facility.
First aid, skin		Immediately flush skin with plenty of water for at least 15 minutes while removing contaminated clothing and shoes. Obtain medical attention without delay. Wash clothing before reuse. Properly dispose of contaminated leather items, such as shoes, belts, and watchbands. Suitable emergency safety shower facility should be immediately available.
ECOLOGICAL PROPERTIES		
Aquatic toxicity, *Green algae*, 72-h ErC50	mg/l	8.8
Aquatic toxicity, *Daphnia magna*, 48-h LC50	mg/l	81
Aquatic toxicity, *Zebra fish*, 96-h LC50	mg/l	597
Biodegradation probability		Based on stringent OECD test guidelines, this material cannot be considered as readily biodegradable; however, these results do not necessarily mean that the material is not biodegradable under environmental conditions.
Biological oxygen demand, 5 days	g/g	0.23
Chemical oxygen demand	g/g	1.76
Theoretical oxygen demand	g/g	2.39
USE & PERFORMANCE		
Manufacturer		Dow
Outstanding properties		increased wet and dry tensile and flexural, and compressive strengths and modulus to the composite

Dowsil™ Z-6094 Silane

PARAMETER	UNIT	VALUE
Recommended for polymers		natural and nitrile rubber, PA-6, PA-6/6 and polybutyleneterephthalate, phenolic, melamine and epoxy thermosets
Recommended for products		composites, foundry and abrasive composite applications
Alternative product(s)		high purity version of Xiameter™ OFS-6020 Silane

Dowsil Z-6119 Silane

PARAMETER	UNIT	VALUE
GENERAL INFORMATION		
Name		Dowsil Z-6119 Silane
CAS #		23779-32-0 or 116912-64-2, 67-56-1, 64-17-5, 57-13-6
EC number	-	245-876-7
Composition		ureidopropyltrialkoxy, 40-50% methanol, 20-30% ethanol, 1-<10% urea
Common synonym		3-ureidopropyltriethoxysilane
Empirical formula		C10H24N2O4Si
Formula		

PARAMETER	UNIT	VALUE
Molecular mass	daltons	246.39
Chemical class	silane	
Mixture	-	yes
Functional organic group		ureido/ethoxy
Non=volatile content	%	45-55
PHYSICAL PROPERTIES		
State	-	liquid
Odor	-	alcohol-like
Color	-	colorless
Boiling point	°C	65
Kinematic viscosity at 25°C	cSt	2.2
Specific gravity at 25°C	-	0.92
HEALTH & SAFETY		
NFPA classification	Flammability	3
	Health	2
	Reactivity	0
HMIS classification	Flammability	3
	Health	4
	Reactivity	0
Carcinogenicity		IARC, OSHA, NTP: no ingredient of this product present at levels greater than or equal to 0.1% is identified as probable, possible or confirmed human carcinogen
TDG class		ALCOHOLS, N.O.S., (Ethanol, Methanol) 3,II

Dowsil Z-6119 Silane

PARAMETER	UNIT	VALUE
ICAO/IATA class	Alcohols, n.o.s., (Ethanol, Methanol), Flammable Liquids, 3,II	
IMDG class	ALCOHOLS, N.O.S, (Ethanol, Methanol),3,II	
Flash point	°C	12
Flash point method	-	TCC
Hazardous combustion products	Carbon oxides, SiO, NOx, formaldehyde, sulfur oxides	
Hazardous ingredients, labelling	Ethanol, Methanol	
Hazardous products of hydrolysis	methanol	
Animal testing, acute toxicity, Rat oral LD50	mg/kg	>5000/ethanol; 300/methanol
Animal testing, acute toxicity, Rat inhalation, LC50	mg/m³	124700/4H / ethanol; 3000/4h/ methanol
Effect of exposure, eye (human)	Avoid contact with eyes. Direct contact may cause severe irritation.	
Effect of exposure, inhalation (human)	Toxic if inhaled.	
Effect of exposure, skin (human)	Toxic in contact with skin. Acute toxicity estimate: 909.09 mg/kg	
Effect of exposure, swallowing (human)	Harmful if swallowed.	
Effect of repeated or overexposure (human)	May cause damage to organs through prolonged or repeated exposure if swallowed.	
Exposure, personal protection	Safety glasses, protective clothing based on chemical resistance data, chemical-resistant gloves, general and local exhaust ventilation.	
First aid, eye	Flush eyes with plenty of water for at least 15 minutes. Remove contact lens, if worn. Get medical attention if irritation develops and persists	
First aid, inhalation	Remove to fresh air. If not breathing, give artificial respiration. If breathing is difficult, give oxygen. Get medical attention.	
First aid, skin	Flush skin with soap and plenty of water for at least for 15 min. and removing contaminated clothing and shoes. Get medical attention.	
Specific target organ	eyes, central nervous system/methanol	
NIOSH, REL	mg/m³	ST325/methanol; 1900/ethanol

Dowsil Z-6119 Silane

PARAMETER	UNIT	VALUE
OSHA, PEL	mg/m^3	260/methanol; 1900/ethanol
ACGIH, TLV	ppm	200/methanol; 1000/ethanol
NIOSH, REL	ppm	ST250/methanol
OSHA, PEL	ppm	200/methanol; 1000/ethanol
UN risk phrases, R	R20/21/22	
US safety phrases, S	S26,S36/37/39	
UN/NA class	-	1986
ECOLOGICAL PROPERTIES		
Aquatic toxicity, *Green algae*, 96-h EC50	mg/l	22000/96H/methanol, 275/72H/ethanol
Aquatic toxicity, *Bluegill sunfish*, 96-h LC50	mg/l	15400/methanol
Aquatic toxicity, *Daphnia magna*, 48-h LC50	mg/l	>10000/methanol; >10000/24H/urea
Aquatic toxicity *Fathead minnow*, 96-h LC50	mg/l	>1000/ethanol
Biodegradation probability		95%/20d/methanol/readily biodegradable; 84%/20d/ethanol/readily biodegradable; 96%/16d/urea/readily biodegradable
Partition coefficient, log K_{oc}	-	-1.73/urea; -0.35/ethanol; -0.77/methanol
USE & PERFORMANCE		
Manufacturer	Dow	
Outstanding properties	adhesion promoter for plastics, coating. Ethyl carbamate free. No alkyltin compounds	
Processing methods	mix well with filler/solvent slurry and dry for filler treatment.	
Concentrations used	0.01-1 wt%/for better adhesion with plastics and metal; 0.1-5 wt%/paint formulation	

Dowsil Z-6120 Silane

PARAMETER	UNIT	VALUE
GENERAL INFORMATION		
Name		Dowsil Z-6120 Silane
CAS #		23843-64-3, 67-56-1, 57-13.6
Composition		3-ureidopropyltrimethoxysilane, 10-20% methanol, 1-5% urea
Common synonym		[3-(trimethoxysilyl)propyl]urea
Empirical formula		C7H18N2O4Si
Formula		
Molecular mass	daltons	222.31
Chemical class	silane	
Functional organic group		ureido/methoxy
Non-volatile content	%	85-95
PHYSICAL PROPERTIES		
State	-	liquid
Odor	-	alcohol-like
Color	-	colorless
Boiling point	°C	65
Kinematic viscosity at 25°C	cSt	3.5
Specific gravity at 25°C	-	1.02
HEALTH & SAFETY		
NFPA classification	Flammability	3
	Health	2
	Reactivity	0
HMIS classification	Flammability	3
	Health	4
	Reactivity	0
Carcinogenicity		IARC, OSHA, NTP: no ingredient of this product present at levels greater than or equal to 0.1% is identified as probable, possible or confirmed human carcinogen
TDG class		FLAMMABLE LIQUID, N.O.S. (methanol)3,III
ICAO/IATA class		Flammable liquids. n.o.s., (Methanol), 3,III
IMDG class		FLAMMABLE LIQUID, N.O.S., (methanol), 3.III

Dowsil Z-6120 Silane

PARAMETER	UNIT	VALUE
Flash point	°C	29
Flash point method	-	CC
Hazardous combustion products	Carbon oxides, SiO, NOx, formaldehyde, Sulfur oxides	
Hazardous ingredients, labelling	methanol	
Hazardous products of hydrolysis	methanol	
Animal testing, acute toxicity, Rat oral LD50	mg/kg	300/methanol
Animal testing, acute toxicity, Rat inhalation, LC50	mg/m^3	3000/4h/ methanol
Effect of exposure, eye (human)	Avoid contact with eyes. Direct contact may cause severe irritation.	
Effect of exposure, inhalation (human)	Toxic if inhaled.	
Effect of exposure, skin (human)	Toxic in contact with skin.	
Effect of exposure, swallowing (human)	Harmful if swallowed.	
Effect of repeated or overexposure (human)	May cause damage to organs through prolonged or repeated exposure if swallowed.	
Exposure, personal protection	Safety glasses, protective clothing based on chemical resistance data, chemical-resistant gloves, general and local exhaust ventilation.	
First aid, eye	Flush eyes with plenty of water for at least 15 minutes. Remove contact lens, if worn. Get medical attention if irritation develops and persists	
First aid, inhalation	Remove to fresh air. If not breathing, give artificial respiration. If breathing is difficult, give oxygen. Get medical attention.	
First aid, skin	Flush skin with soap and plenty of water for at least for 15 min. and removing contaminated clothing and shoes. Get medical attention.	
Specific target organ	eyes, central nervous system/methanol	
NIOSH, REL	mg/m^3	ST325/methanol; 1900/ethanol
OSHA, PEL	mg/m^3	260/methanol; 1900/ethanol
ACGIH, TLV	ppm	200/methanol; 1000/ethanol
NIOSH, REL	ppm	ST250/methanol
OSHA, PEL	ppm	200/methanol; 1000/ethanol

Dowsil Z-6120 Silane

PARAMETER	UNIT	VALUE
UN risk phrases, R	R20/21/22	
US safety phrases, S	S26,S36/37/39	
UN/NA class	-	1993
ECOLOGICAL PROPERTIES		
Aquatic toxicity, *Green algae*, 96-h EC50	mg/l	22000/96H/methanol
Aquatic toxicity, *Bluegill sunfish*, 96-h LC50	mg/l	15400/methanol
Aquatic toxicity, *Daphnia magna*, 48-h LC50	mg/l	>10000/methanol
Biodegradation probability	95%/20d/methanol/readily biodegradable; 96%/16d/urea/readily biodegradable;	
Partition coefficient, log K_{oc}	-	-1.73/urea; -0.35/ethanol
USE & PERFORMANCE		
Manufacturer	Dow	
Outstanding properties	adhesion promoter for plastics and coatings. Ethyl carbamate free. No alkyltin compounds	
Processing methods	mixed well with filler/solvent slurry and dried for filler treatment.	
Concentrations used	0.01-1 wt%/for better adhesion with plastics and metal; 0.1–5 wt%/paint formulation	

Dowsil Z-6121 Silane

PARAMETER	UNIT	VALUE
GENERAL INFORMATION		
Name		Dowsil Z-6121 Silane
CAS #		1760-24-3, 71-36-3, 67-56-1, (Impurity: 68845-16-9 /74956-86-8,107-15-3)
Name		aminoethylaminopropyltrimethoxysilane
Composition		33-49% N-(3-(Trimethoxysilyl)propyl) ethylenediamine , 33-49% Butan-1-ol, 8-<11 methanol, (Impurity: 3-4% N, N-Bis(3-(Trimethoxysilyl)propyl)-1,2-eth-anediamine, 0.3-0.4% ethylenediamine)
Formula		
Chemical class	-	silane
Mixture	-	yes
Active matter	wt%	62
Functional organic group	-	diamino
Solids content	wt%	34
PHYSICAL PROPERTIES		
State	-	liquid
Odor	-	alcohol-like
Color	colorless to pale yellow	
Color, Platinum-cobalt scale	-	400
Boiling point	°C	>65
Solubility (diluents)	alcohols, water	
Specific gravity at 25°C	-	0.91
Viscosity SUS at 25°C	cSt	5
HEALTH & SAFETY		
NFPA classification	Flammability	3
	Health	3
	Instability	0
HMIS classification	Flammability	3
	Health	3
	Physical hazard	0
Carcinogenicity		IARC, OSHA, NTP: no ingredient of this product present at levels greater than or equal to 0.1% is identified as probable, possible or confirmed human carcinogen

Dowsil Z-6121 Silane

PARAMETER	UNIT	VALUE
TDG class		FLAMMABLE LIQUID.N.O.S., (Butan-1-ol, Methanol), 3,III
ICAO/IATA class		Flammable liquids. n.o.s, (Butan-1-ol, Methanol) 3, III
IMDG class		FLAMMABLE LIQUID.N.O.S., (Butan-1-ol, Methanol) 3,III
Flash point	°C	26.6
Flash point method	-	CC
Hazardous ingredients, labelling	Butan-1-ol, Methanol	
Animal testing, acute toxicity, Rat oral LD50	mg/kg	2292/N-(3-(Trimethoxysilyl) propyl)ethylene-diamine, 790/ butanol-1-ol, 866/ ethylenediamine
Animal testing, acute toxicity, Rabbit dermal LD50	mg/kg	2000/N-(3-(Trimethoxysilyl) propyl)ethylenedi-amine, 3438/butan-1-ol; 560/ ethylenediamine
Animal testing, acute toxicity, Rat dermal LD50	mg/kg	2277/calculated
Animal testing, acute toxicity, Rat inhalation, LC50	mg/m³	17760/4H/ butan-1-ol
Effect of exposure, eye (human)	Causes serious eye damage.	
Effect of exposure, inhalation (human)	May cause allergy or asthma symptoms or breathing difficulties if inhaled. May cause respiratory irritation, drowsiness or dizziness or organs damage. Estimate acute toxicity 4.55 mg/l/4H	
Effect of exposure, skin (human)	Causes skin irritation. May cause an allergic skin reaction. Estimate acute toxicity 454.55 mg/kg	
Effect of exposure, swallowing (human)	Harmful if swallowed. Estimated acute toxicity 454.55 mg/kg	
Exposure, personal protection	Safety glasses, protective clothing based on chemical resistance data, chemical-resistant gloves, general and local exhaust ventilation.	
First aid, eye	Flush eyes with plenty of water for at least 15 minutes. Remove contact lens, if worn. Get immediately medical attention.	

Dowsil Z-6121 Silane

PARAMETER	UNIT	VALUE
First aid, inhalation		Remove to fresh air. If not breathing, give artificial respiration. If breathing is difficult, give oxygen. Get medical attention.
First aid, skin		Flush with plenty of water and treat as a caustic burn removing contaminated clothing and shoes
Specific target organ		eyes, central nervous system/methanol
NIOSH, REL	mg/m^3	150/butan-1-ol; ST325/methanol; TWA25/ ethylenediamine
OSHA, PEL	mg/m^3	300/butan-1-ol; 260/methanol; 25/ ethylenediamine
ACGIH, TLV	ppm	20/butan-1-ol; 200/ methanol;10/ ethylenediamine
NIOSH, REL	ppm	50/butan-1-ol; ST250/methanol; TWA10/ ethylenediamine
OSHA, PEL	ppm	100/butan-1-ol; 200/methanol; 10/ ethylenediamine
UN/NA class	-	1993
ECOLOGICAL PROPERTIES		
Aquatic toxicity, *Green algae*, 96-h EC50	mg/l	225/N-(3-(trimethoxysilyl)propyl)ethylenediamine
Aquatic toxicity, *Bluegill sunfish*, 96-h LC50	mg/l	15400/methanol
Aquatic toxicity, *Daphnia magna*, 48-h LC50	mg/l	1328/N-(3-(trimethoxysilyl)propyl)ethylene-diamine; 16.7/ ethylenediamine; 10000/methanol
Aquatic toxicity *Fathead minnow*, 96-h LC50	mg/l	1376
Biodegradation probability		92%/20d/butan-1-ol; 39%/N-(3-(trimethoxysilyl)propyl)ethylenediamine (not readily biodegradable); 95%/20d/methanol; 95%/28d/ethylenediamine

Dowsil Z-6121 Silane

PARAMETER	UNIT	VALUE
Partition coefficient, log K_{oc}		1/butan-1-ol; -0.3/N-(3-(trimethoxysilyl) propyl)ethylenediamine; -0.77/methanol; -4.42/methylenediamine

USE & PERFORMANCE

Manufacturer	Dow
Outstanding properties	improved adhesion and salt-spray resistance of epoxy coatings, effective over a wide range of concentrations. Eliminates the oven cure normally required to develop optimum properties. Improves adhesion of alkyd finishes to glass
Recommended for polymers	epoxy, alkyd, silicone
Recommended applications	coatings
Concentrations used	epoxy coatings: 4% (total epoxy and curing-agent solids); alkyd coatings 1-2:100 (blended immediately before application)

Dowsil Z-6124 Silane

PARAMETER	UNIT	VALUE
GENERAL INFORMATION		
Name		Dowsil Z-6124 Silane
CAS #	-	2996-92-1
EC number		221-066-9
Composition		94-98% Trimethoxyphenylsilane, (impurities: 0.8-3% methanol, 1.4-3% cyclohexyltrimethoxysilane, 0.13-0.17% 1,2-bis(trimethoxysilyl) ethane)
Empirical formula		C13H27N3O3Si
Formula		
Molecular mass	daltons	301.46
Chemical class	-	silane
Mixture	-	yes
Purity	wt%	>94
PHYSICAL PROPERTIES		
State	-	liquid
Color	-	colorless
	-	alcohol-like
Boiling point	°C	>65
Kinematic viscosity at 25°C	cSt	1.7
Refractive index at 20°C	-	1.47
Solubility in water at 25°C		hydrolytic decomposition releasing methanol
Specific gravity at 25°C	-	1.05
Thermal decomposition products		benzene, formaldehyde
HEALTH & SAFETY		
NFPA classification	Flammability	3
	Health	2
	Instability	0
HMIS classification	Flammability	3
	Health	3
	Physical hazard	1
Carcinogenicity		IARC, OSHA, NTP: no ingredient of this product present at levels greater than or equal to 0.1% is identified as probable, possible or confirmed human carcinogen

Dowsil Z-6124 Silane

PARAMETER	UNIT	VALUE
Mutagenicity	In vitro genetic toxicity studies were negative for component(s) tested. Contains component(s) which were negative in some animal genetic toxicity studies and positive in others	
TDG class	Flammable liquid, N.O.S. (Methanol, Alkoxysilane), 3,III	
ICAO/IATA class	Flammable liquid, n.o.s. (Methanol, Alkoxysilane) 3,III	
IMDG class	Flammable liquid, n.o.s. (Methanol, Alkoxysilane) 3,III	
Flash point	°C	29.4
Flash point method	-	CC
Hazardous combustion products	Carbon oxides, NOx, SiOx, formaldehyde.	
Hazardous ingredients, labelling	methanol, alkoxysilane	
Animal testing, acute toxicity, Rat oral LD50	mg/kg	1049/trimethoxy-phenylsilane,300/methanol
Animal testing, acute toxicity, Rat dermal LD50	mg/kg	2471
Animal testing, acute toxicity, Rat inhalation, LC50	mg/m^3	2 ppm/4H/1,2-bis(trimethoxysilyl)ethane
Effect of exposure, eye (human)	Avoid contact with eyes.	
Effect of exposure, inhalation (human)	Toxic if inhaled. Do not breathe vapors or spray mist. Acute toxicity estimated:17.65 mg//4HI	
Effect of exposure, skin (human)	Acute toxicity estimated: > 5000 mg/kg	
Effect of exposure, swallowing (human)	Harmful if swallowed. Acute toxicity estimated:1029 mg/kg	
Effect of repeated or overexposure (human)	May cause damage to organs (bladder, kidney) through prolonged or repeated exposure if swallowed.	
Exposure, personal protection	Safety glasses, protective clothing based on chemical resistance data, chemical-resistant gloves, general and local exhaust ventilation.	
First aid, eye	Flush eyes with plenty of water for at least 15 minutes. Remove contact lens, if worn. Get medical attention if irritation develops and persists	

Dowsil Z-6124 Silane

PARAMETER	UNIT	VALUE
First aid, inhalation	Remove to fresh air. If not breathing, give artificial respiration. If breathing is difficult, give oxygen. Get medical attention.	
First aid, skin	Flush skin with soap and plenty of water for at least for 15 min. and removing contaminated clothing and shoes. Get medical attention.	
TWA Dow IHG	ppm	5
NIOSH, REL	mg/m^3	ST325/methanol
OSHA, PEL	mg/m^3	260/methanol
ACGIH, TLV	ppm	200/methanol
NIOSH, REL	ppm	ST250/methanol
OSHA, PEL	ppm	200/methanol
UN/NA class	-	1993
ECOLOGICAL PROPERTIES		
Aquatic toxicity, *Green algae*, 72-h EC50	mg/l	>0.17, 22000/96H/methanol,
Aquatic toxicity, *Bluegill sunfish*, 96-h LC50	mg/l	15400/methanol
Aquatic toxicity, *Daphnia magna*, 48-h LC50	mg/l	0.0029
Aquatic toxicity, *Rainbow trout*, 96-h LC50	mg/l	>100
Bioconcentration factor	BCF	<10/methanol/ *Leuciscus idus* (*Golden orfe*)
Biodegradation probability	1%/28d/trimethoxyphenylsilane/not readily biodegradable, 95%/20d/methanol/readily biodegradable	
Partition coefficient, log K_{oc}	-	-0.77 methanol
Manufacturer	Dow	
Recommended for polymers	mineral-filled polymers	
Recommended applications	used to modify surface of inorganic fillers such as wollastonite and aluminum trihydroxide it makes surface of inorganic fillers more hydrophobic. Building block to prepare silicone polymer.	
Guidelines for use	The silane can be reacted in the presence of catalyst such as alkaline, acid, metal salt and can be converted to silicone polymer.	
Alternative product(s)	Xiameter OFS-6124, Silane Dynasylan 9165 CP 0330, SiSiB PC8131, KBM-103	

Dowsil Z-6137 Silane

PARAMETER	UNIT	VALUE
GENERAL INFORMATION		
Name		Dowsil Z-6137 Silane
CAS #	-	68400-09-9
General description		Aminofunctional silane homopolymer adhesion promoter. Low methanol content
Composition		Aminoethylaminopropylsilane triol homo-polymer in water
Common synonym		[3-[(2-Aminoethyl)amino]propyl]silan-etriol homopolymer
Empirical formula		H2NC2H4NHC3H6-Si(OH)3
Chemical class		silane
Mixture	-	yes
Active matter	wt%	22
Neutralization equivalent	-	346
Functional organic group		primary amine
PHYSICAL PROPERTIES		
State	-	liquid
Color	-	pale yellow
Odor	-	amine-like
Boiling point	°C	100
Melting point	°C	<0
Specific gravity	-	1
Refractive index	-	1.3765
Kinematic viscosity at 25°C	cSt	1-5
Solubility in water at 25°C		hydrolytic decomposition releasing methanol
Thermal decomposition products		formaldehyde
Vapor pressure at 20°C	kPa	1.7
HEALTH & SAFETY		
NFPA classification	Flammability	1
	Health	0
	Instability	0
HMIS classification	Flammability	1
	Health	0
	Physical hazard	0

Dowsil Z-6137 Silane

PARAMETER	UNIT	VALUE
Carcinogenicity		IARC, OSHA, NTP: no ingredient of this product present at levels greater than or equal to 0.1% is identified as probable, possible or confirmed human carcinogen
Mutagenicity		not classified based on available information.
DOT class		not regulated as a dangerous goods
TDG class		not regulated as a dangerous goods
ICAO/IATA class		not regulated as a dangerous goods
IMDG class		not regulated as a dangerous goods
Flash point	°C	>100
Flash point method	-	CC
Hazardous combustion products		Carbon oxides, SiO, NOx
Animal testing, acute toxicity, Rabbit dermal LD50	mg/kg	no skin irritation
Animal testing, acute toxicity, Rat inhalation, LC50	mg/m^3	>5300/4H
Effect of exposure, eye (human)		Avoid contact with eyes. Direct contact may cause severe irritation.
Effect of exposure, inhalation (human)		Avoid inhalation of vapor or mist. Acute inhalation toxicity estimate: > 200 mg/l/4H
Effect of exposure, skin (human)		Acute dermal toxicity estimated: >5000 mg/kg
Effect of exposure, swallowing (human)		Harmful if swallowed. Acute oral toxicity estimated: > 5,000 mg/kg
Effect of repeated or overexposure (human)		Damage to health by prolonged exposure through inhalation and in contact with skin and if swallowed.
Exposure, personal protection		Safety glasses, protective clothing based on chemical resistance data, chemical-resistant gloves, general and local exhaust ventilation.
First aid, eye		Flush eyes with plenty of water for at least 15 minutes. Remove contact lens, if worn.
First aid, inhalation		Remove to fresh air. If not breathing, give artificial respiration. If breathing is difficult, give oxygen. Get medical attention.

Dowsil Z-6137 Silane

PARAMETER	UNIT	VALUE
First aid, skin	Flush with plenty of water for at least for 15 min. and removing contaminated clothing and shoes. Get medical attention.	
Specific target organ	eyes, central nervous system/methanol	
NIOSH, REL	mg/m^3	ST325/methanol
OSHA, PEL	mg/m^3	260/methanol
ACGIH, TLV	ppm	200/methanol
NIOSH, REL	ppm	ST250/methanol
OSHA, PEL	ppm	200/methanol
UN risk phrases, R	R20/21/22, R36/37/38	
US safety phrases, S	-	S26,S36/37/39
USE & PERFORMANCE		
Manufacturer	Dow	
Outstanding properties	improved adhesion of the organic polymer to inorganic substrate or filler, improved wet and dry physical properties of the composite, improved mixing and compatibility of the filled system. Extremely stable aqueous solution due to the chemistry of amino-functional silanes. Low methanol content	
Recommended for polymers	acrylic, PA, epoxy, phenolics, PVC, urethanes, melamines, nitrile rubber	
Recommended applications	fiberglass, fillers, coatings, adhesives, sealants, foundry, rubber, plastic, resin	
Guidlines for use	as a primer, dilute to 10% solids in water or IPA and apply by dipping or brushing. Dowsil™ Z-6137 Silane should be added during the final thinning stage at a concentration of 0.5-3.0 wt%. If fillers, silicas, or pigments are present higher levels may be needed as the silane may preferentially absorb onto these surfaces.	
Concentrations used	0.1 - 5.0%	

Dowsil Z-6269 Silane

PARAMETER	UNIT	VALUE
GENERAL INFORMATION		
Name		Dowsil Z-6269 Silane
CAS #		171869-89-2, 67-56-1, 1760-24-3
Composition		30-50% 1,2-ethanediamine, N-[3-(trimethoxysilyl)propyl], N'-[(ethenylphenyl)methyl] derivatives., hydrochlorides, 50-70% methanol, <20% N-(3-(trimethoxysilyl)propyl)ethyl-enediamine
Common synonym		N1-(vinylbenzyl)-N2-(3-(trimethoxysilyl) propyl)ethane-1,2-diamine hydrochloride
Empirical formula		(H2C=CHC6H4-CH2-NHC2H4NHC3H6-Si(OCH3)3)•HCl
Formula		
Molecular mass	daltons	374.5
RTECS number	-	DG0875000/ cas171869-89-9; PC1400/cas 67-56-1; KV7400000/ cas1760-24-3
Mixture	-	yes
Active matter	wt%	32
Functional organic group	-	vinylbenzyl-amino
PHYSICAL PROPERTIES		
State	-	liquid
Odor	-	alcohol-like
Color		greenish yellow changing to reddish amber with time
Boiling point	°C	66.8
Kinematic viscosity at 25°C	cSt	2.0
Refractive index at 20°C	-	1.39 - 1.405
Solubility in water at 25°C		hydrolytic decomposition releasing methanol
Specific gravity at 25°C	-	0.90

Dowsil Z-6269 Silane

PARAMETER	UNIT	VALUE
HEALTH & SAFETY		
NFPA classification	Flammability	3
	Health	3
	Reactivity	0
HMIS classification	Flammability	3
	Health	4
	Reactivity	0
Carcinogenicity	IARC, OSHA, NTP: no ingredient of this product present at levels greater than or equal to 0.1% is identified as probable, possible or confirmed human carcinogen	
Mutagenicity	not classified based on available information.	
TDG class	Flammable liquids. METHANOL SOLUTION, 3,II	
ICAO/IATA class	Flammable liquids, n.o.s. Methanol solution, 3, II	
IMDG class	Flammable liquids. METHANOL SOLUTION, 3,II	
Flash point	°C	10.0
Flash point method	-	CC
Animal testing, acute toxicity, Rat oral LD50	mg/kg	2295/N-(3-(Trimethoxysilyl)propyl)ethylenediamine
Animal testing, acute toxicity, Rabbit dermal LD50	mg/kg	>2000/N-(3-(Trimethoxysilyl)propyl)ethylenediamine
Animal testing, acute toxicity, Rat inhalation, LC50	mg/m^3	1.49/4H/dust/mist/N-(3-(Trimethoxysilyl)propyl)ethylenediamine
Effect of exposure, eye (human)	Causes serious eye damage.	
Effect of exposure, inhalation (human)	May cause allergy or asthma symptoms or breathing difficulties if inhaled. May cause respiratory irritation, drowsiness or dizziness or organs damage. Estimate acute toxicity 4.55 mg/l/4H	
Effect of exposure, skin (human)	Causes skin irritation. May cause an allergic skin reaction. Estimate acute toxicity 454.55 mg/kg	

Dowsil Z-6269 Silane

PARAMETER	UNIT	VALUE
Effect of exposure, swallowing (human)	Harmful if swallowed. Estimated acute toxicity 454.55 mg/kg	
Exposure, personal protection	Safety glasses, protective clothing based on chemical resistance data, chemical-resistant gloves, general and local exhaust ventilation.	
First aid, eye	Flush eyes with plenty of water for at least 15 minutes. Remove contact lens, if worn.	
First aid, inhalation	Remove to fresh air. If not breathing, give artificial respiration. If breathing is difficult, give oxygen. Get medical attention.	
First aid, skin	Flush with plenty of water for at least for 15 min. and removing contaminated clothing and shoes. Get medical attention.	
Specific target organ	eyes, central nervous system/methanol	
NIOSH, REL	mg/m^3	ST325/methanol
OSHA, PEL	mg/m^3	260/methanol
ACGIH, TLV	ppm	200/methanol
NIOSH, REL	ppm	ST250/methanol
OSHA, PEL	ppm	200/methanol
UN/NA class	-	1223
ECOLOGICAL PROPERTIES		
Aquatic toxicity, *Green algae*, 96-h EC50	mg/l	22000/96H/methanol, 8.8/72H/N-(3-(Trimethoxysilyl)propyl)ethylenediamine
Aquatic toxicity, *Bluegill sunfish*, 96-h LC50	mg/l	15400/methanol
Aquatic toxicity, *Daphnia magna*, 48-h LC50	mg/l	10000/methanol
Bioconcentration factor	BCF	<10/methanol/ *Leuciscus idus* (*Golden orfe*)
Biodegradation probability	95%/20d/methanol/readily biodegradable; 39%/N-(3-(trimethoxysilyl)propyl) ethylenediamine (not readily biodegradable)	
Partition coefficient, log K_{oc}	-0.77/methanol; -0.3/N-(3-(trimethoxysilyl)propyl)ethylenediamine;	

Dowsil Z-6269 Silane

PARAMETER	UNIT	VALUE
USE & PERFORMANCE		
Manufacturer		Dow
Outstanding properties		improved adhesion of the organic polymer to inorganic substrate or filler, improved wet and dry physical properties of composite, improved mixing and compatibility of the filled system
Recommended for polymers		epoxies for PCBs, polyolefins, all polymer types
Recommended applications		coupling agent for many resin systems; especially useful for fiberglass-reinforced printed circuit boards

Dowsil Z-6341 Silane

PARAMETER	UNIT	VALUE
GENERAL INFORMATION		
Name		Dowsil Z-6341 Silane
CAS #	-	2943-75-1
EC number	-	220-941-2
Composition		88-100% N-Octyltriethoxysilane, <2.0 branched octyltriethoxysilanes
Common synonym		octyltriethoxysilane
Acronym	-	OCTEO
Empirical formula	-	C14H32O3Si
Formula		
Molecular mass	daltons	276.48
Chemical class	-	silane
PHYSICAL PROPERTIES		
State	-	liquid
Odor	-	alcohol-like
Color		colorless to pale yellow
Boiling point	°C	259.6
Melting point	°C	-75
Kinematic viscosity at 25°C	cSt	2.4
Solubility (diluents)		alcohols, chlorinated solvents, aliphatic solvents and low molecular weight cyclic polydimethylsiloxane
Solubility in water at 25°C		hydrolytic decomposition releasing methanol
Specific gravity at 25°C	-	0.877
Thermal decomposition products	-	formaldehyde
Vapor pressure at 20°C	kPa	0.0011
HEALTH & SAFETY		
Carcinogenicity		IARC, OSHA, NTP: no ingredient of this product present at levels greater than or equal to 0.1% is identified as probable, possible or confirmed human carcinogen
Mutagenicity		negative
DOT class		Combustible Liquid, n.o.s. (Ethanol/ alkoxysilane) III. Applies only to containers over 119 gallons or 450 liters.
TDG class		not regulated as dangerous goods

Dowsil Z-6341 Silane

PARAMETER	UNIT	VALUE
ICAO/IATA class	not regulated as a dangerous goods	
IMDG class	not regulated as a dangerous goods	
Autoignition temperature	°C	225.00
Flash point	°C	65
Hazardous ingredients, labelling	Combustible liquid, n.o.s. Ethanol/alk-oxysilane	
Animal testing, acute toxicity, Rat oral LD50	mg/kg	5110
Animal testing, acute toxicity, Rat dermal LD50	mg/kg	6730
Effect of exposure, skin (human)	Causes skin irritation	
Effect of exposure, swallowing (human)	No significant health effects observed in animals at concentrations of 100 mg/kg	
Exposure, personal protection	Safety glasses, protective clothing based on chemical resistance data, chemical-resistant gloves, general and local exhaust ventilation.	
First aid, eye	Flush eyes with plenty of water for at least 15 minutes. Remove contact lens, if worn. Get medical attention if irritation develops and persists	
First aid, inhalation	Remove to fresh air. If not breathing, give artificial respiration. If breathing is difficult, give oxygen. Get medical attention.	
First aid, skin	Flush skin with soap and plenty of water for at least for 15 min. and removing contaminated clothing and shoes. Get medical attention.	
UN/NA class	-	1993
ECOLOGICAL PROPERTIES		
Aquatic toxicity, *Green algae*, 96-h EC50	mg/l	>0.13/72H
Aquatic toxicity, *Daphnia magna*, 48-h LC50	mg/l	>0.049
Aquatic toxicity, *Rainbow trout*, 96-h LC50	mg/l	>0.055
Biodegradation probability	31.5%/not readily biodegradable	
Partition coefficient, log K_{oc}	-	6.41
USE & PERFORMANCE		
Manufacturer	Dow	
Outstanding properties	improved the compatibility of mineral fillers or pigments in polyolefins or to ease their dispersion in nonpolar matrices	
Recommended applications	commercial buildings, parking decks/garages, highways, bridge structures, filler modification	

Dowsil Z-6341 Silane

PARAMETER	UNIT	VALUE
Concentrations used		typical dilution levels are 20% and 40%
Alternative product(s)		Dynasylan OCTEO VP Si208, SiSiB PC5902

Dowsil Z-6376 Silane

PARAMETER	UNIT	VALUE
GENERAL INFORMATION		
Name		Dowsil Z-6376 Silane
CAS #		5089-70-3
EC number	-	225-805-6
Common synonym		γ-chloropropyltriethoxysilane
Empirical formula	-	C9H21O3ClSi
Formula		
Molecular mass	daltons	240.5
Chemical class	-	silane
Active matter	wt%	99
Functionality	-	chloropropyl
Purity	%	>97
PHYSICAL PROPERTIES		
State	-	liquid
Color	-	colorless
Odor	-	alcohol-like
Boiling point	°C	223
Refractive index at 20°C	-	1.42
Specific gravity at 25°C	-	1.09
Kinematic viscosity at 25°C	cSt	1.42
Thermal decomposition products	formaldehyde	
HEALTH & SAFETY		
NFPA classification	Flammability	2
	Health	0
	Instability	0
HMIS classification	Flammability	2
	Health	0
	Physical hazard	0
Carcinogenicity	IARC, OSHA, NTP: no ingredient of this product present at levels greater than or equal to 0.1% is identified as probable, possible or confirmed human carcinogen	
Mutagenicity	negative	
TDG class	FLAMMABLE LIQUIDS, N.O.S. (Ethanol, Chloropropyl triethoxysilane) 3,III	

Dowsil Z-6376 Silane

PARAMETER	UNIT	VALUE
ICAO/IATA class	Flammable liquids. n.o.s. (Ethanol, Chloropropyl triethoxysilane), 3,III	
IMDG class	FLAMMABLE LIQUID, N.O.S. (Ethanol, Chloropropyl triethoxysilane), 3,III	
Flash point	°C	38
Hazardous combustion products	carbon oxides, SiO, NO_x, formaldehyde, sulfur oxides	
Hazardous ingredients, labelling	ethanol, chloropropyl triethoxysilane	
Hazardous products of hydrolysis	methanol	
Animal testing, acute toxicity, Rat oral LD50	mg/kg	>5000
Animal testing, acute toxicity, Rabbit dermal LD50	mg/kg	>2000
Effect of exposure, eye (human)	Avoid contact with eyes. Based on available information not classified as a eye irritant. Rabbit irritation to eyes, reversing within 21 days.	
Effect of exposure, inhalation (human)	Not classified as a irritant based on available information.	
Effect of exposure, skin (human)	Does not cause skin sensitization.	
Effect of exposure, swallowing (human)	Harmful if swallowed.	
Exposure, personal protection	Safety glasses, protective clothing based on chemical resistance data, chemical-resistant gloves, general and local exhaust ventilation.	
First aid, eye	Flush eyes with plenty of water for at least 15 minutes. Remove contact lens, if worn. Get medical attention if irritation develops and persists	
First aid, inhalation	Remove to fresh air. If not breathing, give artificial respiration. If breathing is difficult, give oxygen. Get medical attention.	
First aid, skin	Flush skin with soap and plenty of water for at least for 15 min. and removing contaminated clothing and shoes. Get medical attention.	
Specific target organ	eyes, central nervous system/methanol	
NIOSH, REL	mg/m^3	1900/ethanol
OSHA, PEL	mg/m^3	1900/ethanol
ACGIH, TLV	ppm	1000/ethanol
NIOSH, REL	ppm	1000/ethanol
OSHA, PEL	ppm	1000/ethanol

Dowsil Z-6376 Silane

PARAMETER	UNIT	VALUE
UN/NA class	-	1993
ECOLOGICAL PROPERTIES		
Aquatic toxicity, *Green algae*, 96-h EC50	mg/l	>819/72H/chloro-propyl triethoxysilane
Aquatic toxicity, *Daphnia magna*, 48-h LC50	mg/l	21.2/chloropropyl triethoxysilane, >1000/ethanol
Biodegradation probability		84%/20d/ethanol/readily biodegradable; 46% not readily biodegradable
Partition coefficient, log K_{oc}	-	3.13/chloropropyl triethoxysilane; -0.35/ ethanol
USE & PERFORMANCE		
Manufacturer	Dow	
Outstanding properties	improved chemical bonding of resins to inorganic materials, which improved their physical-mechanical properties, a precursor for functional silane, sol-gel coating ingredient, adhesion promoter, filler treatment	
Recommended for polymers	chlorobutadiene rubber, chlorinated butyl rubber, chlorohydrin rubber, chlorosulfonated polyethylene, etc.	
Recommended applications	rubber-processing aid, used to treat and couple the inorganic fillers in halogenated rubber	

Dowsil Z-6675 Silane

PARAMETER	UNIT	VALUE
GENERAL INFORMATION		
Name		Dowsil Z-6675 Silane
CAS #		23779-32-0 or 116912-64-2
EC number	-	245-876-7
Composition		ureidopropyltrialkoxy
Common synonym		3-ureidopropyltriethoxysilane
Empirical formula	-	C10H24N2O4Si
Formula		

PARAMETER	UNIT	VALUE
Molecular mass	daltons	246.39
Chemical class	-	silane
Mixture	-	yes
Functional organic group	ureido/ethoxy	
Non-volatile content	%	80-90
PHYSICAL PROPERTIES		
State	-	liquid
Color	-	colorless
Odor	-	alkohol-like
Boiling point	°C	65
Density	kg/m³	1030
Kinematic viscosity at 25°C	cSt	35
HEALTH & SAFETY		
NFPA classification	Flammability	3
	Health	2
	Instability	0
HMIS classification	Flammability	3
	Health	2
	Physical hazard	0
Carcinogenicity	IARC, OSHA, NTP: no ingredient of this product present at levels greater than or equal to 0.1% is identified as probable, possible or confirmed human carcinogen	
ICAO/IATA class	Alcohols, n.o.s.,(Ethanol, Methanol), Flammable Liquids, 3,II	
IMDG class	ALCOHOLS, N.O.S, (Ethanol, Methanol),3,II	
Flash point	°C	27

Dowsil Z-6675 Silane

PARAMETER	UNIT	VALUE
Flash point method	-	SCC
Hazardous combustion products	Carbon oxides, SiO, NO$_x$, formaldehyde, Sulfur oxides	
Hazardous ingredients, labelling	ethanol, methanol	
Hazardous products of hydrolysis	methanol	
Animal testing, acute toxicity, Rat oral LD50	mg/kg	1479-2665
Effect of exposure, eye (human)	Avoid contact with eyes. Direct contact may cause severe irritation.	
Effect of exposure, inhalation (human)	Toxic if inhaled.	
Effect of exposure, skin (human)	Toxic in contact with skin.	
Effect of exposure, swallowing (human)	Harmful if swallowed.	
Effect of repeated or overexposure (human)	May cause damage to organs through prolonged or repeated exposure if swallowed.	
Exposure, personal protection	Safety glasses, protective clothing based on chemical resistance data, chemical-resistant gloves, general and local exhaust ventilation.	
First aid, eye	Flush eyes with plenty of water for at least 15 minutes. Remove contact lens, if worn. Get medical attention if irritation develops and persists	
First aid, inhalation	Remove to fresh air. If not breathing, give artificial respiration. If breathing is difficult, give oxygen. Get medical attention.	
First aid, skin	Flush skin with soap and plenty of water for at least for 15 min. and removing contaminated clothing and shoes. Get medical attention.	
NIOSH, REL	mg/m^3	ST325/methanol
ACGIH, TLV	ppm	200/methanol; 1000/ethanol
NIOSH, REL	ppm	ST250/methanol
OSHA, PEL	ppm	200/methanol; 1000/ethanol
UN risk phrases, R	R20/21/22	
US safety phrases, S	S26,S36/37/39	

Dowsil Z-6675 Silane

PARAMETER	UNIT	VALUE
UN/NA class	-	1993
ECOLOGICAL PROPERTIES		
Aquatic toxicity, *Green algae*, 72-h ErC50	mg/l	>1000
Aquatic toxicity, *Zebra fish*, 96-h LC50	mg/l	>934
Aquatic toxicity, *Daphnia magna*, 48-h EC50	mg/l	331
Biodegradation probability	based on stringent OECD test guidelines, this material cannot be considered as readily biodegradable; however, these results do not necessarily mean that the material is not biodegradable under environmental conditions.	
USE & PERFORMANCE		
Manufacturer	Dow	
Outstanding properties	adhesion promoter for plastics, coating. ethyl carbamate free. No alkyltin compounds	
Processing methods	mixed well with filler/solvent slurry and dried for filler treatment.	
Concentrations used	0.01-1 wt%/for better adhesion with plastics and metal; 0.1–5 wt%/paint formulation	

Dowsil Z-6883 Silane

PARAMETER	UNIT	VALUE
GENERAL INFORMATION		
Name		Dowsil Z-6883 Silane
CAS #	-	3068-76-6
EC number	-	221-328-2
Composition		90-100% Phenylaminopropyltrime-thoxysilane
Common synonym		3-(Phenylamino) propyltrimethoxysilane
Empirical formula		PhNHC3H6Si(OCH3)3
Formula		
Molecular mass	daltons	255.39
Chemical class	-	silane
Functional organic group	-	aminophenyl
Purity	%	>96
PHYSICAL PROPERTIES		
State	-	liquid
Odor	-	slight
Color	-	yellowish brown
Boiling point	°C	>35
Kinematic viscosity at 25°C	cSt	9
Refractive index at 20°C	-	1.506
Solubility in water at 25°C		hydrolytic decomposition releasing methanol
Specific gravity at 25°C	-	1.067
Thermal decomposition products	-	formaldehyde
HEALTH & SAFETY		
NFPA classification	Flammability	1
	Health	1
	Instability	0
HMIS classification	Flammability	1
	Health	0
	Physical hazard	0
Carcinogenicity		IARC, OSHA, NTP: no ingredient of this product present at levels greater than or equal to 0.1% is identified as probable, possible or confirmed human carcinogen

Dowsil Z-6883 Silane

PARAMETER	UNIT	VALUE
Mutagenicity	not classified based on available information.	
DOT class	not regulated as a dangerous goods	
TDG class	not regulated as a dangerous goods	
ICAO/IATA class	not regulated as a dangerous goods	
IMDG class	not regulated as a dangerous goods	
Flash point	°C	>100
Flash point method	-	CC
Animal testing, acute toxicity, Rabbit dermal LD50	mg/kg	>2000
Animal testing, acute toxicity, Rat oral LD50	mg/kg	809
Effect of exposure, eye (human)	Avoid contact with eyes. Direct contact may cause severe irritation.	
Effect of exposure, inhalation (human)	Avoid inhalation of vapor or mist. Acute inhalation toxicity estimate: > 200 mg/l/4H	
Effect of exposure, skin (human)	May cause burns. Avoid prolonged or repeated contact with skin. Acute dermal toxicity estimate: > 5,000 mg/kg	
Effect of exposure, swallowing (human)	Harmful if swallowed. Acute oral toxicity estimate: 804.66 mg/kg	
Effect of repeated or overexposure (human)	Damage to health by prolonged exposure through inhalation and in contact with skin and if swallowed.	
Exposure, personal protection	Safety glasses, protective clothing based on chemical resistance data, chemical-resistant gloves, general and local exhaust ventilation.	
First aid, eye	Flush eyes with plenty of water for at least 15 minutes. Remove contact lens, if worn.	
First aid, inhalation	Remove to fresh air. If not breathing, give artificial respiration. If breathing is difficult, give oxygen. Get medical attention.	
First aid, skin	Flush with plenty of water for at least for 15 min. and removing contaminated clothing and shoes. Get medical attention.	
Specific target organ	eyes, central nervous system/methanol	
NIOSH, REL	mg/m³	ST325/methanol
OSHA, PEL	mg/m³	260/methanol
ACGIH, TLV	ppm	200/methanol

Dowsil Z-6883 Silane

PARAMETER	UNIT	VALUE
NIOSH, REL	ppm	ST250/methanol
OSHA, PEL	ppm	200/methanol
UN risk phrases, R	R34,R40,R48/20/21/22,R52/53	
US safety phrases, S	-	S26,S36/37/39,S45
UN/NA class	-	3267
ECOLOGICAL PROPERTIES		
Aquatic toxicity, *Green algae*, 96-h EC50	mg/l	22000/96H/methanol
Aquatic toxicity, *Bluegill sunfish*, 96-h LC50	mg/l	15400/methanol
Aquatic toxicity, *Daphnia magna*, 48-h LC50	mg/l	10000/methanol
Bioconcentration factor	BCF	<10/methanol/*Leuciscus idus* (*Golden orfe*)
Biodegradation probability	95%/20d/methanol/readily biodegradable	
Partition coefficient, log P_{ow}	-	1.1
USE & PERFORMANCE		
Manufacturer	Dow	
Outstanding properties	Improved adhesion between metal and resin, and inorganic fillers and resin	
Recommended for polymers	EMC	
Recommended applications	improves adhesion between metal and EMC	

Dowsil™ Z-8090

PARAMETER	UNIT	VALUE
GENERAL INFORMATION		
Name		Dowsil™ Z-8090
CAS #	-	23410-40-4
Common synonym		aminoethylaminoisobutylmethyldimethoxysilane
Chemical name		N-[3-(dimethoxymethylsilyl)-2-methylpropyl]ethylenediamine
Active matter	wt%	89-100
PHYSICAL PROPERTIES		
State	-	liquid
Odor	-	fishy
Color	-	colorless
Boiling point	°C	>150
Kinematic visccsity at 25°C	cSt	4.2
Specific gravity at 25°C	-	0.96
HEALTH & SAFETY		
NFPA classification	Flammability	1
	Health	3
	Instability	0
HMIS classification	Flammability	1
	Health	3
	Physical hazard	0
Mutagenicity	*In vitro* genetic toxicity studies were negative.	
Teratogenicity	Did not cause birth defects or any other fetal effects in laboratory animals.	
Flash point	°C	106
Flash point method	-	PMCC
Animal testing, acute toxicity, Rat oral LD50	mg/kg	653
Animal testing, acute toxicity, Rabbit dermal LD50	mg/kg	>2000
Animal testing, acute toxicity, Rat inhalation, LC50	mg/l	0.6/4H

Dowsil™ Z-8090

PARAMETER	UNIT	VALUE
First aid, eye	Wash immediately and continuously with flowing water for at least 30 minutes. Remove contact lenses after the first 5 minutes and continue washing. Obtain prompt medical consultation, preferably from an ophthalmologist. Suitable emergency eye wash facility should be immediately available.	
First aid, inhalation	Move person to fresh air and keep comfortable for breathing; consult a physician.	
First aid, skin	Remove material from skin immediately by washing with soap and plenty of water. Remove contaminated clothing and shoes while washing. Seek medical attention if irritation or rash occurs. Wash clothing before reuse. Discard items which cannot be decontaminated, including leather articles such as shoes, belts and watchbands. Suitable emergency safety shower facility should be available in work area.	
UN/NA class	-	3082
ECOLOGICAL PROPERTIES		
Aquatic toxicity, *Green algae*, 72-h EC50	mg/l	8.8
Aquatic toxicity, *Daphnia magna*, 48-h LC50	mg/l	81
Bioaccumulative potential	Bioconcentration potential is low	
Bioconcentration factor	BCF	<100
Biodegradation probability	Based on stringent OECD test guidelines, this material cannot be considered as readily biodegradable; however, these results do not necessarily mean that the material is not biodegradable under environmental conditions.	
Partition coefficient, log P_{ow}	-	1.4
USE & PERFORMANCE		
Manufacturer	Dow	
Recommended for products	paints, inks and coatings	

Dowsil Z-9805 Silane

PARAMETER	UNIT	VALUE
GENERAL INFORMATION		
Name		Dowsil Z-9805 Silane
CAS #	-	780-69-8
EC number		212-305-0
Composition		90-100% triethoxy(phenyl)silane, 1.3-1.7% ethanol
Common synonym		Phenyltriethoxysilane
Empirical formula		(C6H5-Si(OC2H5)3
Formula		
Molecular mass	daltons	240.00
RTECS number	-	VV4900000
Chemical class	-	silane
Functional organic group	-	alkoxyl
Purity	wt%	>96
PHYSICAL PROPERTIES		
State	-	liquid
Odor	-	characteristic
Color	colorless to pale yellow	
Boiling point	°C	235
Kinematic viscosity at 25°C	cSt	1.66
Specific gravity at 25°C	-	1.1
Thermal decomposition products	formaldehyde	
HEALTH & SAFETY		
NFPA classification	Flammability	3
	Health	0
	Reactivity	0
HMIS classification	Flammability	3
	Health	2
	Reactivity	1
Carcinogenicity	IARC, OSHA, NTP: no ingredient of this product present at levels greater than or equal to 0.1% is identified as probable, possible or confirmed human carcinogen	
Mutagenicity	not classified based on available information	

Dowsil Z-9805 Silane

PARAMETER	UNIT	VALUE
TDG class	FLAMMABLE LIQUID, N.O.S. (Ethanol, Cyclohexyltriethoxysilane) 3,III	
ICAO/IATA class	Flammable liquid, n.o.s. (Ethanol, Cyclohexyltriethoxysilane) 3,III	
IMDG class	FLAMMABLE LIQUID, N.O.S. (Ethanol, Cyclohexyltriethoxysilane) 3,III	
Flash point	°C	35
Flash point method	-	SCC
Hazardous combustion products	Carbon oxides, SiOx, formaldehyde.	
Hazardous ingredients, labelling	Ethanol, Cyclohexyltriethoxysilane	
Animal testing, acute toxicity, Rat oral LD50	mg/kg	2830/ triethoxy(phenyl) silane, >7000 ethanol
Animal testing, acute toxicity, Rabbit dermal LD50	mg/kg	3180/ triethoxy(phenyl) silane, >15,800 ethanol
Effect of exposure, eye (human)	Direct contact may cause severe irritation. Vapor may cause eye irritation	
Effect of exposure, skin (human)	Does not cause skin sensitization. Acute toxicity estimated: 3265 mg/kg	
Effect of exposure, swallowing (human)	Low ingestion hazard in normal use. Acute toxicity estimated: 2906 mg/kg	
Effect of repeated or overexposure (human)	May cause damage to organs (bladder, kidney) through prolonged or repeated exposure if swallowed.	
Exposure, personal protection	Safety glasses, protective clothing based on chemical resistance data, chemical-resistant gloves, general and local exhaust ventilation.	
First aid, eye	Flush eyes with plenty of water for at least 15 minutes. Remove contact lens, if worn. Get medical attention if irritation develops and persists	
First aid, inhalation	Remove to fresh air. If not breathing, give artificial respiration. If breathing is difficult, give oxygen. Get medical attention.	
First aid, skin	Flush skin with soap and plenty of water for at least for 15 min. and removing contaminated clothing and shoes. Get medical attention.	
NIOSH, REL	mg/m^3	1900/ethanol

Dowsil Z-9805 Silane

PARAMETER	UNIT	VALUE
OSHA, PEL	mg/m^3	1900/ethanol
ACGIH, TLV	ppm	1000/ethanol
NIOSH, REL	ppm	1000/ethanol
OSHA, PEL	ppm	1000/ethanol
UN/NA class	-	1993
ECOLOGICAL PROPERTIES		
Aquatic toxicity, *Green algae*, 96-h EC50	mg/l	>100/72H/ triethoxy(phenyl) silane, 275/72H/ ethanol
Aquatic toxicity, *Daphnia magna*, 48-h LC50	mg/l	>100/ triethoxy(phenyl) silane
Partition coefficient, log K_{oc}	-	-0.35 ethanol
Stability in water (half-life)	-	1.4 h/pH 7
USE & PERFORMANCE		
Manufacturer	Dow	
Outstanding properties	hydrophobing agent and high-temperature additive for other coupling agents, makes inorganic surfaces hydrophobic	
Recommended for polymers	epoxy, polyester, mineral-filled polymers	
Recommended applications	used to modify the surface of inorganic fillers such as wollastonite and aluminum trihydroxide; it makes the surface of inorganic fillers more hydrophobic	
Alternative product(s)	P-Triethoxy, Dynasylan 9165 CP 032, SiSiB PC8132, KBE-103	
Guidelines for use	the tendency toward self-condensation can be controlled by using fresh solutions, alcoholic solvents, dilution, and careful selection on pH ranges. Silane-triols are most stable at pH 3–4 but condense rapidly at pH 7–9.	

Dynasylan 1122

PARAMETER	UNIT	VALUE
GENERAL INFORMATION		
Name		Dynasylan 1122
CAS #	-	13497-18-2
EC number	-	236-818-1
Composition		bis(3-triethoxysilylpropyl) amine
Common synonym		1-propanamine, 3-(triethoxysilyl)-N-[3-(triethoxysilyl)propyl]-
Empirical formula		C18H43NO6Si2
Formula		
Molecular mass	daltons	425.71
Functional organic group		amino
PHYSICAL PROPERTIES		
State	-	liquid
Odor	-	amine-like
Color		clear colorless to yellowish
Boiling point	°C	408
Melting point	°C	-38
Density at 20°C	kg/m³	970
Refractive index at 20°C	-	1.4265
Solubility (diluents)		alcohols, aliphatic or aromatic hydrocarbons
Specific gravity at 25°C	-	0.97
Vapor pressure at 20°C	kPa	<0.001
Viscosity at 20°C	mPas	5.5
HEALTH & SAFETY		
NFPA classification	Flammability	1
	Health	2
	Reactivity	1
HMIS classification	Flammability	1
	Health	2
	Reactivity	1
DOT class	not regulated	
TDG class	not regulated	
ICAO/IATA class	not regulated	
IMDG class	not regulated	

Dynasylan 1122

PARAMETER	UNIT	VALUE
Autoignition temperature	°C	255
Flash point	°C	95
Flash point method	-	PMCC
Hazardous combustion products		Carbon oxides, Nitrogen oxides (NOx)
Agency rating, listed		EINECS Europe, ECL Korea, ENCS Japan, IECSC China, DSL Canada, PICCS Philippines, TSCA USA
Animal testing, acute toxicity, Rat oral LD50	mg/kg	4580
Effect of exposure, eye (human)		Causes serious eye irritation.
Effect of exposure, inhalation (human)		Avoid inhalation aerosol of mists.
Effect of exposure, skin (human)		Causes skin irritation.
Exposure, personal protection		Safety glasses, protective clothing based on chemical resistance data, chemical-resistant gloves, general and local exhaust ventilation.
First aid, eye		Rinse cautiously with water for several minutes. Remove contact lenses, if present and easy to do. Continue rinsing. If eye irritation persists: Get medical advice/attention.
First aid, inhalation		Remove to fresh air. If not breathing, give artificial respiration. If breathing is difficult, give oxygen. If irritation persists, obtain medical advice.
First aid, skin		Immediately wipe away excess material. Use a waterless hand cleaner to remove as much of the remaining material as possible. Wash with soap and water.
UN risk phrases, R	-	R34
US safety phrases, S		S26,S28,S36/37/39,S45
ECOLOGICAL PROPERTIES		
Aquatic toxicity, *Green algae*, 96-h EC50	mg/l	118/72H
Biodegradation probability		64.5%/28d/not readily biodegradable
USE & PERFORMANCE		
Manufacturer		Evonik Industries
Recommended for polymers		amino resin, acrylics, acrylic copolymers, phenol formaldehydes, PU, silicone, furane and melamine resins
Recommended for products		adhesives and sealants, composites, foundry resins, HFFR cables, paints and coatings

Dynasylan 1122

PARAMETER	UNIT	VALUE
Recommended applications		adhesion promoter to both inorganic materials (for example, glass, metals, and fillers) and organic polymers (thermosets, thermoplastics, and elastomers) as a surface modifier and can be used for the chemical modification of substances. Examples of applications: glass fiber/glass fabric composites: as size constituent or finish, mineral fiber insulating materials, abrasives: as an additive to phenolic resin binders, mineral-filled polymers (composites), or HFFR cables: for pretreatment of fillers and pigments.

Dynasylan 1124

PARAMETER	UNIT	VALUE
GENERAL INFORMATION		
Name	Dynasylan 1124	
CAS #	-	82985-35-1
EC number	-	280-084-5
Composition	bis(3-trimethoxysilylpropyl) amine (impurity <0.3% methanol)	
Empirical formula	-	C12H31NO6Si2
Formula		
Molecular mass	daltons	341.55
RTECS number	-	TX2101000
Chemical class	-	silane
Functional organic group	amino	
PHYSICAL PROPERTIES		
State	-	liquid
Odor	-	amine-like
Color	-	slightly yellowish
Boiling point	°C	285-288
Melting point	°C	-38
Density at 20°C	kg/m^3	1,040
Kinematic viscosity at 20°C	cSt	5.7
Solubility in water at 25°C	hydrolytic decomposition releasing methanol	
Specific gravity at 25°C	-	1.04
Vapor pressure at 20°C	kPa	<0.001
Viscosity at 20°C	mPas	6.5
HEALTH & SAFETY		
NFPA classification	Flammability	1
	Health	3
	Reactivity	1
HMIS classification	Flammability	1
	Health	2
	Reactivity	1
Carcinogenicity	IARC, OSHA, NTP: no ingredient of this product present at levels greater than or equal to 0.1% is identified as probable, possible or confirmed human carcinogen	

Dynasylan 1124

PARAMETER	UNIT	VALUE
DOT class		Environmentally hazardous substance, liquid, n.o.s.(bis(trimethoxysilylpropyl)amine) 9,III
TDG class		Environmentally hazardous substance, liquid, n.o.s.(bis(trimethoxysilylpropyl)amine) 9,III
ICAO/IATA class		Environmentally hazardous substance, liquid, n.o.s.(bis(trimethoxysilylpropyl)amine) 9,III
IMDG class		ENVIRONMENTALLY HAZARDOUS SUBSTANCE, LIQUID, N.O.S. (bis(trimethoxysilylpropyl)amine) 9,III
Flash point	°C	>100
Flash point method	-	PMCC
Hazardous combustion products		Carbon oxides, Nitrogen oxides (NOx)
Agency rating, listed		EINECS Europe, ECL Korea, ENCS Japan, IECSC China, DSL Canada, PICCS Philippines, TSCA USA
Hazardous ingredients, labelling		Environmentally hazardous substance, liquid, n.o.s.(bis(trimethoxysilylpropyl)amine)
Animal testing, acute toxicity, Rat oral LD50	mg/kg	>2000
Animal testing, acute toxicity, Rat dermal LD50	mg/kg	16640/male, 11752/female
Animal testing, acute toxicity, Rat inhalation, LC50	mg/m^3	>40000/4H
Effect of exposure, eye (human)		Causes serious eye damage.
Effect of exposure, inhalation (human)		Avoid inhalation aerosol of mists..
Exposure, personal protection		Safety glasses, protective clothing based on chemical resistance data, chemical-resistant gloves, general and local exhaust ventilation.
First aid, eye		Rinse cautiously with water for several minutes. Remove contact lenses, if present and easy to do. Continue rinsing. If eye irritation persists: Get medical advice/attention.
First aid, inhalation		Remove to fresh air. If not breathing, give artificial respiration. If breathing is difficult, give oxygen. If irritation persists, obtain medical advice.

Dynasylan 1124

PARAMETER	UNIT	VALUE
First aid, skin		Immediately wipe away excess material. Use a waterless hand cleaner to remove as much of the remaining material as possible. Wash with soap and water.
NIOSH, REL	mg/m³	ST325/methanol
OSHA, PEL	mg/m³	260/methanol
UN risk phrases, R	-	R20/22,R36/37/38
US safety phrases, S	-	S26,S36/37/39
UN/NA class	-	3082
ECOLOGICAL PROPERTIES		
Aquatic toxicity, *Green algae*, 96-h EC50	mg/l	>1000/72H
Aquatic toxicity, *Daphnia magna*, 48-h LC50	mg/l	>100
Aquatic toxicity, *Rainbow trout*, 96-h LC50	mg/l	130
Bioaccumulative potential	not bioaccumulative	
Biodegradation probability	64.5%/28d/not readily biodegradable	
USE & PERFORMANCE		
Manufacturer	Evonik Industries	
Outstanding properties	improved flexural strength, tensile strength, impact strength, modulus of elasticity, moisture and corrosion resistance. Offers benefits such as improvement in wet-out, homogeneous distribution of inorganic fillers in polymer matrices, rheological behavior like reduction in viscosity and Newtonian behavior.	
Recommended for polymers	amino resin, acrylics, acrylic copolymers, phenol formaldehyde, PVC, PU, silicone, furan and melamine resins	
Recommended for products	abrasives, insulating materials, foundry resins, adhesives and sealants, composites, paints, and coatings	
Recommended applications	adhesion promoter to both inorganic materials (for example glass, metals, and fillers) and organic polymers (thermosets, thermoplastics, and elastomers), especially suitable for amine hardeners for bonding pastes. High crosslinking potential.	

Dynasylan 1146

PARAMETER	UNIT	VALUE
GENERAL INFORMATION		
Name		Dynasylan 1146
General description		functional oligosiloxane
Composition		Alkylpolysiloxanes, aminoalkyl groups modified + (impurity: max.10% Trimethoxypropylsilane, methanol max 1%)
PHYSICAL PROPERTIES		
State	-	liquid
Odor	-	amino-like
Color		clear to slightly opaque, colorless to yellowish
Boiling point	°C	280-290
Density at 20°C	kg/m^3	1060
pH	10-11 at 20°C (20 g/l H$_2$O)	
Solubility (diluents)	organic solvents such as methanol, ethanol, MEK	
Solubility in water at 25°C	hydrolytic decomposition releasing methanol	
Vapor pressure at 20°C	kPa	<0.1
Viscosity at 25°C	mPas	35
HEALTH & SAFETY		
NFPA classification	Flammability	2
	Health	1
	Reactivity	1
HMIS classification	Flammability	2
	Health	2
	Reactivity	1
DOT class	not regulated in packages 450 liters or less	
TDG class	not regulated in packages 450 liters or less	
ICAO/IATA class	not regulated	
IMDG class	not regulated	
Flash point	°C	>60.1
Flash point method	-	PMCC
Agency rating, listed	EINECS Europe, ECL Korea, IECSC China, DSL Canada, TSCA USA	
Hazardous ingredients, labelling	Combustible liquid, n.o.s.(trimethoxypropylsilane, methanol)	

Dynasylan 1146

PARAMETER	UNIT	VALUE
Effect of exposure, eye (human)	Avoid contact with eyes.	
Effect of exposure, inhalation (human)	May cause respiratory irritation. Do not inhale vapors or aerosols	
Effect of exposure, skin (human)	Avoid contact with skin.	
Effect of exposure, swallowing (human)	Not expected in industrial use. May be harmful if swallowed.	
Exposure, personal protection	Safety glasses, protective clothing based on chemical resistance data, chemical-resistant gloves, general and local exhaust ventilation.	
First aid, eye	Rinse cautiously with water for several minutes. Remove contact lenses, if present and easy to do. Continue rinsing. If eye irritation persists: Get medical advice/attention.	
First aid, inhalation	Remove to fresh air. If not breathing, give artificial respiration. If breathing is difficult, give oxygen. If irritation persists, obtain medical advice.	
First aid, skin	Immediately wipe away excess material. Use a waterless hand cleaner to remove as much of the remaining material as possible. Wash with soap and water.	
NIOSH, REL	mg/m^3	ST325/methanol
OSHA, PEL	mg/m^3	260/methanol
ACGIH, TLV	ppm	200/methanol
OSHA, PEL	ppm	200/methanol
UN/NA class	-	1993
ECOLOGICAL PROPERTIES		
Biodegradation probability	12.8%/28d/not readily biodegradable	
USE & PERFORMANCE		
Manufacturer	Evonik Industries	
Outstanding properties	improved the crosslinking densities of bonding pastes and imparted outstanding hydrophobicity. Innovative silane due to reduced VOC. Low viscosity and low volatility and low odor level during application. Wet adhesion to inorganic and organic substrates	
Recommended for polymers	RTV-silicone, 2K PU, silylated urethanes, two-part epoxy (EP)	

Dynasylan 1146

PARAMETER	UNIT	VALUE
Recommended applications		adhesives, sealants and coatings to improve adhesion of amino-reactive resin to inorganic surface, plastics and inorganic fillers
Concentrations used		0.5-1.5 wt% of total formulation

Dynasylan 1161 EQ

PARAMETER	UNIT	VALUE
GENERAL INFORMATION		
Name	Dynasylan 1161 EQ	
CAS #	42965-91-3, 67-56-1	
EC number	-	256-023-3
Composition	50% 1,2-Ethanediamine,N-(phenylmethyl)-N'-[3-(trimethoxysilyl)propyl]-, monohydrochloride (9CI) + >50% methanol, + 0.1-<1% benzyl chloride	
Common synonym	N-Benzyl-N'-[3-(trimethoxysilyl)propyl] ethylenediamine monohydrochloride	
Empirical formula	$C_{15}H_{28}N_2O_3Si.ClH$	
Formula		
Molecular mass	daltons	348.94
Chemical class	-	silane
Mixture	-	yes
Functional organic group	amino/benzyl	
PHYSICAL PROPERTIES		
State	-	liquid
Odor	-	amino-like
Color	clear colorless to yellowish	
Boiling point	°C	65/methanol
Density at 20°C	kg/m³	943
pH	7.4 at 20°C (10 g/l H_2O)	
Solubility (diluents)	alcohols, aliphatic or aromatic hydrocarbons	
Specific gravity at 25°C	-	0.943
Vapor pressure at 20°C	kPa	11.4
Viscosity at 20°C	mPas	3.45
HEALTH & SAFETY		
NFPA classification	Flammability	3
	Health	2
	Reactivity	0
HMIS classification	Flammability	3
	Health	2
	Reactivity	0
Carcinogenicity	Benzyl chloride/1B carcinogenicity.	

Dynasylan 1161 EQ

PARAMETER	UNIT	VALUE
DOT class	Methanol(solution) 3(6.1) II	
ICAO/IATA class	Methanol(solution) 3(6.1) II	
Autoignition temperature	°C	375
Flash point	°C	8
Flash point method	-	CC
Explosive LEL	wt%	5.5/methanol
Explosive UEL	wt%	44/methanol
Hazardous combustion products	Carbon oxides, chlorine compounds, NO_x	
Hazardous products of hydrolysis	methanol	
Animal testing, acute toxicity, Rat oral LD50	mg/kg	1250/ cas 42965-91-3
Animal testing, acute toxicity, Rat dermal LD50	mg/kg	>2000
Effect of exposure, eye (human)	Causes serious eye damage.	
Effect of exposure, skin (human)	May cause an allergic skin reaction.	
Exposure, personal protection	Safety glasses, protective clothing based on chemical resistance data, chemical-resistant gloves, general and local exhaust ventilation.	
First aid, eye	Rinse cautiously with water for several minutes. Remove contact lenses, if present and easy to do. Continue rinsing. If eye irritation persists: Get medical advice/attention.	
First aid, inhalation	Remove to fresh air. If not breathing, give artificial respiration. If breathing is difficult, give oxygen. If irritation persists, obtain medical advice.	
First aid, skin	Immediately wipe away excess material. Use a waterless hand cleaner to remove as much of the remaining material as possible. Wash with soap and water.	
NIOSH, REL	mg/m³	ST325/methanol
OSHA, PEL	mg/m³	260/methanol, 5/ benzyl chloride
ACGIH, TLV	ppm	200/methanol, 1/ benzyl chloride
NIOSH, REL	ppm	ST250/methanol
OSHA, PEL	ppm	200/methanol, 1/ benzyl chloride
UN/NA class	-	1230

Dynasylan 1161 EQ

PARAMETER	UNIT	VALUE
ECOLOGICAL PROPERTIES		
Aquatic toxicity, *Bluegill sunfish*, 96-h LC50	mg/l	15400/methanol
Aquatic toxicity, *Daphnia magna*, 48-h LC50	mg/l	74
Aquatic toxicity, *Rainbow trout*, 96-h LC50	mg/l	5-16/28d
Biodegradation probability		29%(concentration 10 mgl)/28d/not readily biodegradable, 19.9%(concentration 20 mg/)/28d/not readily biodegradable
Partition coefficient, log K_{oc}	-	-0.77
USE & PERFORMANCE		
Manufacturer		Evonik Industries
Outstanding properties		improved mechanical properties like: flexural strength, tensile strength, impact toughness, modulus of elasticity, and electrical properties like dielectric constant and specific volume resistance. It contains no carcinogenic benzyl chloride and exhibits better polymer wetting
Recommended for polymers		acrylics, PA, PBT, PC, EVA, modified PP, PVB, PVAc
Recommended for products		primers, sealants and adhesives, paints and varnishes
Recommended applications		adhesion promoter

Dynasylan 1175

PARAMETER	UNIT	VALUE
GENERAL INFORMATION		
Name		Dynasylan 1175
CAS #		68092-72-8, 67-56-1
EC number	-	268-470-1
Composition		40% N-vinylbenzyl-N´-aminoethyl-3-aminopropylpolysiloxane, hydrochloride, >20% methanol
Empirical formula	-	C17H30N2O3Si
Formula		
Chemical class	-	silane
Mixture	-	yes
Active matter	wt%	40
Functional organic group	-	amino
Functionality, average	inorganic methoxysilyl	
PHYSICAL PROPERTIES		
State	-	liquid
Color	dark yellow to brown	
Boiling point	°C	65/methanol
Density at 20°C	kg/m³	940
pH	6-7 at 20°C (1:1 in H₂O)	
Saponification value	mg KOH/g	
Solubility (diluents)	alcohols, aliphatic or aromatic hydrocarbons	
Solubility in water at 25°C	miscible with water with hydrolytic decomposition releasing methanol	
Specific gravity at 25°C	-	0.94
Vapor pressure at 20°C	kPa	13
Viscosity at 20°C	mPas	15
HEALTH & SAFETY		
NFPA classification	Flammability	3
	Health	2
	Reactivity	0
HMIS classification	Flammability	3
	Health	2
	Reactivity	0

Dynasylan 1175

PARAMETER	UNIT	VALUE
Carcinogenicity	IARC, OSHA, NTP: no ingredient of this product present at levels greater than or equal to 0.1% is identified as probable, possible or confirmed human carcinogen	
DOT class	Methanol Solution 3(6.1) II	
TDG class	Methanol Solution 3(6.1) II	
ICAO/IATA class	Methanol Solution 3(6.1) II	
IMDG class	METHANOL SOLUTION, 3(6.1) II	
Flash point	°C	9
Flash point method	-	PMCC
Explosive LEL	wt%	5.5/methanol
Explosive UEL	wt%	44/methanol
Hazardous combustion products	Carbon oxides, sulfur oxides	
Agency rating, listed	EINECS Europe, DSL Canada, ECSC China, TSCA USA	
Hazardous ingredients, labelling	Methanol Solution	
Hazardous products of hydrolysis	ethanol	
Effect of exposure, eye (human)	Causes eye irritation.	
Effect of exposure, inhalation (human)	Harmful if inhaled.	
Effect of exposure, skin (human)	Causes skin irritation. Can be absorbed through the skin.	
Effect of exposure, swallowing (human)	Harmful if swallowed.	
Effect of repeated or overexposure (human)	Causes damage to organs	
Exposure, personal protection	Safety glasses, protective clothing based on chemical resistance data, chemical-resistant gloves, general and local exhaust ventilation.	
First aid, eye	Rinse cautiously with water for several minutes. Remove contact lenses, if present and easy to do. Continue rinsing. If eye irritation persists: Get medical advice/attention.	
First aid, inhalation	Remove to fresh air. If not breathing, give artificial respiration. If breathing is difficult, give oxygen. If irritation persists, obtain medical advice.	
First aid, skin	Immediately wipe away excess material. Use a waterless hand cleaner to remove as much of the remaining material as possible. Wash with soap and water.	
NIOSH, REL	mg/m³	ST325/methanol
OSHA, PEL	mg/m³	260/methanol

Dynasylan 1175

PARAMETER	UNIT	VALUE
ACGIH, TLV	ppm	200/methanol
NIOSH, REL	ppm	ST250/methanol
OSHA, PEL	ppm	200/methanol
ECOLOGICAL PROPERTIES		
Aquatic toxicity, *Bluegill sunfish*, 96-h LC50	mg/l	15400/methanol
Aquatic toxicity, *Daphnia magna*, 48-h LC50	mg/l	10000 /methanol
Biological oxygen demand, 20 days	g/g	1.260/methanol
Partition coefficient, log K_{ow}	-	0.77
USE & PERFORMANCE		
Manufacturer	Evonik Industries	
Outstanding properties	improved mechanical properties e.g., flexural strength, tensile strength, impact toughness, modulus of elasticity, and improved electrical properties e.g., dielectric constant, specific volume resistance	
Recommended for polymers	acrylics, PA, EP, PBT, PC, EVA, modified PP, PVB, PVAc, phenolic	
Recommended for products	composites, sealants and adhesives, paint and varnishes	
Recommended applications	adhesion promoter and surface modifier. Dynasylan 1175 binds chemically to both inorganic materials (e.g., glass, metals, fillers) and organic polymers (e.g. thermosets, thermoplastics, elastomers). It can be used in many applications: as a size constituent or finish for glass fiber/ glass fabric composites, as a primer or additive for sealants and adhesives, and for paints and varnishes to improve adhesion to the substrate, for the pretreatment of fillers and pigments for mineral-filled composites.	
Guidelines for use	hydrolysis is preferably carried out in the presence of acetic acid (pH=4)	

Dynasylan 1189

PARAMETER	UNIT	VALUE
GENERAL INFORMATION		
Name		Dynasylan 1189
CAS #		31024-56-3
EC number	-	250-437-8
General description	bifunctional	
Chemical name	N-(n-butyl)-3-aminopropyltrimethoxysi-lane	
Empirical formula	-	C10H25NO2Si
Formula		

H$_3$CO–Si–OCH$_3$ (OCH$_3$), –N–H, butyl chain

PARAMETER	UNIT	VALUE
Molecular mass	daltons	235.40
Chemical class	-	silane
Functional organic group	-	diamino
Functionality, average	inorganic methoxysilyl	
PHYSICAL PROPERTIES		
State	-	liquid
Odor	-	slightly ammoniacal
Color	colorless to yellowish	
Boiling point	°C	238
Melting point	°C	< -38
Density at 20°C	kg/m³	947
pH	10.9 at 20°C (20 g/l H$_2$O)	
Refractive index at 20°C	-	1.4246
Solubility (diluents)	alcohols, aliphatic or aromatic hydrocarbons	
Solubility in water at 25°C	hydrolytic decomposition releasing methanol	
Specific gravity at 25°C	-	0.947
Vapor pressure at 20°C	kPa	0.001
Viscosity at 20°C	mPas	2.5
HEALTH & SAFETY		
NFPA classification	Flammability	1
	Health	2
	Reactivity	0
HMIS classification	Flammability	1
	Health	3
	Reactivity	0

Dynasylan 1189

PARAMETER	UNIT	VALUE
Carcinogenicity	IARC, OSHA, NTP: no ingredient of this product present at levels greater than or equal to 0.1% is identified as probable, possible or confirmed human carcinogen	
DOT class	not regulated	
TDG class	not regulated	
ICAO/IATA class	not regulated	
IMDG class	not regulated	
Autoignition temperature	°C	260
Flash point	°C	>95
Flash point method	-	PMCC
Hazardous combustion products	Carbon oxides, sulfur oxides	
Agency rating, listed	EINECS Europe, ECL Korea, TSCA USA	
Hazardous products of hydrolysis	methanol	
Animal testing, acute toxicity, Rat oral LD50	mg/kg	12825/acute, >500/repeated
Animal testing, acute toxicity, Rabbit dermal LD50	mg/kg	15200
Effect of exposure, eye (human)	Causes serious eye damage.	
Effect of exposure, inhalation (human)	May cause respiratory irritation.	
Effect of exposure, skin (human)	Causes skin irritation. Can be absorbed through the skin.	
Effect of exposure, swallowing (human)	Harmful if swallowed.	
Effect of repeated or overexposure (human)	Liver	
Exposure, personal protection	Safety glasses, protective clothing based on chemical resistance data, chemical-resistant gloves, general and local exhaust ventilation.	
First aid, eye	Rinse cautiously with water for several minutes. Remove contact lenses, if present and easy to do. Continue rinsing. If eye irritation persists: Get medical advice/attention.	
First aid, inhalation	Remove to fresh air. If not breathing, give artificial respiration. If breathing is difficult, give oxygen. If irritation persists, obtain medical advice.	
First aid, skin	Immediately wipe away excess material. Use a waterless hand cleaner to remove as much of the remaining material as possible. Wash with soap and water.	

Dynasylan 1189

PARAMETER	UNIT	VALUE
NIOSH, REL	mg/m³	ST325/methanol
OSHA, PEL	mg/m³	260/methanol
ACGIH, TLV	ppm	200/methanol
NIOSH, REL	ppm	ST250/methanol
OSHA, PEL	ppm	200/methanol
ECOLOGICAL PROPERTIES		
Aquatic toxicity, *Green algae*, 96-h EC50	mg/l	>100/72H
Aquatic toxicity, *Daphnia magna*, 48-h LC50	mg/l	>100
Aquatic toxicity, *Zebra fish*, 96-h LC50	mg/l	>100
Biodegradation probability		24.7%/28d/not readily biodegradable
Partition coefficient, log K_{ow}	-	2.2
USE & PERFORMANCE		
Manufacturer		Evonik Industries
Outstanding properties		improved flexural strength, tensile strength, impact strength, and modulus of elasticity improved moisture and corrosion resistance.
Recommended for polymers		EVA, EMA, epoxies (EP), PA, PU
Recommended for products		composites, foundry resins, adhesives and sealants, paints and coatings, SMP systems
Recommended applications		adhesion promoter and surface modifier. Dynasylan 1189 binds chemically to both inorganic materials (e.g., glass, metals, fillers) and organic polymers (e.g., thermosets, thermoplastics, elastomers). It can be used in many applications: as a size constituent

Dynasylan® 1401

PARAMETER	UNIT	VALUE
GENERAL INFORMATION		
Name		Dynasylan® 1401
Common synonym		N-(2-aminoethyl)-3-aminopropylmethyl-dimethoxysilane
Functional organic group	-	amine
Functionality	-	bifunctional
PHYSICAL PROPERTIES		
State	-	liquid
Odor	-	amine-like
Color	-	clear, yellow
Boiling point	°C	254-271
Density at 20°C	kg/m^3	970
Solubility (diluents)	alcohols, aromatic and aliphatic hydrocarbons	
HEALTH & SAFETY		
Flash point	°C	90
USE & PERFORMANCE		
Manufacturer	Dow	
Outstanding properties	improved mechanical strength, moisture and corrosion resistance, dielectric constant, and volume resistivity	
Recommended for polymers	phenolic and furan resins	
Recommended for products	composites, foundry resins, primers, paints, adhesives, sealants	

Dynasylan 1505

PARAMETER	UNIT	VALUE
GENERAL INFORMATION		
Name		Dynasylan 1505
CAS #	-	3179-76-8
EC number	-	221-660-8
Composition	3-aminopropylmethyldiethoxysilane	
Empirical formula	-	C8H21NO2Si
Formula		
Molecular mass	daltons	191.34
RTECS number	-	UI0700000
Chemical class	-	silane
Functional organic group		amino
PHYSICAL PROPERTIES		
State	-	liquid
Odor	-	amine-like
Color	-	slightly yellowish
Boiling point	°C	202
Melting point	°C	-38
Freezing point	°C	<-20
Density at 20°C	kg/m³	920
Refractive index at 20°C	-	1.427
Solubility (diluents)	alcohols, aliphatic or aromatic hydrocarbons	
Specific gravity at 25°C	-	0.92
Vapor pressure at 20°C	kPa	<0.001
Viscosity at 20°C	mPas	2.0
HEALTH & SAFETY		
HMIS classification	Flammability	2
	Health	3
	Reactivity	1
Flash point	°C	88
Flash point method	-	PMCC
Hazardous combustion products	Carbon oxides, nitrogen oxides	
Agency rating, listed	EINECS Europe, ECL Korea, ENCS Japan, IECSC China, DSL Canada, PICCS Philippines, TSCA USA	
Hazardous ingredients, labelling	Combustible liquid, n.o.s.	

Dynasylan 1505

PARAMETER	UNIT	VALUE
Animal testing, acute toxicity, Rat inhalation, LC50	mg/m^3	>40000/4H
Effect of exposure, eye (human)	Causes eye irritation.	
Effect of exposure, inhalation (human)	May cause respiratory irritation.	
Effect of exposure, skin (human)	Causes skin irritation.	
Exposure, personal protection	Safety glasses, protective clothing based on chemical resistance data, chemical-resistant gloves, general and local exhaust ventilation.	
First aid, eye	Rinse cautiously with water for several minutes. Remove contact lenses, if present and easy to do. Continue rinsing. If eye irritation persists: Get medical advice/attention.	
First aid, inhalation	Remove to fresh air. If not breathing, give artificial respiration. If breathing is difficult, give oxygen. If irritation persists, obtain medical advice.	
First aid, skin	Immediately wipe away excess material. Use a waterless hand cleaner to remove as much of the remaining material as possible. Wash with soap and water.	
UN risk phrases, R	-	R34
US safety phrases, S	-	S26,S36/37/39,S45
UN/NA class	-	3267
USE & PERFORMANCE		
Manufacturer	Evonik Industries	
Outstanding properties	improved flexural strength, tensile strength, impact strength, modulus of elasticity, moisture and corrosion resistance.	
Recommended for polymers	phenolic and furan resins	
Recommended for products	foundry resins, primers	
Recommended applications	adhesion promoter to both inorganic materials (for example, glass, metals, and fillers) and organic polymers (thermosets, thermoplastics, and elastomers) and as a surface modifier. Examples of applications: additive in cold-curing phenolic and furan foundry resins to improve the flexural strength of sand/resin elements with a very long shelf-life of the resins.	

Dynasylan 2201 EQ

PARAMETER	UNIT	VALUE
GENERAL INFORMATION		
Name		Dynasylan 2201 EQ
CAS #		116912-64-2, 67-56-1
EC number	-	245-876-7
General description		bifunctional
Composition		3-ureidopropyltriethoxysilane, 50% in methanol, ethylcarbamate-free
Empirical formula	-	C10H24N2O4Si
Formula		
Molecular mass	daltons	264.40
Chemical class	-	silane
Functional organic group	-	ureido
PHYSICAL PROPERTIES		
State	-	liquid
Odor	-	alcoholic
Boiling point	°C	73.4
Acid number	mg KOH/g	0
Density at 20°C	kg/m³	920
Refractive index	-	1.39
Solubility in water at 25°C		partly miscible partial hydrolytic decomposition releasing methanol
Specific gravity at 25°C	-	0.92
Vapor pressure at 20°C	kPa	12
Viscosity at 20°C	mPas	2.5
HEALTH & SAFETY		
NFPA classification	Flammability	3
	Health	2
	Reactivity	1
HMIS classification	Flammability	3
	Health	2
	Reactivity	1
Carcinogenicity		contains no carcinogenic by-products. Dynasylan 2201 EQ is therefore not classified as carcinogenic, class 2
DOT class		Methanol Solution 3(6.1) II

Dynasylan 2201 EQ

PARAMETER	UNIT	VALUE
TDG class	Methanol Solution 3(6.1) II	
ICAO/IATA class	Methanol Solution, 3(6.1) II	
IMDG class	METHANOL SOLUTION, 3(6.1) II	
Autoignition temperature	°C	425/methanol
Flash point	°C	13
Flash point method	-	PMCC
Explosive LEL	wt%	5.5/methanol
Explosive UEL	wt%	44/methanol
Agency rating, listed	EINECS Europe, ECL Korea, ENCS Japan, AICS Australia, IECSC China, DSL Canada, PICCS Philippines, TSCA USA	
Hazardous ingredients, labelling	Methanol Solution	
Hazardous products of hydrolysis	methanol	
Animal testing, acute toxicity, Rat oral LD50	mg/kg	
Effect of exposure, eye (human)	Causes eye irritation.	
Effect of exposure, inhalation (human)	Causes respiratory irritation. Acute toxicity estimated 3 mg/m^3/vapor .	
Effect of exposure, skin (human)	Causes skin irritation. Hazardous by absorption through the skin. Acute toxicity estimated 300 mg/kg	
Effect of exposure, swallowing (human)	Harmful if swallowed. Acute toxicity estimated 100 mg/kg.	
Effect of repeated or overexposure (human)	May cause liver and kidney injuries.	
Exposure, personal protection	Safety glasses, protective clothing based on chemical resistance data, chemical-resistant gloves, general and local exhaust ventilation.	
First aid, eye	Rinse cautiously with water for several minutes. Remove contact lenses, if present and easy to do. Continue rinsing. If eye irritation persists: Get medical advice/attention.	
First aid, inhalation	Remove to fresh air. If not breathing, give artificial respiration. If breathing is difficult, give oxygen. If irritation persists, obtain medical advice.	
First aid, skin	Immediately wipe away excess material. Use a waterless hand cleaner to remove as much of the remaining material as possible. Wash with soap and water.	
NIOSH, REL	mg/m^3	ST325/methanol
OSHA, PEL	mg/m^3	260/methanol

Dynasylan 2201 EQ

PARAMETER	UNIT	VALUE
ACGIH, TLV	ppm	200/methanol
NIOSH, REL	ppm	ST250/methanol
OSHA, PEL	ppm	200/methanol
UN/NA class	-	1230
ECOLOGICAL PROPERTIES		
Aquatic toxicity, *Bluegill sunfish*, 96-h LC50	mg/l	15400/methanol
Aquatic toxicity, *Daphnia magna*, 48-h LC50	mg/l	10000 /methanol
Aquatic toxic ty, *Zebra fish*, 96-h LC50	mg/l	>100
Biological oxygen demand, 20 days	g/g	1.260/methanol
USE & PERFORMANCE		
Manufacturer	Evonik Industries	
Outstanding properties	improved mechanical properties (flexural strength, tensile strength, impact toughness, and modulus of elasticity), improved resistance to moisture and corrosion, and higher temperature of deflection under load	
Recommended for polymers	thermosets, thermoplastics, phenolic, furan, and melamine	
Recommended for products	composites, foundry resins, sealants and adhesives, paints and varnishes	
Recommended applications	adhesion promoter to both inorganic materials (e.g., glass, metals, fillers) and organic polymers (e.g., thermosets, thermoplastics, elastomers), thus functioning as a surface modifier	

Dynasylan® 4150

PARAMETER	UNIT	VALUE
GENERAL INFORMATION		
Name		Dynasylan® 4150
Common synonym		non-ionic silane
PHYSICAL PROPERTIES		
State	-	liquid
Color	-	colorless
Freezing point	°C	-10
Density at 20°C	kg/m³	1100
Solubility (diluents)	most organic solvents and water	
Viscosity at 20°C	mPas	10-20
HEALTH & SAFETY		
Flash point	°C	>95
USE & PERFORMANCE		
Manufacturer	Dow	
Outstanding properties	imparts outstanding wetting and dispersibility to fillers and pigments used in the cosmetic industry (for example, titanium dioxide and clay). It can considerably reduce the dispersion viscosity of several fillers and pigments, consequently optimizing the processability of formulations.	
Recommended for products	cosmetics (treatment of many common cosmetic fillers, including talc, mica, and silicates, as well as metal oxide pigments, such as titanium dioxide, zinc oxide, and iron oxide)	
Recommended applications	color cosmetics mineral make-up (eyeshadow, blushes), the skin covering powder, fluid foundation	
Processing methods	pre-treatment of the filler or pigment is best achieved in a high-speed mixer by spraying Dynasylan® 4150 into a well-agitated filler or pigment bed.	
Guidelines for use	because of the excellent wetting and dispersion, high filler or pigment loads at a given viscosity can be realized with Dynasylan® 4150.	
Concentration used	wt%	1

Dynasylan 6490

PARAMETER	UNIT	VALUE
GENERAL INFORMATION		
Name		Dynasylan 6490
CAS #	-	131298-48-1
General description		functional oligosiloxane
Composition		Vinylmethoxysiloxane
Chemical class		silane
Functional organic group		methoxy
SiO_2 content	wt%	54
PHYSICAL PROPERTIES		
State	-	liquid
Odor	-	nearly odorless
Color	-	colorless
Boiling point	°C	222
Melting point	°C	-70
Density at 20°C	kg/m³	1,000
Solubility in water at 25°C		hydrolytic decomposition releasing methanol
Sulfur oxide (SO_2) content	wt%	54/SO_2
Vapor pressure at 20°C	kPa	0.2
Viscosity at 20°C	mPas	2-4
HEALTH & SAFETY		
NFPA classification	Flammability	2
	Health	1
	Reactivity	1
HMIS classification	Flammability	2
	Health	1
	Reactivity	1
DOT class	not regulated in packages 450 liters or less	
TDG class	not regulated in packages 450 liters or less	
ICAO/IATA class	not regulated	
IMDG class	not regulated	
Flash point	°C	87
Flash point method	-	PMCC
Agency rating, listed	EINECS Europe, ECL Korea, IECSC China, DSL Canada, TSCA USA	
Effect of exposure, eye (human)	Avoid contact with eyes.	

Dynasylan 6490

PARAMETER	UNIT	VALUE
Effect of exposure, inhalation (human)		May cause respiratory irritation. Do not inhale vapors or aerosols
Effect of exposure, skin (human)		Avoid contact with skin.
Effect of exposure, swallowing (human)		Not expected in industrial use. May be harmful if swallowed.
Exposure, personal protection		Safety glasses, protective clothing based on chemical resistance data, chemical-resistant gloves, general and local exhaust ventilation.
First aid, eye		Rinse cautiously with water for several minutes. Remove contact lenses, if present and easy to do. Continue rinsing. If eye irritation persists: Get medical advice/attention.
First aid, inhalation		Remove to fresh air. If not breathing, give artificial respiration. If breathing is difficult, give oxygen. If irritation persists, obtain medical advice.
First aid, skin		Immediately wipe away excess material. Use a waterless hand cleaner to remove as much of the remaining material as possible. Wash with soap and water.
UN/NA class	-	1993
USE & PERFORMANCE		
Manufacturer		Evonik Industries
Outstanding properties		Improved filler dispersion, good processability, environmentally friendly (low VOC), safety and handling during processing, excellent balance between tensile strength and elongation at break at a high level, improved chemical resistance, strongly reduced tendency to stress cracking, higher impact strength and abrasion resistance
Recommended for polymers		EPDM, EVA, PE
Recommended for products		cables (also flame retardant)
Recommended applications		compatibilizer between inorganic fillers (e.g. kaolin, MDH, ATH) and organic polymers, major field of application for mineral-filled compounds is the cable industry. In addition, it can be used in many other applications such as filler and pigment coatings, dispersions, etc.

Dynasylan 6498

PARAMETER	UNIT	VALUE
GENERAL INFORMATION		
Name	Dynasylan 6498	
CAS #	-	29434-25-1
Composition	vinylethoxysiloxane	
Chemical class	-	silane
Active matter	wt%	>= 98
Functional organic group	-	ethoxy
PHYSICAL PROPERTIES		
State	-	liquid
Odor	-	nearly odorless
Color	colorless to yellowish	
Boiling point	°C	245
Density at 20°C	kg/m³	1,000
Solubility in water at 25°C	not miscible, partial decomposition by hydrolysis	
Sulfur oxide (SO$_2$) content	wt%	45/SO$_2$
Viscosity at 20°C	mPas	3-7
HEALTH & SAFETY		
NFPA classification	Flammability	2
	Health	1
	Reactivity	0
HMIS classification	Flammability	2
	Health	1
	Reactivity	0
DOT class	not regulated in packages 450 liters or less	
TDG class	not regulated in packages 450 liters or less	
ICAO/IATA class	not regulated	
IMDG class	not regulated	
Flash point	°C	75
Flash point method	-	PMCC
Agency rating, listed	EINECS Europe, ECL Korea, IECSC China, DSL Canada, TSCA USA	
Effect of exposure, eye (human)	Avoid contact with eyes.	
Effect of exposure, inhalation (human)	May cause respiratory irritation. Do not inhale vapors or aerosols	
Effect of exposure, skin (human)	Avoid contact with skin.	

Dynasylan 6498

PARAMETER	UNIT	VALUE
Effect of exposure, swallowing (human)		Not expected in industrial use. May be harmful if swallowed.
Exposure, personal protection		Safety glasses, protective clothing based on chemical resistance data, chemical-resistant gloves, general and local exhaust ventilation.
First aid, eye		Rinse cautiously with water for several minutes. Remove contact lenses, if present and easy to do. Continue rinsing. If eye irritation persists: Get medical advice/attention.
First aid, inhalation		Remove to fresh air. If not breathing, give artificial respiration. If breathing is difficult, give oxygen. If irritation persists, obtain medical advice.
First aid, skin		Immediately wipe away excess material. Use a waterless hand cleaner to remove as much of the remaining material as possible. Wash with soap and water.
UN/NA class	-	1993
USE & PERFORMANCE		
Manufacturer	Evonik Industries	
Outstanding properties	improved filler dispersion, good processability, environmentally friendly (low VOC), safety and handling during processing, excellent balance between tensile strength and elongation at break at a high level, improved chemical resistance, strongly reduced tendency to stress cracking, higher impact strength and abrasion resistance	
Recommended for polymers	EPDM, EVA, EMA, PE	
Recommended for products	cables, HFFR cables	
Recommended applications	adhesion promoter, dispersion and hydrophobic agent in mineral filled compounds	

Dynasylan 6598

PARAMETER	UNIT	VALUE
GENERAL INFORMATION		
Name	Dynasylan 6598	
CAS #	-	201615-10-3
Composition	Vinyl-alkyl siloxane oligomer	
Acronym	-	VTMOEO
Chemical class	-	silane
Functional organic group	-	ethoxy
SiO_2 content	wt%	42
PHYSICAL PROPERTIES		
State	-	liquid
Odor	-	aromatic
Color	-	colorless
Boiling point	°C	255
Freezing point	°C	-55
Density at 20°C	kg/m³	1,000
Viscosity at 20°C	mPas	3-7
HEALTH & SAFETY		
NFPA classification	Flammability	2
	Health	1
	Reactivity	1
HMIS classification	Flammability	2
	Health	1
	Reactivity	1
DOT class	not regulated in packages 450 liters or less	
TDG class	not regulated in packages 450 liters or less	
ICAO/IATA class	not regulated	
IMDG class	not regulated	
Flash point	°C	70
Flash point method	-	PMCC
Effect of exposure, eye (human)	Avoid contact with eyes.	
Effect of exposure, inhalation (human)	May cause respiratory irritation. Do not inhale vapors or aerosols	
Effect of exposure, skin (human)	May cause skin irritation.	
Exposure, personal protection	Safety glasses, protective clothing based on chemical resistance data, chemical-resistant gloves, general and local exhaust ventilation.	

Dynasylan 6598

PARAMETER	UNIT	VALUE
First aid, eye		Rinse cautiously with water for several minutes. Remove contact lenses, if present and easy to do. Continue rinsing. If eye irritation persists: Get medical advice/attention.
First aid, inhalation		Remove to fresh air. If not breathing, give artificial respiration. If breathing is difficult, give oxygen. If irritation persists, obtain medical advice.
First aid, skin		Immediately wipe away excess material. Use a waterless hand cleaner to remove as much of the remaining material as possible. Wash with soap and water.
ECOLOGICAL PROPERTIES		
Biodegradation probability		12.6%/28d/not readily biodegradable
USE & PERFORMANCE		
Manufacturer		Evonik Industries
Outstanding properties		improved electrical properties of the filled compounds, provides the highest level of available stability under wet conditions. Lower VOC than monomeric silanes
Recommended for polymers		EVA, PE, EPDM
Recommended for products		cable, HFFR compounds
Recommended applications		adhesion promoter in mineral-filled, peroxide-crosslinked compounds. A major field of application for mineral-filled compounds is the cable industry. EPDM and kaolin can be processed into cable compounds through the adhesion promoting and hydrophobic effects of Dynasylan 6598. It can also be used in the manufacture of halogen-free, non-toxic, environmentally-friendly flame retardant compounds based on EVA or PE and ATH or MDH.

Dynasylan® 9116

PARAMETER	UNIT	VALUE
GENERAL INFORMATION		
Name		Dynasylan® 9116
Common synonym		monomeric long-chain alkylfunctional silane
PHYSICAL PROPERTIES		
State	-	liquid
Color	-	clear, colorless
Boiling point	°C	180
Density at 20°C	kg/m^3	890
Solubility (diluents)	petroleum ether, toluene	
Viscosity at 20°C	mPas	7
HEALTH & SAFETY		
Flash point	°C	165
USE & PERFORMANCE		
Manufacturer	Dow	
Outstanding properties	the long-chain alkyl functionality results in unique compound properties (better dispersion, high filler loadings, enhanced mechanical properties, reduced water uptake) when Dynasylan® 9116 treated fillers are incorporated into polymers (e.g., polypropylene). It exhibits superior hydrophobicity on substrates and forms weather- and moisture-resistant bonds to substrates	
Recommended for polymers	polypropylene	
Recommended for products	treatment of mineral fillers and pigments or inorganic surfaces in general (e.g., concrete, ATH, MDH); waterproofing by impregnation when used as water-repellent improvements in toughness in wollastonite-filled polypropylene, allowing the formulation of highly-filled compounds.	

Dynasylan® 9896

PARAMETER	UNIT	VALUE
GENERAL INFORMATION		
Name		Dynasylan® 9896
Common synonym		oligomeric short-chain alkylfunctional silane
PHYSICAL PROPERTIES		
State	-	liquid
Odor	-	odorless
Color	-	clear, colorless
Boiling point	°C	355
Density at 20°C	kg/m³	1040
pH (500 g/l)	-	3-4
Solubility	petroleum ether, toluene, alcohol	
Viscosity at 20°C	mPas	60
HEALTH & SAFETY		
Flash point	°C	63-90
USE & PERFORMANCE		
Manufacturer	Dow	
Outstanding properties	improved filler dispersion good process-ability significantly reduced water-uptake	
Recommended for polymers	polyethylene, polypropylene	
Recommended for products	coatings, printablility of mineral-filled plastics; surface modifier to generate hydrophobicity (e.g., on inorganic pigments, mineral fillers)	

Dynasylan AMEO

PARAMETER	UNIT	VALUE
GENERAL INFORMATION		
Name	Dynasylan AMEO	
CAS #	-	919-30-2
EC number	-	213-048-4
Composition	3-Aminopropyltriethoxysilane	
Common synonym	3-Triethoxysilylpropylamine	
Acronym		AMEO/APTES
Empirical formula	C9H23NO3Si/(OC2H5)3Si-(CH2)3-NH2	
Formula		

Molecular mass	daltons	221.37
RTECS number	-	TX2100000
Chemical class	-	silane
Functional organic group	-	primary amine
PHYSICAL PROPERTIES		
State	-	liquid
Odor	-	amine-like
Color	-	colorless
Boiling point	°C	220
Melting point	°C	-70
Density at 20°C	kg/m³	950
Refractive index at 20°C	-	1.422
Solubility (diluents)	alcohols, aliphatic or aromatic hydrocarbons	
Solubility in water at 25°C	5.4 g/l, hydrolytic decomposition releasing methanol	
Specific gravity at 25°C	-	0.95
Vapor pressure at 20°C	kPa	0.0002
Viscosity at 20°C	mPas	1.85
HEALTH & SAFETY		
Carcinogenicity	no evidence of carcinogenicity	
Mutagenicity	negative	
Teratogenicity	no evidence of reproduction toxicity	
Autoignition temperature	°C	200
Flash point	°C	80-90
Flash point method	-	PMCC
Hazardous combustion products	Carbon oxides, nitrogen oxides	

Dynasylan AMEO

PARAMETER	UNIT	VALUE
Animal testing, acute toxicity, Rat oral LD50	mg/kg	>4000
Effect of exposure, eye (human)	Causes severe eyes burns.	
Effect of exposure, skin (human)	Causes severe skin burns. May cause an allergic skin reaction.	
Effect of exposure, swallowing (human)	Harmful if swallowed.	
Exposure, personal protection	Safety glasses, protective clothing based on chemical resistance data, chemical-resistant gloves, general and local exhaust ventilation.	
First aid, eye	Rinse cautiously with water for several minutes. Remove contact lenses, if present and easy to do. Continue rinsing. If eye irritation persists: Get medical advice/attention.	
First aid, inhalation	Remove to fresh air. If not breathing, give artificial respiration. If breathing is difficult, give oxygen. If irritation persists, obtain medical advice.	
First aid, skin	Immediately wipe away excess material. Use a waterless hand cleaner to remove as much of the remaining material as possible. Wash with soap and water.	
ECOLOGICAL PROPERTIES		
Bioaccumulative potential	not bioaccumulative	
Biodegradation probability	partially biodegradable	
Partition coefficient, log K_{oc}	-	1.7
USE & PERFORMANCE		
Manufacturer	Evonik Industries	
Recommended for polymers	PA, PBT, PC, EVA, PVB, PVAc, PVC, PS, PU, epoxy resin, unsaturated polyester, PP, phenolic, furan, and melamine resins, NBR, acrylates, silicone	
Recommended for products	composites, abrasives, insulating materials, sealants and adhesives, cables, paints and coatings	
Guidelines for use	it can undergo reactions with ketone or ester solvents. Silane or silanized substrates can react with carbon dioxide to form carbonates and/or carbamates.	

Dynasylan AMEO

PARAMETER	UNIT	VALUE
Recommended applications		adhesion promoter, surface modifier, and as a reactant for product modification. Dynasylan AMEO binds chemically to both inorganic materials (e.g., glass, metals, fillers) and organic polymers (e.g., thermosets, thermoplastics, and elastomers). Examples of applications: mineral-filled polymers (composites) or HFFR cables for pre-treatment of fillers and pigments or as an additive, paints and coatings as an additive, and primer for improving adhesion to the substrate, mineral fiber insulating materials, abrasives as an additive to phenolic resin binders

Dynasylan AMEO-T

PARAMETER	UNIT	VALUE
GENERAL INFORMATION		
Name		Dynasylan AMEO-T
CAS #	-	919-30-2
Composition		90% 3-aminopropyltriethoxysilane, blend of primary and secondary aminofunctional silanes
Common synonym		3-triethoxysilylpropylamine
Acronym	-	AMEO/APTES
Formula		
Chemical class	-	silane
Mixture	-	blend
Functional organic group		primary amino/diamino
PHYSICAL PROPERTIES		
State	-	liquid
Odor	-	amine-like
Color	-	yellowish
Solubility (diluents)		alcohols, aliphatic or aromatic hydrocarbons
Density at 20°C	kg/m^3	950
Boiling point	°C	>68
Refractive index at 20°C	-	1.425
Viscosity at 20°C	mPas	2
HEALTH & SAFETY		
Flash point	°C	80-90
Teratogenicity		no evidence of reproduction toxicity
Hazardous combustion products		Carbon oxides, nitrogen oxides
Animal testing, acute toxicity, Rat oral LD50	mg/kg	>4000
Effect of exposure, eye (human)		Causes severe eyes burns.
Effect of exposure, skin (human)		Causes severe skin burns. May cause an allergic skin reaction.
Effect of exposure, swallowing (human)		Harmful if swallowed.
Exposure, personal protection		Safety glasses, protective clothing based on chemical resistance data, chemical-resistant gloves, general and local exhaust ventilation.

Dynasylan AMEO-T

PARAMETER	UNIT	VALUE
First aid, eye	Rinse cautiously with water for several minutes. Remove contact lenses, if present and easy to do. Continue rinsing. If eye irritation persists: Get medical advice/attention.	
First aid, inhalation	Remove to fresh air. If not breathing, give artificial respiration. If breathing is difficult, give oxygen. If irritation persists, obtain medical advice.	
First aid, skin	Immediately wipe away excess material. Use a waterless hand cleaner to remove as much of the remaining material as possible. Wash with soap and water.	
ECOLOGICAL PROPERTIES		
Aquatic toxicity, *Daphnia magna*, 48-h LC50	mg/l	331
Bioaccumulative potential	not bioaccumulative	
Biodegradation probability	partially biodegradable	
USE & PERFORMANCE		
Manufacturer	Evonik Industries	
Outstanding properties	reduction of viscosity, Newtonian behavior, increased filler loading	
Recommended for polymers	acrylics, acrylic copolymers, PA, PBT, PC, EVA, modified PP, PVB, PVAc, PVC, phenolic, furan, and melamine resins, silicones	
Recommended for products	insulating materials, foundry resins, sealants and adhesives, cables, paints and coatings	
Recommended applications	adhesion promoter, surface modifier, and as a reactant for product modification. Dynasylan AMEO-T binds chemically to both inorganic materials (e.g., glass, metals, fillers) and organic polymers (e.g., thermosets, thermoplastics, and elastomers). Examples of applications: mineral-filled polymers (composites) and HFFR cables for pre-treatment of fillers and pigments or as an additive, paints and coatings as an additive, and primer for improving adhesion to the substrate, mineral fiber insulating materials, abrasives as an additive to phenolic resin binders.	

Dynasylan AMMO

PARAMETER	UNIT	VALUE
GENERAL INFORMATION		
Name	Dynasylan AMMO	
CAS #	-	13822-56-5
EC number	-	237-511-5
Composition	3-aminopropyltrimethoxy silane, (impurity 0.3% methanol)	
Common synonym	3-(trimethoxysilyl)-1-propanamin	
Empirical formula	-	C6H17NO3Si
Formula		
Molecular mass	daltons	179.29
Chemical class	-	silane
Functional organic group	-	primary amino
PHYSICAL PROPERTIES		
State	-	liquid
Odor	-	amine-like
Color	-	colorless
Density at 20°C	kg/m³	1,020
Boiling point	°C	194
Kinematic viscosity at 20°C	cSt	1.95
pH	>9.0 at 20°C (20 g/l H₂O)	
Refractive index at 20°C	-	1.425
Solubility (diluents)	alcohols, aliphatic or aromatic hydrocarbons	
Viscosity at 20°C	mPas	2.0
HEALTH & SAFETY		
NFPA classification	Flammability	2
	Health	3
	Reactivity	1
HMIS classification	Flammability	2
	Health	3
	Reactivity	1
Carcinogenicity	IARC, OSHA, NTP: no ingredient of this product present at levels greater than or equal to 0.1% is identified as probable, possible or confirmed human carcinogen	
Mutagenicity	negative	

Dynasylan AMMO

PARAMETER	UNIT	VALUE
DOT class	Combustible liquid, n.o.s. (3-Aminopropyltrimethoxy) 8,II	
ICAO/IATA class	not regulated	
IMDG class	not regulated	
Autoignition temperature	°C	295
Flash point	°C	90
Flash point method	-	PMCC
Hazardous combustion products	Carbon oxides, nitrogen oxides	
Hazardous ingredients, labelling	Combustible liquid, n.o.s. (3-Aminopropyltrimethoxy)	
Hazardous products of hydrolysis	methanol	
Animal testing, acute toxicity, Rat oral LD50	mg/kg	>2000
Animal testing, acute toxicity, Rabbit dermal LD50	mg/kg	>10000
Animal testing, acute toxicity, Rat inhalation, LC50	mg/m^3	>40000/4H
Effect of exposure, eye (human)	Causes severe eyes burns.	
Effect of exposure, skin (human)	Causes severe skin burns. May cause an allergic skin reaction.	
Exposure, personal protection	Safety glasses, protective clothing based on chemical resistance data, chemical-resistant gloves, general and local exhaust ventilation.	
First aid, eye	Rinse cautiously with water for several minutes. Remove contact lenses, if present and easy to do. Continue rinsing. If eye irritation persists: Get medical advice/attention.	
First aid, inhalation	Remove to fresh air. If not breathing, give artificial respiration. If breathing is difficult, give oxygen. If irritation persists, obtain medical advice.	
First aid, skin	Immediately wipe away excess material. Use a waterless hand cleaner to remove as much of the remaining material as possible. Wash with soap and water.	
NIOSH, REL	mg/m^3	ST325/methanol
OSHA, PEL	mg/m^3	260/methanol
ACGIH, TLV	ppm	200/methanol
NIOSH, REL	ppm	ST250/methanol
OSHA, PEL	ppm	200/methanol
UN risk phrases, R	-	R36/38,R34

Dynasylan AMMO

PARAMETER	UNIT	VALUE
US safety phrases, S	S26,S36/37/39,S45,S25	
UN/NA class	-	1993
ECOLOGICAL PROPERTIES		
Aquatic toxicity, *Green algae*, 96-h EC50	mg/l	>1000/72H
Aquatic toxicity, *Zebra fish*, 96-h LC50	mg/l	>934
Bioaccumulative potential	not bioaccumulative	
Biodegradation probability	67%/28d/not readily biodegradable	
Partition coefficient, log K_{oc}	-	0.2
USE & PERFORMANCE		
Manufacturer	Evonik Industries	
Outstanding properties	improved mechanical properties, for example, flexural strength, tensile strength, impact strength and modulus of elasticity, improved moisture, corrosion resistance, and electrical properties: for example, dielectric constant, volume resistivity.	
Recommended for polymers	acrylics, acrylic copolymers, phenolic, furan, and melamine resins, PA, PBT, PC, EVA, modified PP, PVB, PVAc, phenoplasts, PVC, silicones	
Recommended for products	foundry resins, primers, composites, sealants and adhesives, paints and coatings	
Recommended applications	adhesion promoter, surface modifier, and as a reactant for product modification. Dynasylan AMMO binds chemically to both inorganic materials (e.g., glass, metals, fillers) and organic polymers (e.g., thermosets, thermoplastics, and elastomers). Examples of application: especially suitable for amine hardeners, mineral-filled composites for pretreatment of fillers and pigments, paints and coatings as an additive and primer for improving adhesion to the substrate, glass fiber/glass fabric composites as a size ingredient, or finish glass and metal primers.	

Dynasylan BDAC

PARAMETER	UNIT	VALUE
GENERAL INFORMATION		
Name	Dynasylan BDAC	
CAS #	-	13170-23-5
EC number	-	236-112-3
Composition	97% di-tert-butoxydiacetoxysilane	
Common synonym	triacetoxy-tert-butoxysilane	
Acronym	-	BDAC
Empirical formula	-	C12H24O6Si
Chemical class	-	silane
Functional organic group	-	acetoxy
PHYSICAL PROPERTIES		
Odor	-	acetic acid
Color	clear colorless to yellowish	
Boiling point	°C	102
Density at 20°C	kg/m³	1,030
Solubility in water at 25°C	not miscible, hydrolytic decomposition releasing methanol	
Specific gravity at 20°C	-	1.091
Viscosity at 20°C	mPas	7
HEALTH & SAFETY		
NFPA classification	Flammability	1
	Health	3
	Reactivity	1
HMIS classification	Flammability	1
	Health	3
	Reactivity	1
ICAO/IATA class	Corrosive liquid, acidic, organic, n.o.s.(diacetoxydi-tertbutoxysilane) 8,II	
IMDG class	CORROSIVE LIQUID, ACIDIC, ORGANIC, N.O.S.(diacetoxyditert-butoxysilane) 8,II	
Flash point	°C	95
Flash point method	-	PMCC
Agency rating, listed	EINECS Europe, ECL Korea, ENCS Japan, AICS Australia, IECSC China, DSL Canada, PICCS Philippines, TSCA USA	
Hazardous products of hydrolysis	acetic acid	
Effect of exposure, eye (human)	Causes severe eyes burns. Causes serious eye damage	

Dynasylan BDAC

PARAMETER	UNIT	VALUE
Effect of exposure, inhalation (human)	Avoid inhalation aerosol of mists. If aerosol or mists are inhaled, take affected persons out into the fresh air. Possible discomforts include severe irritation of mucus lining (nose, throat, eyes), cough, sneezing and flow of tears.	
Effect of exposure, skin (human)	Causes severe skin burns.	
Effect of exposure, swallowing (human)	Harmful if swallowed.	
Exposure, personal protection	Safety glasses, protective clothing based on chemical resistance data, chemical-resistant gloves, general and local exhaust ventilation.	
First aid, eye	Rinse cautiously with water for several minutes. Remove contact lenses, if present and easy to do. Continue rinsing. If eye irritation persists: Get medical advice/attention.	
First aid, inhalation	Remove to fresh air. If not breathing, give artificial respiration. If breathing is difficult, give oxygen. If irritation persists, obtain medical advice.	
First aid, skin	Immediately wipe away excess material. Use a waterless hand cleaner to remove as much of the remaining material as possible. Wash with soap and water.	
ACGIH, TLV	mg/m^3	10/acetic acid
UN risk phrases, R	-	R34,R36/37/38
US safety phrases, S	S26,S36/37/39,S45,S28	
UN/NA class	-	3265
USE & PERFORMANCE		
Manufacturer	Evonik Industries	
Recommended applications	adhesion promoter for room-temperature vulcanizing silicone sealants	

Dynasylan® BTSE

PARAMETER	UNIT	VALUE
GENERAL INFORMATION		
Name		Dynasylan® BTSE
Common synonym		bis(triethoxysilyl)ethane
Functionality	-	dipodal alkylsilane
Functional groups	-	ethoxy (six)
Acidity (H$_2$SO$_4$)	wt%	0.07-0.14
PHYSICAL PROPERTIES		
State	-	liquid
Odor	-	odorless
Color	-	milky
Boiling point	°C	315 (initial)
Density at 20°C	kg/m^3	950
Gel time	min	3-10
Solvent	-	ethanol
Viscosity at 20°C	mPas	4
HEALTH & SAFETY		
Flash point	°C	113
USE & PERFORMANCE		
Manufacturer		Dow
Outstanding properties		hydrophobic and high crosslink density silane layers
Recommended for polymers		silicone
Recommended for products		scratch-resistant hardcoats, corrosion protection primers, sealants, sol-gel systems, electronic materials
Recommended applications		its main applications are: silicones: adhesion promoter and crosslinking agent for silicone sealants and putty compounds; sol-gel systems: due to its high crosslinking density, it is used as an additive in sol-gel systems. It can increase the chemical as well as mechanical stability of such coatings and primers.
Processing		PVD-precursor for barrier coatings on plastics (plasma-activated), silane barriers against water and oxygen for plastic foils or films
Guideline for use		usually requires acid- or alkali-catalyzed hydrolysis

Dynasylan DAMO

PARAMETER	UNIT	VALUE
GENERAL INFORMATION		
Name	Dynasylan DAMO	
CAS #	-	1760-24-3
General description	bifunctional	
Composition	60-100% N-[3-(trimethoxysilyl)propyl] ethylenediamine, 5-10% N,N'-bis[3-(trimethoxysilyl)propyl]ethylenediamine, (impurities: 1-5% N,N-bis[3-(trimethoxysilyl)propyl]ethylenediamine, 0.1-1% methanol)	
Acronym	-	AEAPTMS
Empirical formula	C8H22N2O3Si	
Formula		
Molecular mass	daltons	222.36
Chemical class	-	silane
Functional organic group	-	diamino
PHYSICAL PROPERTIES		
State	-	liquid
Odor	-	amine-like
Color	colorless to slightly yellowish	
Boiling point	°C	140 at 20 hPa
Density at 20°C	kg/m³	1,030
pH	10.2 at 20°C (10 g/l H$_2$O)	
Refractive index at 20°C	-	1.447
Solubility (diluents)	alcohols, aliphatic or aromatic hydrocarbons	
Solubility in water at 25°C	hydrolytic decomposition releasing methanol	
Specific gravity at 25°C	-	1.03
Vapor pressure at 20°C	kPa	0.6
Viscosity at 20°C	mPas	6.0
HEALTH & SAFETY		
NFPA classification	Flammability	1
	Health	2
	Reactivity	1

Dynasylan DAMO

PARAMETER	UNIT	VALUE
HMIS classification	Flammability	1
	Health	3
	Reactivity	1
Carcinogenicity	IARC, OSHA, NTP: no ingredient of this product present at levels greater than or equal to 0.1% is identified as probable, possible or confirmed human carcinogen	
Mutagenicity	no evidence	
Teratogenicity	no evidence	
DOT class	not regulated	
ICAO/IATA class	not regulated	
IMDG class	not regulated	
Flash point	°C	>120
Flash point method	-	PMCC
Hazardous combustion products	Carbon oxides, nitrogen oxides, formaldehyde	
Agency rating, listed	EINECS Europe, ECL Korea, ENCS Japan, AICS Australia, IECSC China, DSL Canada, PICCS Philippines, TSCA USA	
Hazardous ingredients, labelling	Combustible liquid, n.o.s.	
Hazardous products of hydrolysis	methanol	
Animal testing, acute toxicity, Rat oral LD50	mg/kg	>2000/acute, >500/repeated
Animal testing, acute toxicity, Rat inhalation, LC50	mg/m³	>1490-2440/dust/mist
Effect of exposure, eye (human)	Causes serious eye damage.	
Effect of exposure, inhalation (human)	Avoid breathing fume, mist, vapors, spray.	
Effect of exposure, skin (human)	May cause an allergic skin reaction	
Effect of exposure, swallowing (human)	Harmful if swallowed.	
Exposure, personal protection	Safety glasses, protective clothing based on chemical resistance data, chemical-resistant gloves, general and local exhaust ventilation.	
First aid, eye	Rinse cautiously with water for several minutes. Remove contact lenses, if present and easy to do. Continue rinsing. If eye irritation persists: Get medical advice/attention.	

Dynasylan DAMO

PARAMETER	UNIT	VALUE
First aid, inhalation		Remove to fresh air. If not breathing, give artificial respiration. If breathing is difficult, give oxygen. If irritation persists, obtain medical advice.
First aid, skin		Immediately wipe away excess material. Use a waterless hand cleaner to remove as much of the remaining material as possible. Wash with soap and water.
NIOSH, REL	mg/m³	ST325/methanol
OSHA, PEL	mg/m³	260/methanol
ACGIH, TLV	ppm	260/methanol
NIOSH, REL	ppm	ST250/methanol
OSHA, PEL	ppm	260/methanol
ECOLOGICAL PROPERTIES		
Aquatic toxicity, *Green algae*, 96-h EC50	mg/l	126/72H
Aquatic toxicity, *Daphnia magna*, 48-h LC50	mg/l	81
Biodegradation probability		39%/28d/not readily biodegradable
USE & PERFORMANCE		
Manufacturer		Evonik Industries
Outstanding properties		improved properties, such as flexural strength, tensile strength, impact strength, and modulus of elasticity as well improved moisture and corrosion resistance.
Recommended for polymers		acrylics and acrylic copolymers, EVA, EMA, EP, PA, PBT, PC, modified PP, PVC, PVAc, PU, silicones
Recommended for products		composites, primers, sealants and adhesives, paints and coatings
Recommended applications		adhesion promoter

Dynasylan DAMO-T

PARAMETER	UNIT	VALUE
GENERAL INFORMATION		
Name		Dynasylan DAMO-T
CAS #		1760-24-3
General description		bifunctional
Composition		N-(2-aminoethyl)-3-aminopropyltrimethoxysilane
Acronym	-	AEAPTMS
Empirical formula	-	C8H22N2O3Si
Formula		
Molecular mass	daltons	222.36
Chemical class	-	silane
Functional organic group	diamino/inorganic methoxysilyl	
PHYSICAL PROPERTIES		
State	-	liquid
Odor	-	amine-like
Color	colorless to slightly yellowish	
Density at 20°C	kg/m^3	1030
Kinematic viscosity at 20°C	cSt	6.8
Refractive index at 20°C	-	1.444
Solubility (diluents)	alcohols, aliphatic or aromatic hydrocarbons	
Solubility in water at 25°C	hydrolytic decomposition releasing methanol	
Specific gravity at 25°C	-	1.03
Viscosity at 20°C	mPas	7.0
HEALTH & SAFETY		
NFPA classification	Flammability	3
	Health	2
	Reactivity	0
HMIS classification	Flammability	3
	Health	2
	Reactivity	0
Hazardous combustion products	Carbon oxides, nitrogen oxides, formaldehyde	
Flash point	°C	120

Dynasylan DAMO-T

PARAMETER	UNIT	VALUE
Exposure, personal protection		Safety glasses, protective clothing based on chemical resistance data, chemical-resistant gloves, general and local exhaust ventilation.
First aid, eye		Rinse cautiously with water for several minutes. Remove contact lenses, if present and easy to do. Continue rinsing. If eye irritation persists: Get medical advice/attention.
First aid, inhalation		Remove to fresh air. If not breathing, give artificial respiration. If breathing is difficult, give oxygen. If irritation persists, obtain medical advice.
First aid, skin		Immediately wipe away excess material. Use a waterless hand cleaner to remove as much of the remaining material as possible. Wash with soap and water.
USE & PERFORMANCE		
Manufacturer		Evonik Industries
Outstanding properties		adjusted reactivity. Improves properties, such as flexural strength, tensile strength, modulus of elasticity and moisture and corrosion resistance.
Recommended for polymers		phenolic resin, furan resin, PA, PBT, PC, EVA, modified PP, PVAc, PVC, acrylates, and silicones
Recommended for products		composites, primers, foundry resins, sealants and adhesives, paints and coatings
Recommended applications		adhesion promoter and coupling agent to both inorganic materials (for example, glass, metals and fillers) and organic polymers (thermosets, thermoplastics, and elastomers).

Dynasylan GLYEO

PARAMETER	UNIT	VALUE
GENERAL INFORMATION		
Name	Dynasylan GLYEO	
CAS #	-	2602-34-8
EC number	-	220-011-6
Composition	94% [3-(2,3-epoxypropoxy)propyl] triethoxysilane	
Common synonym	3-glycidyloxypropyltriethoxysilane	
Empirical formula	-	C12H26O5Si
Formula		
Chemical class	-	silane
Mixture	-	yes
Functional organic group	-	ethoxy
PHYSICAL PROPERTIES		
State	-	liquid
Color	-	colorless
Boiling point	°C	270
Melting point	°C	-70
Density at 20°C	kg/m^3	1,006
pH	3.5-4.0 at 20°C (1000 g/l)	
Solubility (diluents)	alcohols, aliphatic or aromatic hydrocarbons, ketones	
Solubility in water at 25°C	hydrolytic decomposition releasing methanol	
Specific gravity at 25°C	-	1.006
Vapor pressure at 20°C	kPa	0.11
Viscosity at 20°C	mPas	3.35
HEALTH & SAFETY		
NFPA classification	Flammability	1
	Health	2
	Reactivity	1
HMIS classification	Flammability	1
	Health	2
	Reactivity	1

Dynasylan GLYEO

PARAMETER	UNIT	VALUE
Carcinogenicity	IARC, OSHA, NTP: no ingredient of this product present at levels greater than or equal to 0.1% is identified as probable, possible or confirmed human carcinogen	
Mutagenicity	no evidence of mutagenic effects	
DOT class	not regulated	
ICAO/IATA class	not regulated	
IMDG class	not regulated	
Flash point	°C	125
Flash point method	-	PMCC
Hazardous combustion products	carbon oxides	
Hazardous products of hydrolysis	ethanol	
Animal testing, acute toxicity, Rat oral LD50	mg/kg	>2000/acute, >1000/repeated
Animal testing, acute toxicity, Rabbit dermal LD50	mg/kg	>2000
Animal testing, acute toxicity, Rat inhalation, LC50	mg/m^3	>5200/dust/mist
Effect of exposure, eye (human)	Not a hazardous substance.	
Effect of exposure, inhalation (human)	Not a hazardous substance.	
Effect of exposure, skin (human)	Not irritating to skin.	
Effect of repeated or overexposure (human)	Not classified as specific target organ toxicant, repeated exposure	
Exposure, personal protection	Safety glasses, protective clothing based on chemical resistance data, chemical-resistant gloves, general and local exhaust ventilation.	
First aid, eye	Rinse cautiously with water for several minutes. Remove contact lenses, if present and easy to do. Continue rinsing. If eye irritation persists: Get medical advice/attention.	
First aid, inhalation	Remove to fresh air. If not breathing, give artificial respiration. If breathing is difficult, give oxygen. If irritation persists, obtain medical advice.	
First aid, skin	Immediately wipe away excess material. Use a waterless hand cleaner to remove as much of the remaining material as possible. Wash with soap and water.	
UN risk phrases, R	R36/37/38,R20	
US safety phrases, S	S26	

Dynasylan GLYEO

PARAMETER	UNIT	VALUE
ECOLOGICAL PROPERTIES		
Aquatic toxicity, *Green algae*, 96-h EC50	mg/l	>100/72H
Aquatic toxicity, *Daphnia magna*, 48-h LC50	mg/l	>100
Aquatic toxicity, *Zebra fish*, 96-h LC50	mg/l	>100
Bioaccumulative potential	low bioaccumulation	
Biodegradation probability	53%/28d/partially biodegradable	
Partition coefficient, log K_{ow}	-	2
USE & PERFORMANCE		
Manufacturer	Evonik Industries	
Outstanding properties	improved mechanical properties, such as flexural strength, tensile strength, impact strength and modulus of elasticity, improved moisture and corrosion resistance, and improved electrical properties, for example, dielectric constant, volume resistivity. Improved following processing properties: better filler dispersion, rheological behavior (i.e., viscosity reduction) Newtonian behavior, increased filler loading	
Recommended for polymers	acrylates, ABS, BR, EPDM, PBT, PS, PVAc, PVC, polysulfides, epoxy, phenolic, and melamine resins	
Recommended for products	adhesives and sealants, composites, foundry resins, paints and coatings, primers	
Recommended applications	used as a coupling agent, crosslinking agents, surface modifier, as an additive, and as a primer in coatings improving adhesion to the substrate, especially glass and metal, used as a pretreatment of fillers and pigments	

Dynasylan GLYMO

PARAMETER	UNIT	VALUE
GENERAL INFORMATION		
Name		Dynasylan GLYMO
CAS #	-	2530-83-8
EC number	-	219-784-2
Composition		98% [3-(2,3-epoxypropoxy)propyl] trimethoxysilane
Common synonym		3-glycidyloxypropyltrimethoxysilane
Acronym		TMSPGE/GLYMO
Empirical formula	-	C9H20O5Si
Formula		
Molecular mass	daltons	236.34
RTECS number	-	VV4025000
Chemical class	-	silane
Functional organic group	-	methoxy
PHYSICAL PROPERTIES		
State	-	liquid
Odor	-	terpentine-like
Color	-	colorless
Boiling point	°C	90
Melting point	°C	-70
Density at 20°C	kg/m³	1,070
Refractive index at 20°C	-	1.429
Kinematic viscosity at 20°C	cSt	2.7
Solubility (diluents)		alcohols, aliphatic or aromatic hydrocarbons
Solubility in water at 25°C		36.5 g/l at 20°C, hydrolytic decomposition
Specific gravity at 25°C	-	1.07
Vapor pressure at 20°C	kPa	0.01
Viscosity at 20°C	mPas	3.65
HEALTH & SAFETY		
NFPA classification	Flammability	1
	Health	2
	Reactivity	1

Dynasylan GLYMO

PARAMETER	UNIT	VALUE
HMIS classification	Flammability	1
	Health	2
	Reactivity	1
Carcinogenicity	IARC, OSHA, NTP: no ingredient of this product present at levels greater than or equal to 0.1% is identified as probable, possible or confirmed human carcinogen	
Mutagenicity	no evidence of mutagenic effects	
DOT class	not regulated	
ICAO/IATA class	not regulated	
IMDG class	not regulated	
Autoignition temperature	°C	400
Flash point	°C	122
Flash point method	-	PMCC
Explosive LEL	wt%	0.7
Explosive UEL	wt%	13.6
Hazardous combustion products	carbon oxides	
Hazardous products of hydrolysis	methanol	
Animal testing, acute toxicity, Rat oral LD50	mg/kg	8025
Animal testing, acute toxicity, Rabbit dermal LD50	mg/kg	4248
Animal testing, acute toxicity, Rat inhalation, LC50	mg/m³	>5300/dust/mist
Effect of exposure, eye (human)	Causes serious eye irritation	
Effect of exposure, inhalation (human)	Avoid breathing fume, mist, vapors, spray.	
Effect of exposure, skin (human)	Not irritating to skin, not sensitizing	
Effect of repeated or overexposure (human)	Not classified for long-term human health effects or for aquatic toxicity	
Exposure, personal protection	Safety glasses, protective clothing based on chemical resistance data, chemical-resistant gloves, general and local exhaust ventilation.	
First aid, eye	Rinse cautiously with water for several minutes. Remove contact lenses, if present and easy to do. Continue rinsing. If eye irritation persists: Get medical advice/attention.	
First aid, inhalation	Remove to fresh air. If not breathing, give artificial respiration. If breathing is difficult, give oxygen. If irritation persists, obtain medical advice.	

Dynasylan GLYMO

PARAMETER	UNIT	VALUE
First aid, skin		Immediately wipe away excess material. Use a waterless hand cleaner to remove as much of the remaining material as possible. Wash with soap and water.
NIOSH, REL	mg/m³	ST325/methanol
OSHA, PEL	mg/m³	260/methanol
ACGIH, TLV	ppm	200/methanol
NIOSH, REL	ppm	ST250/methanol
OSHA, PEL	ppm	200/methanol
UN risk phrases, R	R36/38,R21,R52,R41	
US safety phrases, S	S28A,S26,S36/37/39	
ECOLOGICAL PROPERTIES		
Aquatic toxicity, *Green algae*, 96-h EC50	mg/l	119/7d
Aquatic toxicity, *Daphnia magna*, 48-h LC50	mg/l	324
Bioaccumulative potential	not bioaccumulation	
Biodegradation probability	37.0%/28d/not readily biodegradable	
Partition coefficient, log K_{ow}	-	0.5
USE & PERFORMANCE		
Manufacturer	Evonik Industries	
Outstanding properties	improved mechanical properties, such as flexural strength, tensile strength, impact strength and modulus of elasticity, improved moisture and corrosion resistance, and improved electrical properties, for example, dielectric constant, volume resistivity. Improved following processing properties: filler dispersion, rheological behavior (i.e., viscosity reduction), Newtonian behavior, increased filler loading, non-yellowing	
Recommended for polymers	acrylates, PA, PU, PVAc, epoxy resin, phenolic resin, polysulfides, unsaturated polyester, PP	
Recommended for products	composites, foundry resins, paints and coatings, primers	

Dynasylan GLYMO

PARAMETER	UNIT	VALUE
Recommended applications		adhesion promoter-can be formulated into the resin part of 2K-EP, and 2K-PU, crosslinking agents, surface modifier, used as an additive and as a primer in coating improved adhesion to the substrate, especially glass and metal, improved shelf life over aminosilanes in polyurethanes, used as a pretreatment of fillers and pigments.

Dynasylan HYDROSIL 1151

PARAMETER	UNIT	VALUE
GENERAL INFORMATION		
Name		Dynasylan HYDROSIL 1151
Composition		aqueous 3-aminopropylsilane hydrolysate
Chemical class	-	silane
Mixture	-	yes
Active matter	wt%	40
Functional organic group	-	primary amine
PHYSICAL PROPERTIES		
State		liquid/water solution
Odor	-	amine-like
Color		colorless to slightly yellowish
Boiling point	°C	100-150
Melting point	°C	-3.0
Density at 20°C	kg/m^3	1,060
pH	10-12 at 20°C (1:1 in H$_2$O)	
Refractive index at 20°C	-	1.371
Solubility (diluents)	-	alcohols
Solubility in water at 25°C	g/l	miscible
Specific gravity at 20°C	-	1.06
Vapor pressure at 20°C	kPa	0.001
Viscosity at 20°C	mPas	3.7
HEALTH & SAFETY		
NFPA classification	Flammability	1
	Health	1
	Reactivity	0
HMIS classification	Flammability	1
	Health	1
	Reactivity	0
DOT class	not regulated	
TDG class	not regulated	
ICAO/IATA class	not regulated	
IMDG class	not regulated	
Flash point	°C	80
Flash point method	-	PMCC
Agency rating, listed	EINECS Europe, ECL Korea, ENCS Japan, AICS Australia, IECSC China, DSL Canada, PICCS Philippines, TSCA USA	

Dynasylan HYDROSIL 1151

PARAMETER	UNIT	VALUE
Animal testing, acute toxicity, Rat oral LD50	mg/kg	>2000
Animal testing, acute toxicity, Rabbit dermal LD50	mg/kg	no skin irritation
Effect of exposure, eye (human)	Avoid contact with eyes.	
Effect of exposure, inhalation (human)	Avoid inhalation aerosol of mists. If aerosol or mists are inhaled, take affected persons out into the fresh air. Possible discomforts include severe irritation of mucus lining (nose, throat, eyes), cough, sneezing and flow of tears.	
Effect of exposure, skin (human)	Avoid contact with skin.	
Exposure, personal protection	Safety glasses, protective clothing based on chemical resistance data, chemical-resistant gloves, general and local exhaust ventilation.	
First aid, eye	Rinse cautiously with water for several minutes. Remove contact lenses, if present and easy to do. Continue rinsing. If eye irritation persists: Get medical advice/attention.	
First aid, inhalation	Remove to fresh air. If not breathing, give artificial respiration. If breathing is difficult, give oxygen. If irritation persists, obtain medical advice.	
First aid, skin	Immediately wipe away excess material. Use a waterless hand cleaner to remove as much of the remaining material as possible. Wash with soap and water.	

ECOLOGICAL PROPERTIES

Aquatic toxicity, *Green algae*, 96-h EC50	mg/l	>1000/72H
Aquatic toxicity, *Daphnia magna*, 48-h LC50	mg/l	331
Aquatic toxicity, *Zebra fish*, 96-h LC50	mg/l	>934
Biodegradation probability	8.00%/28d/not readily biodegradable	
Partition coefficient, log K_{oc}	-	-0.5/calculated

USE & PERFORMANCE

Manufacturer	Evonik Industries

Dynasylan HYDROSIL 1151

PARAMETER	UNIT	VALUE
Outstanding properties		improved mechanical properties: flexural strength, tensile strength, impact strength and modulus of elasticity, moisture and corrosion resistance, superior adhesion, better filler dispersion, high flash point, UV light stable, temperature and chemical-resistance. Almost solvent free, waterborne, odorless, reduced film thickness results in lower weight, cost-saving, chrome-free formulation.
Recommended for polymers		acrylics and acrylic copolymers, epoxies, PU, PVAc, EVA, EMA, PA, PBT, PC, PP, PVC, silicones
Recommended for products		composites, foundry resins, insulating materials, paints and coatings, primers, sealants and adhesives
Recommended applications		Dynasylan Hydrosil products are completely hydrolyzed multifunctional silane oligomers in water. These products provide barrier protection as well as adhesion promotion to a subsequent topcoat. The combination with the appropriate coating acts as an effective metal protection system against corrosion. For certain systems, the corrosion protection properties are on par with traditional pre-treatment systems, e. g., iron or zinc phosphate and chromate.

Dynasylan HYDROSIL 2627

PARAMETER	UNIT	VALUE
GENERAL INFORMATION		
Name		Dynasylan HYDROSIL 2627
Composition		aminofunctional oligomeric siloxane, carrier water
Chemical class		silane
Mixture	-	yes
Active matter	wt%	20
Functional organic group		amino/alkyl
PHYSICAL PROPERTIES		
State		liquid/water solution
Odor	-	slight
Color		clear to slightly yellowish
Color, Gardner scale	-	<1
Boiling point	°C	101/water
Density at 20°C	kg/m^3	1,062
pH	11 at 20°C (1:1 in H$_2$O)	
Solubility in water at 25°C	g/l	miscible
Specific gravity at 20°C	-	1.062
Viscosity at 20°C	mPas	3.28
Volatility	-	VOC-free
HEALTH & SAFETY		
NFPA classification	Flammability	1
	Health	1
	Reactivity	0
HMIS classification	Flammability	0
	Health	1
	Reactivity	0
DOT class	not regulated	
TDG class	not regulated	
ICAO/IATA class	not regulated	
IMDG class	not regulated	
Flash point	°C	101
Flash point method	-	PMCC
Agency rating, listed	EINECS Europe, IECSC China, DSL Canada, PICCS Philippines, TSCA USA	
Effect of exposure, eye (human)	Avoid contact with eyes.	

Dynasylan HYDROSIL 2627

PARAMETER	UNIT	VALUE
Effect of exposure, inhalation (human)		Avoid inhalation aerosol of mists. If aerosol or mists are inhaled, take affected persons out into the fresh air. Possible discomforts include severe irritation of mucus lining (nose, throat, eyes), cough, sneezing and flow of tears.
Effect of exposure, skin (human)		Avoid contact with skin.
Exposure, personal protection		Safety glasses, protective clothing based on chemical resistance data, chemical-resistant gloves, general and local exhaust ventilation.
First aid, eye		Rinse cautiously with water for several minutes. Remove contact lenses, if present and easy to do. Continue rinsing. If eye irritation persists: Get medical advice/attention.
First aid, inhalation		Remove to fresh air. If not breathing, give artificial respiration. If breathing is difficult, give oxygen. If irritation persists, obtain medical advice.
First aid, skin		Immediately wipe away excess material. Use a waterless hand cleaner to remove as much of the remaining material as possible. Wash with soap and water.
USE & PERFORMANCE		
Manufacturer		Evonik Industries
Outstanding properties		improved mechanical properties: flexural strength, tensile strength, impact strength and modulus of elasticity, moisture and corrosion resistance, superior adhesion, better filler dispersion, high flash point, UV light stable, temperature, and chemical resistance. Almost solvent-free, waterborne, odorless, reduced film thickness results in lower weight, cost-saving, chrome-free formulation.

Dynasylan HYDROSIL 2627

PARAMETER	UNIT	VALUE
Recommended applications		Dynasylan HYDROSIL products are completely hydrolyzed multifunctional silane oligomers in water. These products provide barrier protection as well as adhesion promotion to a subsequent topcoat. The combination with the appropriate coating acts as an effective metal protection system against corrosion. For certain systems, the corrosion protection properties are on par with traditional pre-treatment systems, e. g. iron or zinc phosphate and chromate.
Concentrations used		0.5-5% of active compounds

Dynasylan HYDROSIL 2776

PARAMETER	UNIT	VALUE
GENERAL INFORMATION		
Name		Dynasylan HYDROSIL 2776
Composition		alkylpolysiloxanes, amine-modified, carrier water
Chemical class	-	silane
Mixture	-	yes
Active matter	wt%	26
Functional organic group	-	diamino
PHYSICAL PROPERTIES		
State		liquid/water solution
Color		colorless to slightly yellowish
Color, Pt-Co scale	-	<=300
Boiling point	°C	101/water
Density at 20°C	kg/m^3	1000-1100
pH		11 at 20°C (1:1 in H$_2$O)
Solubility in water at 25°C	g/l	miscible
Specific gravity at 20°C	-	1.1
Vapor pressure at 20°C	kPa	1.84
Viscosity at 20°C	mPas	4-8
Volatility	-	VOC-free
HEALTH & SAFETY		
NFPA classification	Flammability	0
	Health	1
	Reactivity	0
HMIS classification	Flammability	0
	Health	1
	Reactivity	0
DOT class	not regulated	
TDG class	not regulated	
ICAO/IATA class	not regulated	
IMDG class	not regulated	
Autoignition temperature	°C	>650
Flash point	°C	>93
Flash point method	-	PMCC
Effect of exposure, eye (human)	Avoid contact with eyes.	

Dynasylan HYDROSIL 2776

PARAMETER	UNIT	VALUE
Effect of exposure, inhalation (human)		Avoid inhalation aerosol of mists. If aerosol or mists are inhaled, take affected persons out into the fresh air. Possible discomforts include severe irritation of mucus lining (nose, throat, eyes), cough, sneezing and flow of tears.
Effect of exposure, skin (human)		Avoid contact with skin.
Exposure, personal protection		Safety glasses, protective clothing based on chemical resistance data, chemical-resistant gloves, general and local exhaust ventilation.
First aid, eye		Rinse cautiously with water for several minutes. Remove contact lenses, if present and easy to do. Continue rinsing. If eye irritation persists: Get medical advice/attention.
First aid, inhalation		Remove to fresh air. If not breathing, give artificial respiration. If breathing is difficult, give oxygen. If irritation persists, obtain medical advice.
First aid, skin		Immediately wipe away excess material. Use a waterless hand cleaner to remove as much of the remaining material as possible. Wash with soap and water.
USE & PERFORMANCE		
Manufacturer		Evonik Industries
Outstanding properties		improved mechanical properties: flexural strength, tensile strength, impact strength and modulus of elasticity, moisture and corrosion resistance, superior adhesion, better filler dispersion, high flash point, UV light stable, temperature, and chemical resistance. Almost solvent-free, waterborne, odorless, reduced film thickness results in lower weight, cost-saving, chrome-free formulation

Dynasylan HYDROSIL 2776

PARAMETER	UNIT	VALUE
Recommended applications		Dynasylan HYDROSIL products are completely hydrolyzed multifunctional silane oligomers in water. These products provide barrier protection as well as adhesion promotion to a subsequent topcoat. The combination with the appropriate coating acts as an effective metal protection system against corrosion. For certain systems, the corrosion protection properties are on par with traditional pre-treatment systems, e. g., iron or zinc phosphate and chromate.
Concentrations used		0.5-5% of active compounds

Dynasylan HYDROSIL 2907

PARAMETER	UNIT	VALUE
GENERAL INFORMATION		
Name		Dynasylan HYDROSIL 2907
Composition		amino-vinyl-functional oligomeric siloxane, carrier water + (impurity <1% methanol and <1% formic acid)
Chemical class	-	silane
Mixture	-	yes
Functional organic group	-	amino/vinyl
PHYSICAL PROPERTIES		
State		liquid/water solution
Odor	-	amine-like
Color		colorless to slightly yellowish
Color, Pt-Co scale	-	<100
Boiling point	°C	102
Density at 20°C	kg/m³	1122-1129
pH		4-4.5 at 20°C (1:1 in H_2O)
Solubility in water at 25°C	g/l	miscible
Specific gravity at 20°C	-	1.102-1.103
Viscosity at 20°C	mPas	<20
Volatility	-	VOC-free
HEALTH & SAFETY		
NFPA classification	Flammability	1
	Health	1
	Reactivity	0
HMIS classification	Flammability	1
	Health	1
	Reactivity	0
Carcinogenicity		IARC, OSHA, NTP: no ingredient of this product present at levels greater than or equal to 0.1% is identified as probable, possible or confirmed human carcinogen
TDG class		not regulated
ICAO/IATA class		not regulated
IMDG class		not regulated
Autoignition temperature	°C	>650
Flash point	°C	>80
Flash point method	-	PMCC
Agency rating, listed		EINECS Europe, IECSC China, PICCS Philippines, TSCA USA

Dynasylan HYDROSIL 2907

PARAMETER	UNIT	VALUE
Effect of exposure, eye (human)		Causes eye irritation.
Effect of exposure, inhalation (human)		May cause respiratory irritation. Do not inhale vapors or aerosols. If aerosol or mists are inhaled, take affected persons out into the fresh air. Possible discomforts include severe irritation of mucus lining (nose, throat, eyes), cough, sneezing and flow of tears.
Effect of exposure, skin (human)		May cause skin irritation.
Exposure, personal protection		Safety glasses, protective clothing based on chemical resistance data, chemical-resistant gloves, general and local exhaust ventilation.
First aid, eye		Rinse cautiously with water for several minutes. Remove contact lenses, if present and easy to do. Continue rinsing. If eye irritation persists: Get medical advice/attention.
First aid, inhalation		Remove to fresh air. If not breathing, give artificial respiration. If breathing is difficult, give oxygen. If irritation persists, obtain medical advice.
First aid, skin		Immediately wipe away excess material. Use a waterless hand cleaner to remove as much of the remaining material as possible. Wash with soap and water.
USE & PERFORMANCE		
Manufacturer		Evonik Industries
Outstanding properties		improved mechanical properties: flexural strength, tensile strength, impact strength and modulus of elasticity, moisture and corrosion resistance, superior adhesion, better filler dispersion, high flash point, UV light stable, temperature and chemical-resistance. Almost solvent-free, waterborne, odorless, reduced film thickness results in lower weight, cost-saving, chrome-free formulation

Dynasylan HYDROSIL 2907

PARAMETER	UNIT	VALUE
Recommended applications		Dynasylan HYDROSIL products are completely hydrolyzed multifunctional silane oligomers in water. These products provide barrier protection as well as adhesion promotion to a subsequent topcoat. The combination with the appropriate coating acts as an effective metal protection system against corrosion. For certain systems, the corrosion protection properties are on par with traditional pre-treatment systems, e. g., iron or zinc phosphate and chromate.
Concentrations used		0.5-5% of active compounds

Dynasylan HYDROSIL 2909

PARAMETER	UNIT	VALUE
GENERAL INFORMATION		
Name		Dynasylan HYDROSIL 2909
Composition		organofunctional siloxane oligomer, carrier water (impurity <0.3% methanol)
Chemical class	-	silane
Mixture	-	yes
Active matter	wt%	37
Functional organic group	-	amino
PHYSICAL PROPERTIES		
State	liquid/water solution	
Odor	-	amine-like
Color	colorless to slightly yellowish	
Color, Gardner scale	-	<1
Density at 20°C	kg/m^3	1094
pH	4 at 20°C (1:1 in H$_2$O)	
Solubility in water at 25°C	g/l	miscible
Specific gravity at 20°C	-	1.094
Vapor pressure at 20°C	kPa	2.29
Viscosity at 20°C	mPas	152
Volatility	-	VOC-free
HEALTH & SAFETY		
NFPA classification	Flammability	1
	Health	1
	Reactivity	0
HMIS classification	Flammability	1
	Health	1
	Reactivity	0
Carcinogenicity	IARC, OSHA, NTP: no ingredient of this product present at levels greater than or equal to 0.1% is identified as probable, possible or confirmed human carcinogen	
ICAO/IATA class	not regulated	
IMDG class	not regulated	
Autoignition temperature	°C	470
Flash point	°C	>95
Flash point method	-	PMCC
Effect of exposure, eye (human)	Causes eye irritation.	

Dynasylan HYDROSIL 2909

PARAMETER	UNIT	VALUE
Effect of exposure, inhalation (human)	May cause respiratory irritation. Do not inhale vapors or aerosols. If aerosol or mists are inhaled, take affected persons out into the fresh air. Possible discomforts include severe irritation of mucus lining (nose, throat, eyes), cough, sneezing and flow of tears.	
Effect of exposure, skin (human)	May cause skin irritation.	
Effect of exposure, swallowing (human)	Acute toxicity estimated > 5,000 mg/kg.	
Exposure, personal protection	Safety glasses, protective clothing based on chemical resistance data, chemical-resistant gloves, general and local exhaust ventilation.	
First aid, eye	Rinse cautiously with water for several minutes. Remove contact lenses, if present and easy to do. Continue rinsing. If eye irritation persists: Get medical advice/attention.	
First aid, inhalation	Remove to fresh air. If not breathing, give artificial respiration. If breathing is difficult, give oxygen. If irritation persists, obtain medical advice.	
First aid, skin	Immediately wipe away excess material. Use a waterless hand cleaner to remove as much of the remaining material as possible. Wash with soap and water.	
NIOSH, REL	mg/m^3	ST325/methanol
OSHA, PEL	mg/m^3	260/methanol
ACGIH, TLV	ppm	200/methanol
NIOSH, REL	ppm	ST250/methanol
OSHA, PEL	ppm	200/methanol

USE & PERFORMANCE

Manufacturer	Evonik Industries	
Outstanding properties	Improved mechanical properties: flexural strength, tensile strength, impact strength and modulus of elasticity, moisture and corrosion resistance, superior adhesion, better filler dispersion, high flash point, UV light stable, temperature and chemical-resistance. Almost solvent-free, waterborne, odorless, reduced film thickness results in lower weight, cost-saving, chrome-free formulation	
Recommended for polymers	2K PU	

Dynasylan HYDROSIL 2909

PARAMETER	UNIT	VALUE
Recommended for products		2K PU solvent based coating on electro-galvanized steel and on cold rolled steel
Recommended applications		Dynasylan HYDROSIL products are completely hydrolyzed multifunctional silane oligomers in water. These products provide barrier protection as well as adhesion promotion to a subsequent topcoat. The combination with the appropriate coating acts as an effective metal protection system against corrosion. For certain systems, the corrosion protection properties are on par with traditional pre-treatment systems, e. g. iron or zinc phosphate and chromate.
Concentrations used		0.5-5% of active compounds

Dynasylan HYDROSIL 2926

PARAMETER	UNIT	VALUE
GENERAL INFORMATION		
Name		Dynasylan HYDROSIL 2926
Composition		Epoxy-functional oligomeric siloxane, carrier water
Chemical class	-	silane
Mixture	-	yes
Active matter	wt%	30
Functional organic group	-	hydroxy
PHYSICAL PROPERTIES		
State	liquid/water solution	
Odor	-	slight
Color	-	colorless
Color, Gardner scale	-	<1
Density at 20°C	kg/m^3	1091
pH	3.0 at 20°C (1:1 in H_2O)	
Solubility in water at 25°C	g/l	miscible
Specific gravity at 20°C	-	1.091
Viscosity at 20°C	mPas	6-7
Volatility	-	VOC-free
HEALTH & SAFETY		
Flash point	°C	>98
Exposure, personal protection		Safety glasses, protective clothing based on chemical resistance data, chemical-resistant gloves, general and local exhaust ventilation.
First aid, eye		Rinse cautiously with water for several minutes. Remove contact lenses, if present and easy to do. Continue rinsing. If eye irritation persists: Get medical advice/attention.
First aid, inhalation		Remove to fresh air. If not breathing, give artificial respiration. If breathing is difficult, give oxygen. If irritation persists, obtain medical advice.

Dynasylan HYDROSIL 2926

PARAMETER	UNIT	VALUE
First aid, skin		Immediately wipe away excess material. Use a waterless hand cleaner to remove as much of the remaining material as possible. Wash with soap and water.

USE & PERFORMANCE

Manufacturer		Evonik Industries
Outstanding properties		improved mechanical properties: flexural strength, tensile strength, impact strength and modulus of elasticity, moisture and corrosion resistance, superior adhesion, better filler dispersion, high flash point, UV light stable, temperature and chemical-resistance. Almost solvent-free, waterborne, odorless, reduced film thickness results in lower weight, cost-saving, chrome-free formulation
Recommended applications		Dynasylan HYDROSIL products are completely hydrolyzed multifunctional silane oligomers in water. These products provide barrier protection as well as adhesion promotion to a subsequent topcoat. The combination with the appropriate coating acts as an effective metal protection system against corrosion. For certain systems, the corrosion protection properties are on par with traditional pre-treatment systems, e. g., iron or zinc phosphate and chromate.

Dynasylan MEMO

PARAMETER	UNIT	VALUE
GENERAL INFORMATION		
Name		Dynasylan MEMO
CAS #	-	2530-85-0
General description		bifunctional
Composition		100% 3-methacryloxypropyl trimethoxysilane
Formula		

PARAMETER	UNIT	VALUE
Chemical class	-	silane
Functional organic group	-	methacryl
PHYSICAL PROPERTIES		
State	-	liquid
Odor	-	slightly aromatic
Color		colorless to yellowish
Boiling point	°C	255
Melting point	°C	<-20
Density at 20°C	kg/m³	1,040
Refractive index at 20°C	-	1.432
Solubility (diluents)	alcohols, aliphatic or aromatic hydrocarbons	
Solubility in water at 25°C	hydrolytic decomposition releasing methanol	
Specific gravity at 25°C	-	1.04
Vapor pressure at 20°C	kPa	0.01
Viscosity at 20°C	mPas	2.8
HEALTH & SAFETY		
NFPA classification	Flammability	1
	Health	1
	Reactivity	1
HMIS classification	Flammability	1
	Health	1
	Reactivity	1
DOT class	not regulated	
TDG class	not regulated	
ICAO/IATA class	not regulated	
IMDG class	not regulated	
Flash point	°C	110

Dynasylan MEMO

PARAMETER	UNIT	VALUE
Flash point method	-	PMCC
Explosive LEL	wt%	0.90
Explosive UEL	wt%	5.40
Hazardous combustion products	Carbon oxides, Sulfur oxides	
Agency rating, listed	EINECS Europe, ECL Korea, ENCS Japan, AICS Australia, IECSC China, DSL Canada, PICCS Philippines, TSCA USA	
Hazardous products of hydrolysis	methanol	
Animal testing, acute toxicity, Rat oral LD50	mg/kg	>2000
Animal testing, acute toxicity, Rabbit dermal LD50	mg/kg	no skin irritation
Animal testing, acute toxicity, Rat dermal LD50	mg/kg	>2000
Animal testing, acute toxicity, Rat inhalation, LC50	mg/m^3	2280/4H/aerosol
Effect of exposure, eye (human)	Avoid contact with eyes.	
Effect of exposure, skin (human)	Avoid contact with skin.	
Exposure, personal protection	Safety glasses, protective clothing based on chemical resistance data, chemical-resistant gloves, general and local exhaust ventilation.	
First aid, eye	Rinse cautiously with water for several minutes. Remove contact lenses, if present and easy to do. Continue rinsing. If eye irritation persists: Get medical advice/attention.	
First aid, inhalation	Remove to fresh air. If not breathing, give artificial respiration. If breathing is difficult, give oxygen. If irritation persists, obtain medical advice.	
First aid, skin	Immediately wipe away excess material. Use a waterless hand cleaner to remove as much of the remaining material as possible. Wash with soap and water.	
NIOSH, REL	mg/m^3	ST325/methanol
OSHA, PEL	mg/m^3	260/methanol
ACGIH, TLV	ppm	200/methanol
NIOSH, REL	ppm	ST250/methanol
OSHA, PEL	ppm	200/methanol
ECOLOGICAL PROPERTIES		
Aquatic toxicity, *Green algae*, 96-h EC50	mg/l	>536/72H
Aquatic toxicity, *Daphnia magna*, 48-h LC50	mg/l	> 876
Aquatic toxicity, *Zebra fish*, 96-h LC50	mg/l	>100

Dynasylan MEMO

PARAMETER	UNIT	VALUE
Biodegradation probability	74%/28d/readily biodegradable	
Partition coeff cient, log K_{ow}	-	2.1
USE & PERFORMANCE		
Manufacturer	Evonik Industries	
Outstanding properties	improved dispersion of the fillers and pigments, provided a reduction in the settling of the fillers and pigments and reduction in viscosity, improved flow properties of the resin, and improved mechanical properties of the molded products. Offers benefits such as improved wet adhesion strength, chemical resistance, and mar resistance	
Recommended for polymers	PA, PU, epoxy resin, polyesters, polyolefins, unsaturated polyester, PP	
Recommended applications	adhesion promoter, surface modifier, co-monomer for polymer synthesis and crosslinker.	

Dynasylan® MTES

PARAMETER	UNIT	VALUE
GENERAL INFORMATION		
Name		Dynasylan® MTES
Common synonym		alkyltrialkoxysilane
Composition		partially hydrolyzed to form a preproduct that can be further crosslinked using temperature. This prehydrolysis is often done in conjunction with other orga-nofunctional silanes, silicic acid esters (e.g., Dynasylan® A) or even an aqueous silica sol.
Functionality	-	trifunctional
Functional groups	-	alkoxy
PHYSICAL PROPERTIES		
State	-	liquid
Color	-	colorless
Boiling point	°C	142 (initial)
Density at 20°C	kg/m^3	890
Dilution		miscible with aliphatic and aromatic solvents
Viscosity at 20°C	mPas	0.6
HEALTH & SAFETY		
Flash point	°C	30
USE & PERFORMANCE		
Manufacturer		Dow
Outstanding properties		hydrolysis leads to silanol groups which, in a subsequent condensation reaction, form very stable siloxane bonds (-Si-O-Si-). Condensation occurs parallel to hydrolysis once a certain amount of silanol groups have been formed. The absolute and relative rates of hydrolysis and condensation depend on a number of factors. The most important factors include pH, concentration, solvent, tem-perature, and the catalyst.
Recommended for products		sol-gel systems; mar resistant coatings having a higher UV-stability than tradi-tional organic coatings

Dynasylan MTMO

PARAMETER	UNIT	VALUE
GENERAL INFORMATION		
Name	Dynasylan MTMO	
CAS #	-	4420-74-0
EC number	-	224-588-5
General description	bifunctional	
Composition	3-mercaptopropyltri-methoxysilane	
Acronym		MTMO
Empirical formula	-	C6H16O3SSi
Formula		
Molecular mass	daltons	196.34
Chemical class	-	silane
Functional organic group	-	methacryl
PHYSICAL PROPERTIES		
State	-	liquid
Odor	-	ester-like
Color	clear colorless to yellowish	
Boiling point	°C	85
Melting point	°C	<-50
Density at 20°C	kg/m³	1,060
Refractive index at 20°C	-	1.445
Solubility in water at 25°C	hydrolytic decomposition releasing methanol	
Solvent solubility	alcohols, ketones,, and aliphatic or aromatic hydrocarbons	
Specific gravity at 25°C	-	1.06
Vapor pressure at 20°C	kPa	0.02
Viscosity at 20°C	mPas	2.0
HEALTH & SAFETY		
NFPA classification	Flammability	1
	Health	2
	Reactivity	1
HMIS classification	Flammability	1
	Health	2
	Reactivity	1

Dynasylan MTMO

PARAMETER	UNIT	VALUE
Carcinogenicity		IARC, OSHA, NTP: no ingredient of this product present at levels greater than or equal to 0.1% is identified as probable, possible or confirmed human carcinogen
DOT class		(3-mercaptopropyl-trimethoxysilane) 3, III. Not regulated in packages 450 liter or less.
ICAO/IATA class		Environmentally hazardous substance, liquid, n.o.s.(3mercaptopropyl-trime-thoxysilane) 9, III
IMDG class		ENVIRONMENTALLY HAZARDOUS SUBSTANCE, LIQUID, N.O.S.(3-mer-captopropyl-trimethoxysilane)9, III
Flash point	°C	85
Flash point method	-	PMCC
Hazardous combustion products		Carbon oxides, sulfur oxides
Agency rating, listed		EINECS Europe, ECL Korea, ENCS Japan, AICS Australia, IECSC China, DSL Canada, PICCS Philippines, TSCA USA
Hazardous ingredients, labelling		(3-mercaptopropyl-trimethoxysilane)
Hazardous products of hydrolysis		methanol
Animal testing, acute toxicity, Rat oral LD50	mg/kg	933/male,774/female
Effect of exposure, eye (human)		Avoid contact with eyes.
Effect of exposure, inhalation (human)		Avoid breathing fume, mist, vapors, spray.
Effect of exposure, skin (human)		May cause an allergic skin reaction. Skin sensitisation
Effect of exposure, swallowing (human)		Harmful if swallowed.
Exposure, personal protection		Safety glasses, protective clothing based on chemical resistance data, chemical-resistant gloves, general and local exhaust ventilation.
First aid, eye		Rinse cautiously with water for several minutes. Remove contact lenses, if present and easy to do. Continue rinsing. If eye irritation persists: Get medical advice/attention.
First aid, inhalation		Remove to fresh air. If not breathing, give artificial respiration. If breathing is difficult, give oxygen. If irritation persists, obtain medical advice.

Dynasylan MTMO

PARAMETER	UNIT	VALUE
First aid, skin		Immediately wipe away excess material. Use a waterless hand cleaner to remove as much of the remaining material as possible. Wash with soap and water.
NIOSH, REL	mg/m^3	ST325/methanol
OSHA, PEL	mg/m^3	260/methanol
ACGIH, TLV	ppm	200/methanol
NIOSH, REL	ppm	ST250/methanol
OSHA, PEL	ppm	200/methanol
UN risk phrases, R		R22,R43,R51/53
US safety phrases, S		S24/25,S36/37,S61,S57,S37,S24
UN/NA class	-	1993/DOT, 3082/ IMDG & ICAO/IATA
ECOLOGICAL PROPERTIES		
Aquatic toxicity, *Green algae*, 96-h EC50	mg/l	267/72H
Aquatic toxicity, *Bluegill sunfish*, 96-h LC50	mg/l	439
Aquatic toxicity, *Daphnia magna*, 48-h LC50	mg/l	6.7
Biodegradation probability		51%/28d/not readily biodegradable
USE & PERFORMANCE		
Manufacturer		Evonik Industries
Outstanding properties		improved compression set, mechanical properties, such as flexural strength, tensile strength, impact strength and modulus of elasticity, improved moisture, and corrosion resistance, and improved electrical properties, for example, dielectric constant, volume resistivity, increased thermal resistivity. Improved following processing properties: cure time, filler dispersion, rheological behavior (i.e., viscosity reduction), increased filler loading.
Recommended for polymers		PVC, PVDC, PU, polysulfides, sulfur-cured and metal oxide-cured elastomers
Recommended for products		composites, crosslinkers, primers, sealants and adhesives
Recommended applications		adhesion promoter, crosslinking agent, surface modifier or reactive reagent, Dynasylan MTMO bind to both inorganic materials (e.g. glass, metals, fillers) and organic polymers (e.g. thermosets, thermoplastics, elastomers).

Dynasylan® MTMS

PARAMETER	UNIT	VALUE
GENERAL INFORMATION		
Name		Dynasylan® MTMS
Common synonym		alkyltrialkoxysilane
Common name		methyltrimethoxysilane
Composition		partially hydrolyzed to form a preproduct that can be further crosslinked using temperature. This prehydrolysis is often done in conjunction with other orga-nofunctional silanes, silicic acid esters (e.g., Dynasylan® A) or even a aqueous silica sol.
Functionality	-	trifunctional
Functional groups	-	alkoxy
PHYSICAL PROPERTIES		
State	-	liquid
Color	-	colorless
Boiling point	°C	102 (initial)
Density at 20°C	kg/m^3	960
Viscosity at 20°C	mPas	0.5
HEALTH & SAFETY		
Flash point	°C	9
USE & PERFORMANCE		
Manufacturer		Dow
Outstanding properties		prehydrolysis often is done in conjunc-tion with other organofunctional silanes (e.g., Dynasylan® GLYMO), silicic acid esters, or even an aqueous silica sol. This preproduct can be modified even further by the addition of organic resins or inorganic nanoparticles such as AEROSIL®. It is also possible to con-struct an inorganic/organic network by adding silanes containing organofunc-tional groups (e.g. aminopropyl groups) and organic resins and polymerizing using standard organic methods.
Recommended for products		sol-gel systems; mar resistant coatings having a higher UV-stability than tradi-tional organic coatings

Dynasylan® OCTEO

PARAMETER	UNIT	VALUE
GENERAL INFORMATION		
Name		Dynasylan® OCTEO
Common synonym		monomeric medium-chain alkylfunctional silane
Functionality	-	alkyl
PHYSICAL PROPERTIES		
State	-	liquid
Color	-	clear, colorless
Boiling point	°C	265
Density at 20°C	kg/m³	880
Solvents	petroleum ether, toluene	
Viscosity at 20°C	mPas	2
HEALTH & SAFETY		
Flash point	°C	>93
USE & PERFORMANCE		
Manufacturer	Dow	
Outstanding properties	Dynasylan® OCTEO treated minerals or pigments are incorporated into polymers, e.g., polyethylene or polypropylene.	
Recommended for polymers	polyethylene, polypropylene	
Recommended for products	surface modifier to generate hydrophobicity on inorganic fillers and pigments, e.g., titanium dioxide.	
Concentrations used	wt%	0.5-1.5

Dynasylan® OCTMO

PARAMETER	UNIT	VALUE
GENERAL INFORMATION		
Name		Dynasylan® OCTMO
Common synonym		monomeric medium-chain alkylfunctional silane
Functionality	-	alkyl
PHYSICAL PROPERTIES		
State	-	liquid
Color	-	clear, colorless
Boiling point	°C	246
Density at 20°C	kg/m³	910
Solvents		petroleum ether, toluene
Viscosity at 20°C	mPas	2
HEALTH & SAFETY		
Flash point	°C	102
USE & PERFORMANCE		
Manufacturer		Dow
Outstanding properties		Dynasylan® OCTMO silane can be used as a surface modifier to generate hydrophobicity on inorganic fillers and pigments. The medium-chain alkyl functionality results in unique compound properties when Dynasylan® OCTMO treated minerals or pigments are incorporated into polymers, e.g. polyethylene or polypropylene.
Recommended for polymers		polyethylene, polypropylene
Recommended for products		surface modifier to generate hydrophobicity on inorganic fillers and pigments, e.g. titanium dioxide.
Concentrations used	wt%	0.5-1.5

Dynasylan® PTEO

PARAMETER	UNIT	VALUE
GENERAL INFORMATION		
Name		Dynasylan® PTEO
Common synonym		alkyltrialkoxysilane
Functionality	-	trifunctional
PHYSICAL PROPERTIES		
State	-	liquid
Color	-	colorless
Boiling point	°C	175 (initial)
Density at 20°C	kg/m³	890
Dilution solvent	-	ethanol
Viscosity at 20°C	mPas	1
HEALTH & SAFETY		
Flash point	°C	57
USE & PERFORMANCE		
Manufacturer		Dow
Features		partially hydrolyzed to form a preproduct that can be further crosslinked using temperature. This pre-hydrolysis often is done in conjunction with other organofunctional silanes, silicic acid esters, or even an an aqueous silica sol. This pre-product can be modified even further by the addition of organic resins or inorganic nanoparticles such as Aerosil®. It is also possible to construct an inorganic/organic network by adding silanes containing organofunctional groups (e.g., aminopropyl groups) and organic resins.
Recommended for products		sol-gel systems; mar resistant coatings having a higher UV-stability than tradiional organic coatings
Guidelines for use		Dynasylan® PTEO reacts slower with water than Dynasylan® PTMO, and often, a hydrolysis catalyst (mineral acids or ammonia, or even acetic acid and amines) must be added to hydrolyze at appreciable rates. Hydrolysis can also be furthered by adding a cosolvent such as ethanol.

Dynasylan® SIVO 110

PARAMETER	UNIT	VALUE
GENERAL INFORMATION		
Name		Dynasylan® SIVO 110
Functionality	-	multifunctional
Solid content	wt%	35
PHYSICAL PROPERTIES		
State	-	liquid
Color		opaque to slightly milky, colorless to slightly yellow
Boiling point	°C	96
Density at 20°C	kg/m³	1140
pH	-	4.3
Dilution	-	water
Viscosity at 20°C	mPas	7
HEALTH & SAFETY		
Flash point	°C	>95
USE & PERFORMANCE		
Manufacturer		Dow
Features		VOC-free, a water-borne sol-gel system having high hardness, excellent scratch and mar resistance, superior stability in boiling water, very good adhesion to various substrates, very good adhesion to organic topcoats (e.g., epoxies), flexibility, the very low thickness of formed layers (dry layer of < 2 µm), excellent resistance to solvents and other chemicals
Recommended for products		sol-gel systems; sol-gel-based hybrid coatings, corrosion resistant primer systems, transparent sol-gel topcoats
Guidelines for use		temperature resistance of up to 220°C coatings comprising high hardness can be additionally improved by the introduction of up to 20 wt-% silica

Dynasylan® SIVO 121

PARAMETER	UNIT	VALUE
GENERAL INFORMATION		
Name		Dynasylan® SIVO 121
Functionality	-	multifunctional
PHYSICAL PROPERTIES		
State	-	liquid
Color	yellowish slightly turbid	
Boiling point	°C	96
Density at 20°C	kg/m³	1010
pH	-	4
Dilution	-	water
Viscosity at 20°C	mPas	1
HEALTH & SAFETY		
Flash point	°C	>90
USE & PERFORMANCE		
Manufacturer	Dow	
Features	Dynasylan® SIVO 121 containing wood preservation agents can be painted, rolled or sprayed. In a spray application it is mandatory to minimize the generation of Aerosol emissions by suited procedures (e.g. application of HPLV spray processes, air driven low pressure spray processes. Consecutive application steps should be carried out wet-in-wet as a dried coat of Dynasylan® SIVO 121 will almost immediately exhibit a strong repelling effect. Consequently a second application step would therefore be much less effective.	
Recommended for products	wood surface impregnation agent. It can act most prominently as a hydrophobizing or olephobizing agent for plain wood surfaces or weathered, originally scumble paint treated wood surfaces	
Guidelines for use	For a full range effectiveness and durability (up to 1500 h of QUV-stability in transparent systems) 300 g/m² of undiluted Dynasylan® SIVO 121 are necessary. That amount can be applied by either one-step or multi-step procedures, depending on the absorptiveness of the wood substrate.	

Dynasylan SIVO 202

PARAMETER	UNIT	VALUE
GENERAL INFORMATION		
Name		Dynasylan SIVO 202
Composition		multifunctional aminosilane system
Chemical class		silane
PHYSICAL PROPERTIES		
State	-	liquid
Color	-	light straw
Density at 20°C	kg/m³	1030
Solubility (diluents)		alcohols, aliphatic, aromatic hydrocarbon, can undergo reactions with ketones or ester solvents
Viscosity at 20°C	mPas	4
HEALTH & SAFETY		
Agency rating, listed		EINECS Europe, ECL Korea, IECSC China, PICCS Philippines, TSCA USA
Effect of exposure, eye (human)		Avoid contact with eyes.
Effect of exposure, skin (human)		Avoid contact with skin.
Effect of exposure, swallowing (human)		Not harmful or toxic after single oral ingestion of high concentrations.
Exposure, personal protection		Safety glasses, protective clothing based on chemical resistance data, chemical-resistant gloves, general and local exhaust ventilation.
First aid, eye		Rinse cautiously with water for several minutes. Remove contact lenses, if present and easy to do. Continue rinsing. If eye irritation persists: Get medical advice/attention.
First aid, inhalation		Remove to fresh air. If not breathing, give artificial respiration. If breathing is difficult, give oxygen. If irritation persists, obtain medical advice.
First aid, skin		Immediately wipe away excess material. Use a waterless hand cleaner to remove as much of the remaining material as possible. Wash with soap and water.

Dynasylan SIVO 202

PARAMETER	UNIT	VALUE
USE & PERFORMANCE		
Manufacturer		Evonik Industries
Outstanding properties		Exhibits excellent wetting of substrates and forms weather- and moisture-bonds to substrates that are difficult to adhere to. The main benefits of Dynasylan SIVO 202 in silylated PU, MS polymers adhesives and sealants are improved adhesion to plastics such as PC, PVC, and improved mechanical properties.
Recommended for polymers		epoxies, PC, PU, PVC, silicones
Recommended for products		adhesives and sealants, primers
Recommended applications		wide variety of applications, particularly in adhesives and sealants to improve adhesion of hybrid products (e.g., silylated PU, MS polymers), RTV-silicones, two-part PU, two-part epoxies to inorganic surfaces, and some difficult plastic surface
Concentrations used		0.5-2% of total formulation

Dynasylan SIVO 210

PARAMETER	UNIT	VALUE
GENERAL INFORMATION		
Name		Dynasylan SIVO 210
Composition		proprietary aminosilane composition
Common synonym		3-(trimethoxysilyl)-1-propanamin
Chemical class	-	silane
Functional organic group	-	amino/methoxy
PHYSICAL PROPERTIES		
State	-	liquid
Odor	-	amine-like
Color		clear colorless to yellowish
Boiling point	°C	240
Density at 20°C	kg/m^3	970
pH		11.0 at 20°C (20 g/l H$_2$O)
Solubility (diluents)		alcohols, aliphatic or aromatic hydrocarbons
Specific gravity at 25°C	-	0.97
Viscosity at 20°C	mPas	4-40
HEALTH & SAFETY		
Flash point	°C	>95
Hazardous combustion products		Carbon oxides, nitrogen oxides
Agency rating, listed		EINECS Europe, ECL Korea, ECSC China, PICCS Philippines, TSCA USA
Exposure, personal protection		Safety glasses, protective clothing based on chemical resistance data, chemical-resistant gloves, general and local exhaust ventilation.
First aid, eye		Rinse cautiously with water for several minutes. Remove contact lenses, if present and easy to do. Continue rinsing. If eye irritation persists: Get medical advice/attention.
First aid, inhalation		Remove to fresh air. If not breathing, give artificial respiration. If breathing is difficult, give oxygen. If irritation persists, obtain medical advice.
First aid, skin		Immediately wipe away excess material. Use a waterless hand cleaner to remove as much of the remaining material as possible. Wash with soap and water.

Dynasylan SIVO 210

PARAMETER	UNIT	VALUE
USE & PERFORMANCE		
Manufacturer		Evonik Industries
Outstanding properties		Improved mechanical properties, for example, flexural strength, tensile strength, impact strength, and modulus of elasticity, improved moisture, corrosion resistance, as well improved filler dispersion, rheological behavior (reduction in viscosity), higher filler loading.
Recommended for polymers		acrylics, acrylic copolymers, PA, PBT, PC, EVA, modified PP, PVB, PVAC, PVC, phenolic, furan, and melamine resins
Recommended for products		foundry resins, metal primers, paints and coatings, sealants and adhesives
Recommended applications		adhesion promoter to both inorganic materials (for example, glass, metals and fillers) and organic polymers (thermosets, thermoplastics, and elastomers). Examples of application: mineral-filled polymers (composites) or HFFR cables for pre-treatment of fillers and pigments or as an additive, paints and coatings as an additive and primer for improving adhesion to the substrate, mineral fiber insulating materials, abrasives as an additive to phenolic resin binders

Dynasylan SIVO 214

PARAMETER	UNIT	VALUE
GENERAL INFORMATION		
Name		Dynasylan SIVO 214
Composition		proprietary aminosilane composition
Empirical formula		C10H27N3O3Si
Molecular mass	daltons	265.43
Chemical class	-	silane
Functional organic group	-	amino/methoxy
PHYSICAL PROPERTIES		
State	-	liquid
Odor	-	amine-like
Color		clear colorless to yellowish
Boiling point	°C	>68 (4 hPa)
Density at 20°C	kg/m³	950
pH		11.0 at 20°C (20 g/l H_2O)
Solubility (diluents)		alcohols, aliphatic or aromatic hydrocarbons
Viscosity at 20°C	mPas	2
HEALTH & SAFETY		
Flash point	°C	98
Hazardous combustion products		Carbon oxides, nitrogen oxides
Agency rating, listed		EINECS Europe, ECL Korea, ECSC China, PICCS Philippines, TSCA USA
Hazardous products of hydrolysis		methanol
Exposure, personal protection		Safety glasses, protective clothing based on chemical resistance data, chemical-resistant gloves, general and local exhaust ventilation.
First aid, eye		Rinse cautiously with water for several minutes. Remove contact lenses, if present and easy to do. Continue rinsing. If eye irritation persists: Get medical advice/attention.
First aid, inhalation		Remove to fresh air. If not breathing, give artificial respiration. If breathing is difficult, give oxygen. If irritation persists, obtain medical advice.

Dynasylan SIVO 214

PARAMETER	UNIT	VALUE
First aid, skin		Immediately wipe away excess material. Use a waterless hand cleaner to remove as much of the remaining material as possible. Wash with soap and water.
USE & PERFORMANCE		
Manufacturer		Evonik Industries
Outstanding properties		improved mechanical properties, for example, flexural strength, tensile strength, impact strength and modulus of elasticity, improved moisture, corrosion resistance, as well improved filler dispersion, rheological behavior (reduction in viscosity), higher filler loading. Dynasylan SIVO 214 forms highly cross-linked networks between substrates and inorganic matrices.
Recommended for polymers		acrylics, acrylic copolymers, PA, PBT, PC, EVA, modified PP, PVB, PVAc, PVC, phenolic, furan, melamine resins
Recommended for products		composites, foundry resins, paints and coatings, primers, sealants and adhesives
Recommended applications		adhesion promoter to both inorganic materials (for example, glass, glass fiber, glass wool, mineral wool, silicic acid, quartz, sand, cristobalite, wollastonite, mica, also suitable are aluminum hydroxide, kaolin, talc and metals oxides i metals) and organic polymers (thermosets, thermoplastics, and elastomers). Examples of application: mineral-filled polymers (composites) or HFFR cables for pre-treatment of fillers and pigments or as an additive, paints and coatings as an additive and primer for improving adhesion to the substrate, mineral fiber insulating materials, abrasives as an additive to phenolic resin binders.
Concentrations used		0.5-2% solution

Dynasylan® SIVO 408

PARAMETER	UNIT	VALUE
GENERAL INFORMATION		
Name		Dynasylan® SIVO 408
Composition		oligomeric short-chain alkylfunctional silane
Functionality	-	alkylfunctional
Functional groups	-	ethoxy
PHYSICAL PROPERTIES		
State	-	liquid
Color		clear, colorless to slightly yellow
Boiling point	°C	96
Density at 20°C	kg/m³	1040
pH	-	3-4
Dilution		petroleum ether, toluene, alcohol
Viscosity at 20°C	mPas	35
HEALTH & SAFETY		
Flash point	°C	>25
USE & PERFORMANCE		
Manufacturer		Dow
Features		Dynasylan® SIVO 408 forms covalent bonds to the inorganic surfaces and will not migrate out of the final compound as it will happen to silicone oils used as surface modifiers.
Recommended for polymers		polyethylene, polypropylene
Recommended for products		surface modifier to generate hydrophobicity
Concentrations used	wt%	0.5-1.5

Dynasylan® SIVO 560

PARAMETER	UNIT	VALUE
GENERAL INFORMATION		
Name		Dynasylan® SIVO 560
Composition		multifunctional silane system with a high concentration of unsaturated organic groups
Functionality	-	multilfunctional
Functional groups	-	ethoxy
PHYSICAL PROPERTIES		
State	-	liquid
Color		clear, colorless
Boiling point	°C	142
Density at 20°C	kg/m^3	1010
Refractive index at 20°C	-	1.408-1.414
Dilution		alcohols, aliphatic and aromatic hydrocarbons
Viscosity at 20°C	mPas	1
HEALTH & SAFETY		
Flash point	°C	34
USE & PERFORMANCE		
Manufacturer		Dow
Features		Dynasylan® SIVO 408 forms covalent bonds to the inorganic surfaces and will not migrate out of the final compound as it will happen to silicone oils used as surface modifiers.
Recommended for polymers		polyesters and polyolefins, peroxide-crosslinked elastomers
Recommended for products		additive for casting resins (unsaturated polyester, MMA), especially for the production of engineered stones, surface modifier for pigments and fillers for thermosets (unsaturated polyester, MMA),
Guidelines for use		Dynasylan® SIVO 560 is either applied to the inorganic substrate as a pretreatment by dipping, spraying, or coating or it may be added directly to the resin matrix (additive process).
Concentrations used	wt%	0.5-1.5

Dynasylan® SIVO Clear EC

PARAMETER	UNIT	VALUE
GENERAL INFORMATION		
Name		Dynasylan® SIVO Clear EC
Composition		one-component user-friendly coating system for smooth surfaces such as glass and glazed ceramic
PHYSICAL PROPERTIES		
State	-	liquid
Color		clear, colorless
Density at 20°C	kg/m³	790
Viscosity at 20°C	mPas	2.5
HEALTH & SAFETY		
Flash point	°C	12.5
USE & PERFORMANCE		
Manufacturer		Dow
Recommended for products		treatment of windshields, coating for oil-, water-, and dirt-repellent treatment
Guidelines for use		do not apply the product at temperatures below 10°C

Dynasylan TRIAMO

PARAMETER	UNIT	VALUE
GENERAL INFORMATION		
Name		Dynasylan TRIAMO
CAS #		35141-30-1, 103526-27-8
EC number	-	252-390-9
Composition		>60-<100%N-(2-aminoethyl)-N'-(3-trimetoxysilyl)propyl)ethylenediamine + 5%-<10% N-(2-aminoethyl-N-(3-(3-trimetoxysilyl)-propyl)-1, (impurity 0.1-1% methanol)
Common synonym		4,7,10-triazadecyl-trimethoxysilane
Empirical formula	-	C10H27N3O3Si
Formula		
Molecular mass	daltons	265.43
Chemical class	-	silane
Functional organic group	-	amino
PHYSICAL PROPERTIES		
State	-	liquid
Odor	-	amine-like
Color	-	clear dark yellow
Boiling point	°C	114-168 at 2.7 hPa
Density at 20°C	kg/m^3	1,030
Refractive index at 20°C	-	1.46
Solubility (diluents)	alcohols, aliphatic or aromatic hydrocarbons	
Solubility in water at 25°C	hydrolytic decomposition releasing ethanol	
Specific gravity at 25°C	-	1.04
Vapor pressure at 20°C	kPa	0.2
Viscosity at 20°C	mPas	20
HEALTH & SAFETY		
NFPA classification	Flammability	1
	Health	2
	Reactivity	1
HMIS classification	Flammability	1
	Health	2
	Reactivity	1
DOT class	not regulated	

Dynasylan TRIAMO

PARAMETER	UNIT	VALUE
TDG class	not regulated	
ICAO/IATA class	Environmentally hazardous substance, liquid, n.o.s.(4,7,10 TRIAZADECYL TRIMETHOXY SILANE) 9,III	
IMDG class	ENVIRONMENTALLY HAZARDOUS SUBSTANCE, LIQUID, N.O.S.(4,7,10-TRIAZADECYL TRIMETHOXY SILANE)	
Flash point	°C	137
Hazardous combustion products	Carbon oxides, nitrogen oxides	
Agency rating, listed	EINECS Europe, ECL Korea, AICS Australia, IECSC China, DSL Canada, PICCS Philippines, TSCA USA	
Hazardous products of hydrolysis	methanol	
Animal testing, acute toxicity, Rat oral LD50	mg/kg	7758
Animal testing, acute toxicity, Rat dermal LD50	mg/kg	16640
Animal testing, acute toxicity, Rat inhalation, LC50	mg/m^3	>40000/4H
Effect of exposure, eye (human)	Causes serious eye damage.	
Effect of exposure, inhalation (human)	Avoid inhalation aerosol of mists.	
Effect of exposure, skin (human)	Causes skin irritation. May cause allergic skin reaction.	
Exposure, personal protection	Safety glasses, protective clothing based on chemical resistance data, chemical-resistant gloves, general and local exhaust ventilation.	
First aid, eye	Rinse cautiously with water for several minutes. Remove contact lenses, if present and easy to do. Continue rinsing. If eye irritation persists: Get medical advice/attention.	
First aid, inhalation	Remove to fresh air. If not breathing, give artificial respiration. If breathing is difficult, give oxygen. If irritation persists, obtain medical advice.	
First aid, skin	Immediately wipe away excess material. Use a waterless hand cleaner to remove as much of the remaining material as possible. Wash with soap and water.	
NIOSH, REL	mg/m^3	ST325/methanol
OSHA, PEL	mg/m^3	260/methanol
ACGIH, TLV	ppm	200/methanol
NIOSH, REL	ppm	ST250/methanol
OSHA, PEL	ppm	200/methanol

Dynasylan TRIAMO

PARAMETER	UNIT	VALUE
UN risk phrases, R	-	R21,R34,R43,R14
US safety phrases, S	S26,S36/37/39,S45,S8,S30,S24/25	
UN/NA class	-	3082
USE & PERFORMANCE		
Manufacturer	Evonik Industries	
Outstanding properties	improved adhesion, moisture, and corrosion resistance	
Recommended for polymers	PVC plastisols, silicones	
Recommended for products	foundry resins	
Recommended applications	adhesion promoter to both inorganic materials (e.g., glass, metals, fillers) and organic polymers (e.g., thermosets, thermoplastics, elastomers), used in applications such as PVC-plastisols as an additive (0.5-1%) to improve adhesion between PVC and metals	
Concentrations used	0.5-1.0%	

Dynasylan VTEO

PARAMETER	UNIT	VALUE
GENERAL INFORMATION		
Name	Dynasylan VTEO	
CAS #	-	78-08-0
EC number	-	201-081-7
General description	bifunctional	
Composition	98% triethoxy(vinyl)silane	
Common synonym	vinyltriethoxysilane	
Acronym		VTEO/VTES
Empirical formula	C8H18O3Si	
Formula		
Molecular mass	daltons	190.31
Chemical class	silane	
Functional organic group	vinyl	
Functionality, average	inorganic triethoxysilyl	
PHYSICAL PROPERTIES		
State	-	liquid
Odor	-	aromatic
Color	-	colorless
Boiling point	°C	158
Melting point	°C	-93
Density at 20°C	kg/m^3	900
Refractive index at 20°C	-	1.397
Solubility in water at 25°C	hydrolytic decomposition releasing ethanol	
Specific gravity at 25°C	-	0.9
Vapor pressure at 20°C	kPa	0.7
Vapor pressure at 50°C	kPa	2.0
Viscosity at 20°C	mPas	0.7
HEALTH & SAFETY		
NFPA classification	Flammability	2
	Health	1
	Reactivity	1
HMIS classification	Flammability	2
	Health	1
	Reactivity	1

Dynasylan VTEO

PARAMETER	UNIT	VALUE
Carcinogenicity		IARC, OSHA, NTP: no ingredient of this product present at levels greater than or equal to 0.1% is identified as probable, possible or confirmed human carcinogen
Teratogenicity		no evidence
DOT class		FLAMMABLE LIQUID, N.O.S.(triethoxy(vinyl)silane) 3,III. Not regulated in packages 450 liter or less.
ICAO/IATA class		Flammable liquid, n.o.s.(triethoxy(vinyl) silane) 3, III
IMDG class		FLAMMABLE LIQUID, N.O.S.(triethoxy(vinyl)silane) 3, III
Flash point	°C	38
Flash point method	-	PMCC
Explosive LEL	wt%	0.53
Explosive UEL	wt%	15
Agency rating, listed		EINECS Europe, ECL Korea, ENCS Japan, AICS Australia, IECSC China, DSL Canada, PICCS Philippines, TSCA USA
Hazardous ingredients, labelling		Flammable liquid, n.o.s.(triethoxyvinylsilane)
Hazardous products of hydrolysis		ethanol
Animal testing, acute toxicity, Rat oral LD50	mg/kg	>5000
Animal testing, acute toxicity, Rabbit dermal LD50	mg/kg	no skin irritation
Exposure, personal protection		Safety glasses, protective clothing based on chemical resistance data, chemical-resistant gloves, general and local exhaust ventilation.
First aid, eye		Rinse cautiously with water for several minutes. Remove contact lenses, if present and easy to do. Continue rinsing. If eye irritation persists: Get medical advice/attention.
First aid, inhalation		Remove to fresh air. If not breathing, give artificial respiration. If breathing is difficult, give oxygen. If irritation persists, obtain medical advice.
First aid, skin		Immediately wipe away excess material. Use a waterless hand cleaner to remove as much of the remaining material as possible. Wash with soap and water.
NIOSH, REL	mg/m³	1900/ethanol

Dynasylan VTEO

PARAMETER	UNIT	VALUE
OSHA, PEL	mg/m^3	1900/ethanol
ACGIH, TLV	ppm	1000/ethanol
NIOSH, REL	ppm	1000/ethanol
OSHA, PEL	ppm	1000/ethanol
UN risk phrases, R	R10,R36	
US safety phrases, S	S26,S36	
UN/NA class	-	1993
ECOLOGICAL PROPERTIES		
Aquatic toxicity, *Daphnia magna*, 48-h LC50	mg/l	168.7
Aquatic toxicity, *Zebra fish*, 96-h LC50	mg/l	>100
Biodegradation probability	51%/28d/not partial biodegradable	
Partition coefficient, log K_{ow}	-	3.0
USE & PERFORMANCE		
Manufacturer	Evonik Industries	
Outstanding properties	improved mechanical and electrical properties especially after exposure to moisture. When bonded to an inorganic filler, Dynasylan VTEO hydrophobized the filler surface, improving the compatibility of fillers with polymers, leading to better dispersibility, reduced melt viscosity, and easier processing of filled plastics. Surface coating of glass, metal, or ceramics with Dynasylan VTEO will improve adhesion, especially of acrylic systems, and corrosion or scratch resistance.	
Recommended for polymers	PE, acrylics, SBR, silicones	
Recommended for products	adhesion promoter, cable insulation, composites, moisture scavenger	
Recommended applications	adhesion promoter for various mineral-filled polymers, improving mechanical and electrical properties, especially after exposure to moisture.	

Dynasylan VTMO

PARAMETER	UNIT	VALUE
GENERAL INFORMATION		
Name		Dynasylan VTMO
CAS #	-	2768-02-07
EC number	-	220-449-8
General description	bifunctional	
Composition	>98% triethoxy(vinyl)silanevinyltrimethoxysilane, (impurity 0.2% tetramethyl orthosilicate)	
Empirical formula	C5H12O3Si/(CH3O)3Si-CH=CH2	
Formula		
Molecular mass	daltons	148.23
RTECS number	-	VV6700000
Chemical class	-	silane
Functional organic group	-	vinyl
Functionality, average	inorganic trimethoxysilyl	
PHYSICAL PROPERTIES		
State	-	liquid
Color	-	colorless
Odor	-	aromatic
Boiling point	°C	123
Density at 20°C	kg/m^3	970
Refractive index at 20°C	-	1.39
Solubility in water at 25°C	hydrolytic decomposition	
Specific gravity at 25°C	-	0.97
Viscosity at 20°C	mPas	1.0
HEALTH & SAFETY		
NFPA classification	Flammability	3
	Health	1
	Reactivity	0
HMIS classification	Flammability	3
	Health	1
	Reactivity	0
DOT class	FLAMMABLE LIQUID, N.O.S.(trimethoxyvinylsilane) 3,III	
ICAO/IATA class	Flammable liquid, n.o.s.(trimethoxyvinylsilane) 3, III	

Dynasylan VTMO

PARAMETER	UNIT	VALUE
IMDG class	Flammable liquid, n.o.s.(trimethoxyvinylsilane) 3,III	
Flash point	°C	25
Flash point method	-	PMCC
Agency rating, listed	EINECS Europe, ECL Korea, ENCS Japan, AICS Australia, IECSC China, DSL Canada, PICCS Philippines, TSCA USA	
Hazardous ingredients, labelling	Flammable liquid, n.o.s.(trimethoxyvinylsilane)	
Hazardous products of hydrolysis	ethanol	
Effect of exposure, eye (human)	Serious eye damage (tetramethyl orthosilicate).	
Effect of exposure, inhalation (human)	Harmful if inhaled. May cause respiratory irritation.	
Effect of exposure, skin (human)	Skin irritation (tetramethyl orthosilicate)	
Exposure, personal protection	Safety glasses, protective clothing based on chemical resistance data, chemical-resistant gloves, general and local exhaust ventilation.	
First aid, eye	Rinse cautiously with water for several minutes. Remove contact lenses, if present and easy to do. Continue rinsing. If eye irritation persists: Get medical advice/attention.	
First aid, inhalation	Remove to fresh air. If not breathing, give artificial respiration. If breathing is difficult, give oxygen. If irritation persists, obtain medical advice.	
First aid, skin	Immediately wipe away excess material. Use a waterless hand cleaner to remove as much of the remaining material as possible. Wash with soap and water.	
UN risk phrases, R	R10,R36/37/38,R20	
US safety phrases, S	S26,S36/37/39,S37/39,S16	
UN/NA class	-	1993
ECOLOGICAL PROPERTIES		
Aquatic toxicity, *Green algae*, 96-h EC50	mg/l	210/7d
Aquatic toxicity, *Daphnia magna*, 48-h LC50	mg/l	168.7
Aquatic toxicity, *Rainbow trout*, 96-h LC50	mg/l	191

Dynasylan VTMO

PARAMETER	UNIT	VALUE
Biodegradation probability		51%/28d/not partial biodegradable
USE & PERFORMANCE		
Manufacturer		Evonik Industries
Outstanding properties		Improved tear and crack resistance, chemical resistance, abrasion resistance, and memory effect. Also, improved corrosion resistance and adhesion strength in wet conditions as well as wet scrub resistance
Recommended for polymers		acrylic, SBR, PA, PE, EP, phenolic, silicones
Recommended for products		peroxide-initiated grafting, cable insulation, moisture scavenger, sealants
Recommended applications		adhesion promoter, crosslinking agent, surface modifier, and moisture scavenger.

Dynasylan VTMOEO

PARAMETER	UNIT	VALUE
GENERAL INFORMATION		
Name		Dynasylan VTMOEO
CAS #		1067-53-4, (impurity 109-86-4)
EC number	-	213-934-0
General description		bifunctional organosilane
Composition		tris(2-methoxyethoxy)vinylsilane, 0.2.% 2-methoxyethanol + impurity <0.3% impurity 2-methoxyethanol
Common synonym		vinyltris(2-methoxyethoxy)silane
Acronym		VTMOEO
Formula		
Molecular mass	daltons	280.0
RTECS number	-	VV6826000
Functional organic group	-	vinyl
Functionality, average	2-methoxy-ethoxy-silyl	
PHYSICAL PROPERTIES		
State	-	liquid
Odor	-	aromatic
Color	-	colorless
Boiling point	°C	258 (1013 hPa)
Melting point	°C	-130
Density at 20°C	kg/m^3	1,040
Density at 25°C	kg/m^3	1,030
Refractive index at 20°C	-	1.430
Solubility in water at 20°C	71g/l (decomposition by hydrolysis)	
Specific gravity at 25°C	-	1.03
Sulfur oxide (SO$_2$) content	wt%	42/SO$_2$
Vapor pressure at 20°C	kPa	0.00043
Viscosity at 20°C	mPas	2.7
HEALTH & SAFETY		
NFPA classification	Flammability	2
	Health	1
	Reactivity	0

Dynasylan VTMOEO

PARAMETER	UNIT	VALUE
HMIS classification	Flammability	1
	Health	2
	Reactivity	1
Carcinogenicity	no evidence of carcinogenicity	
DOT class	not regulated in packages 450 liter or less	
TDG class	not regulated in packages 450 liter or less	
ICAO/IATA class	not regulated	
IMDG class	not regulated	
Autoignition temperature	°C	210
Flash point	°C	115
Flash point method	-	PMCC
Agency rating listed	EINECS Europe, ECL Korea, IECSC China, DSL Canada, TSCA USA	
Effect of exposure, eye (human)	Avoid contact with eyes.	
Effect of exposure, inhalation (human)	May cause respiratory irritation. Do not inhale vapors or aerosols	
Effect of exposure, skin (human)	May cause skin irritation. Can be absorbed through the skin (2-methoxy-ethanol).	
Effect of exposure, swallowing (human)	Not harmful or toxic after single oral ingestion of high concentrations.	
Exposure, personal protection	Safety glasses, protective clothing based on chemical resistance data, chemical-resistant gloves, general and local exhaust ventilation.	
First aid, eye	Rinse cautiously with water for several minutes. Remove contact lenses, if present and easy to do. Continue rinsing. If eye irritation persists: Get medical advice/attention.	
First aid, inhalation	Remove to fresh air. If not breathing, give artificial respiration. If breathing is difficult, give oxygen. If irritation persists, obtain medical advice.	
First aid, skin	Immediately wipe away excess material. Use a waterless hand cleaner to remove as much of the remaining material as possible. Wash with soap and water.	
UN/NA class	-	1993

Dynasylan VTMOEO

PARAMETER	UNIT	VALUE
USE & PERFORMANCE		
Manufacturer		Evonik Industries
Outstanding properties		improved filler dispersion, good processability, environmentally friendly (low VOC), safety and handling during processing, excellent balance between tensile strength and elongation at break at a high level, improved chemical resistance, strongly reduced tendency to stress cracking, higher impact strength, and abrasion resistance.
Recommended for polymers		EVA, EMA
Recommended for products		cable
Recommended applications		adhesion promoter, dispersion and hydrophobation agent in mineral filled compounds

Geniosil GF 31

PARAMETER	UNIT	VALUE
GENERAL INFORMATION		
Name		Geniosil GF 31
CAS #	-	2530-85-0
EC number	-	219-785-8
Composition		3-methacryloxypropyltrimethoxysilane
Empirical formula		C10H20O5Si
Formula		
Molecular mass	daltons	248.4
RTECS number	-	UC0230000
Chemical class	silane	
Functional organic group	methacryl	
Purity	wt%	>98
PHYSICAL PROPERTIES		
State	-	liquid
Odor	-	slight
Color	-	colorless
Boiling point	°C	125 at 15 hPa
Melting point	°C	< -50
Density at 25°C	kg/m³	1,050
Hydrolyzable chloride (as HCl)	mg/kg	>30
pH	3-4 5% aq. solution	
Solubility (diluents)	organic solvents, such as alcohols, toluene and acetone	
Solubility in water at 25°C	insoluble in neutral water	
Specific gravity at 25°C	-	1.05
Vapor pressure at 20°C	kPa	>0.2
Viscosity at 25°C	mPas	2.6
HEALTH & SAFETY		
DOT class	not regulated	
TDG class	not regulated	
ICAO/IATA class	not regulated	
IMDG class	not regulated	
Autoignition temperature	°C	265
Flash point	°C	>100
Flash point method	-	PMCC

Geniosil GF 31

PARAMETER	UNIT	VALUE
Hazardous combustion products	Carbon oxides, NOx, SiOx, formaldehyde.	
Agency rating, listed	EINECS Europe, ECL Korea, ENCS Japan, AICS Australia, IECSC China, DSL Canada, PICCS Philippines, TSCA USA, NZIoC-New Zealand, TCSI Taiwan, REACH-EU	
Hazardous products of hydrolysis	methanol	
Animal testing, acute toxicity, Rat oral LD50	mg/kg	>2000
Animal testing, acute toxicity, Rabbit dermal LD50	mg/kg	no skin irritation
Animal testing, acute toxicity, Rat dermal LD50	mg/kg	>2000
Animal testing, acute toxicity, Rat inhalation, LC50	mg/m^3	>2280/4H
Effect of exposure, eye (human)	Avoid contact with eyes.	
Effect of exposure, inhalation (human)	Do not inhale gases/vapors/aerosols.	
Effect of exposure, skin (human)	Avoid contact with skin.	
Effect of exposure, swallowing (human)	Harmful if swallowed.	
Exposure, personal protection	Safety glasses, protective clothing based on chemical resistance data, chemical-resistant gloves, general and local exhaust ventilation.	
First aid, eye	Rinse cautiously with water for several minutes. Remove contact lenses, if present and easy to do. Continue rinsing. If eye irritation persists: Get medical advice/attention.	
First aid, inhalation	Keep the patient calm. Protect against loss of body heat. Seek medical advice and clearly identify substance.	
First aid, skin	Remove contaminated or soaked clothing. Wash off with plenty of water or water and soap immediately for 10-15 minutes. In serious cases, use emergency shower immediately. Seek medical advice and clearly identify substance.	
NIOSH, REL	mg/m^3	ST325/methanol
OSHA, PEL	mg/m^3	260/methanol
ACGIH, TLV	ppm	200/methanol
NIOSH, REL	ppm	ST250/methanol
OSHA, PEL	ppm	200/methanol
UN risk phrases, R	R36/37/R38,R37/38,R22	
US safety phrases, S	-	S26,S28,S37/39

Geniosil GF 31

PARAMETER	UNIT	VALUE
ECOLOGICAL PROPERTIES		
Aquatic toxicity, *Daphnia magna*, 48-h LC50	mg/l	>100
Aquatic toxicity, *Zebra fish*, 96-h LC50	mg/l	>100
Biodegradation probability	74%/28d/readily biodegradable	
USE & PERFORMANCE		
Manufacturer	Wacker Chemie AG	
Outstanding properties	improved filler dispersibility while reducing the filler's sedimentation tendency. It also results in a major reduction in the melt viscosity of casting resins, as well as a marked improvement in the mechanical properties of glass-fiber-reinforced or mineral-filled plastics. In addition, the use of Geniosil GF 31 in these materials leads to a sizable increase in moisture (vapor) resistance as well as greater resistance to acids and bases. Improved mechanical properties and enhanced substrate adhesion.	
Recommended for polymers	thermosetting plastics (e.g., polyacrylates, polyesters), thermoplastics (polyesters, polyolefins), and elastomers	
Recommended for products	heat and/or radiation-curable adhesives and sealants	
Recommended applications	coupling agent in glass fiber or glass fabric-reinforced polyester and polyolefin molded components. In addition, Geniosil GF 31 is used as a surface modifier for fillers and pigments used in thermosetting plastics (e.g., polyacrylates, polyesters), thermoplastics (polyesters, polyolefins) and elastomers, as well as a polymer co,mponent in surface coatings. Further important areas of application are heat and/or radiation-curable adhesives, and sealants	
Conditions to avoid	contact with moisture must be avoided during processing to prevent undesired hydrolysis with moisture.	

Geniosil GF 56

PARAMETER	UNIT	VALUE
GENERAL INFORMATION		
Name		Geniosil GF 56
CAS #	-	78-08-0
EC number	-	201-081-7
Composition	vinyltriethoxysilane	
Common synonym	ethenyltriethyloxy silane	
Acronym		VTEO
Empirical formula		C8H18O3Si
Formula		

$$H_5C_2O-\overset{\displaystyle OC_2H_5}{\underset{\displaystyle OC_2H_5}{Si}}$$

PARAMETER	UNIT	VALUE
Molecular mass	daltons	190.3
RTECS number	-	VV6700000
Chemical class	silane	
Functional organic group	vinyl	
Purity	wt%	>98
PHYSICAL PROPERTIES		
State	-	liquid
Odor	-	aromatic
Color	-	clear, colorless
Boiling point	°C	158
Melting point	°C	< -50
Density at 25°C	kg/m³	910
Hydrolyzable chloride (as HCl)	mg/kg	<10
Hydrolysis half-life	0.9 h/pH 7; 0.1 h/pH 4	
Refractive index at 20°C	-	1.397
Solubility (diluents)	dissolves readily in standard organic solvent	
Solubility in water at 25°C	hydrolytic decomposition releasing methanol	
Specific gravity at 20°C	-	0.91
Vapor pressure at 20°C	kPa	0.3
Vapor pressure at 50°C	kPa	<1.5
Viscosity at 25°C	mPas	0.56
HEALTH & SAFETY		
Mutagenicity	negative	
TDG class	Dangerous Goods, Flammable liquid, n.o.s. (Triethoxy(vinyl)silane) 3,III	

Geniosil GF 56

PARAMETER	UNIT	VALUE
ICAO/IATA class		Dangerous Goods, Flammable liquid, n.o.s. (Triethoxy(vinyl)silane) 3,III
IMDG class		Dangerous Goods, Flammable liquid, n.o.s. (Triethoxy(vinyl)silane) 3,III
Autoignition temperature	°C	265
Flash point	°C	37
Flash point method	-	PMCC
Explosive LEL	wt%	0.7
Explosive UEL	wt%	17.0
Hazardous combustion products		Carbon oxides, NOx, SiOx, formaldehyde.
Agency rating, listed		EINECS Europe, ECL Korea, ENCS Japan, AICS Australia, IECSC China, DSL Canada, PICCS Philippines, TSCA USA, NZIoC-New Zealand, TCSI Taiwan, REACH-EU
Hazardous ingredients, labelling		triethoxy(vinyl)silane
Animal testing, acute toxicity, Rat oral LD50	mg/kg	>5000
Animal testing, acute toxicity, Rabbit dermal LD50	mg/kg	no skin irritation
Animal testing, acute toxicity, Rat dermal LD50	mg/kg	>2000
Effect of exposure, eye (human)		Avoid contact with eyes.
Effect of exposure, skin (human)		Avoid contact with skin.
Effect of exposure, swallowing (human)		Harmful if swallowed.
Exposure, personal protection		Safety glasses, protective clothing based on chemical resistance data, chemical-resistant gloves, general and local exhaust ventilation.
First aid, eye		Rinse cautiously with water for several minutes. Remove contact lenses, if present and easy to do. Continue rinsing. If eye irritation persists: Get medical advice/attention.
First aid, inhalation		Keep the patient calm. Protect against loss of body heat. Seek medical advice and clearly identify substance.
First aid, skin		Remove contaminated or soaked clothing. Wash off with plenty of water or water and soap immediately for 10-15 minutes. In serious cases, use emergency shower immediately. Seek medical advice and clearly identify substance.
UN risk phrases, R	-	R10,R36/37

Geniosil GF 56

PARAMETER	UNIT	VALUE
US safety phrases, S	-	S26,S36
UN/NA class	-	1993
ECOLOGICAL PROPERTIES		
Aquatic toxicity, *Daphnia magna*, 48-h LC50	mg/l	168.7
Bioaccumulative potential	not bioaccumulative	
Biodegradation probability	51%/28d/rapid biological degradation of the organic hydrolysis product	
Partition coefficient, log K_{oc}	-	-0.02
USE & PERFORMANCE		
Manufacturer	Wacker Chemie AG	
Outstanding properties	improved bonding between organic surface coatings and inorganic substrates; as a result, these coatings demonstrate markedly greater scratch resistance and resistance to chemicals.	
Recommended for polymers	polyacrylates, crosslinked polyethylene	
Recommended for products	silane-modified polymers that serve as binders in paints and adhesives, pipes and cables made of silane-crosslinked polyethylene	
Recommended applications	adhesion promoter in primers and coatings, and is used in the production of organically modified fillers for cable insulation	
Conditions to avoid	contact with moisture must be avoided during processing to prevent undesired hydrolysis with moisture	

Geniosil GF 60

PARAMETER	UNIT	VALUE
GENERAL INFORMATION		
Name	Geniosil GF 60	
CAS #	-	23432-62-4
EC number	-	245-659-7
Composition	100% N-ethyl[3-(Trimethoxysilyl) propyl] carbamate +(impurity <0.5% (3-Isocyanatopropyl) trimethoxysilane)	
Empirical formula	(CH3O)3SiC3H6NH-CO-OCH3	
Formula		
Molecular mass	daltons	237.30
Chemical class	silane	
Functional organic group	carbamato	
Purity	wt%	>96
PHYSICAL PROPERTIES		
State	-	liquid
Odor	-	characteristically aromatic
Color	clear, colorless to light yellowish	
Boiling point	°C	102 at 0.1 kPa
Melting point	°C	<-50
Density at 20°C	kg/m^3	1,108.7
Kinematic viscosity at 25°C	cSt	13
Vapor pressure at 50°C	kPa	< 0.001
Vapor pressure at 100°C	kPa	0.011
HEALTH & SAFETY		
DOT class	not regulated	
TDG class	not regulated	
ICAO/IATA class	not regulated	
IMDG class	not regulated	
Autoignition temperature	°C	385
Flash point	°C	99
Flash point method	-	CC
Hazardous combustion products	Carbon oxides, SiOx, formaldehyde.	
Agency rating, listed	EINECS Europe, AICS Australia, IECSC China, DSL Canada, PICCS Philippines, TSCA USA	

Geniosil GF 60

PARAMETER	UNIT	VALUE
Animal testing, acute toxicity, Rat oral LD50	mg/kg	>2000
Effect of exposure, inhalation (human)	May cause allergy or asthma symptoms or breathing difficulties if inhaled.	
Effect of exposure, skin (human)	May cause an allergic skin reaction.	
Exposure, personal protection	Safety glasses, protective clothing based on chemical resistance data, chemical-resistant gloves, general and local exhaust ventilation.	
First aid, eye	Rinse cautiously with water for several minutes. Remove contact lenses, if present and easy to do. Continue rinsing. If eye irritation persists: Get medical advice/attention.	
First aid, inhalation	Keep the patient calm. Protect against loss of body heat. Seek medical advice and clearly identify substance.	
First aid, skin	Remove contaminated or soaked clothing. Wash off with plenty of water or water and soap immediately for 10-15 minutes. In serious cases, use emergency shower immediately. Seek medical advice and clearly identify substance.	
OSHA, PEL	mg/m^3	260/methanol, 30/N,N-dimethylformamide
ACGIH, TLV	ppm	200/methanol
NIOSH, REL	ppm	ST250/methanol
OSHA, PEL	ppm	200/methanol, 10/N,N-dimethylformamide

ECOLOGICAL PROPERTIES

Aquatic toxicity, *Daphnia magna*, 48-h LC50	mg/l	>100
Aquatic toxicity, *Rainbow trout*, 96-h LC50	mg/l	>100

USE & PERFORMANCE

Manufacturer	Wacker Chemie AG
Outstanding properties	used in the production of moisture-crosslinked adhesive and sealant formulations. As a bifunctional compound, Geniosil GF 60 can interact with many different organic polymers to form a molecular bridge between inorganic and organic substrates.

Geniosil GF 60

PARAMETER	UNIT	VALUE
Recommended applications		acts as an adhesion promoter, reactive diluent or water scavenger, used in combination with other adhesion promoters

Geniosil GF 62

PARAMETER	UNIT	VALUE
GENERAL INFORMATION		
Name		Geniosil GF 62
CAS #		4130-08-9, (impurity: 64-19-7, 108-24-7)
EC number	-	223-943-1
Composition		<100% triacetoxy vinylsilane + (impurity: <2% acetic acid, <1% acetic anhydride)
Common synonym		Vinyltreiacetoxysilane
Empirical formula		(CH3CO2)3SiCH=CH2
Formula		
Molecular mass	daltons	232.30
Chemical class	silane	
Functional organic group	vinyl	
Purity	wt%	>90
PHYSICAL PROPERTIES		
State	-	liquid
Odor	-	vinegar-like
Color		colorless to slightly yellow
Boiling point	°C	112 at 1.3 kPa
Melting point	°C	7.0
Density at 20°C	kg/m^3	1,160
Hydrolyzable chloride (as HCl)	mg/kg	<100
Kinematic viscosity at 25°C	cSt	1.53
pH	2 at 25°C (50 g/l H$_2$O)	
Solubility (diluents)	organic solvents, such as methanol, toluene, acetone	
Solubility in water at 25°C	hydrolytic decomposition releasing methanol	
Vapor pressure at 20°C	kPa	0.2
Vapor pressure at 50°C	kPa	0.037
Viscosity at 25°C	mPas	7.0
HEALTH & SAFETY		
Carcinogenicity		IARC, OSHA, NTP: no ingredient of this product present at levels greater than or equal to 0.1% is identified as probable, possible or confirmed human carcinogen

Geniosil GF 62

PARAMETER	UNIT	VALUE
Mutagenicity	no data available	
DOT class	Dangerous Goods, Corrosive liquid, acidic, organic, n.o.s. (Triacetoxyvinylsilane)8,II	
ICAO/IATA class	Dangerous Goods, Corrosive liquid, acidic, organic, n.o.s. (Triacetoxyvinylsilane)8,II	
IMDG class	Dangerous Goods, Corrosive liquid, acidic, organic, n.o.s. (Triacetoxyvinylsilane)8,II, no marine pollutant	
Autoignition temperature	°C	400.00/liquid
Flash point	°C	110
Flash point method	-	CC
Hazardous combustion products	Carbon oxides, SiOx, formaldehyde.	
Agency rating, listed	EINECS Europe, ECL Korea, ENCS Japan, AICS Australia, IECSC China, DSL Canada, PICCS Philippines, TSCA USA	
Hazardous ingredients, labelling	Corrosive liquid, acidic, organic, n.o.s. (Triacetoxyvinylsilane)	
Effect of exposure, eye (human)	Causes serious eye damage.	
Effect of exposure, skin (human)	Causes severe skin burns damage.	
Exposure, personal protection	Safety glasses, protective clothing based on chemical resistance data, chemical-resistant gloves, general and local exhaust ventilation.	
First aid, eye	Rinse cautiously with water for several minutes. Remove contact lenses, if present and easy to do. Continue rinsing. If eye irritation persists: Get medical advice/attention.	
First aid, inhalation	Keep the patient calm. Protect against loss of body heat. Seek medical advice and clearly identify substance.	
First aid, skin	Remove contaminated or soaked clothing. Wash off with plenty of water or water and soap immediately for 10-15 minutes. In serious cases, use emergency shower immediately. Seek medical advice and clearly identify substance.	
OSHA, PEL	mg/m^3	25/acetic acid, 20/ acetic anhydride
ACGIH, TLV	ppm	10/acetic acid, 1/ acetic anhydride

Geniosil GF 62

PARAMETER	UNIT	VALUE
OSHA, PEL	ppm	10/acetic acid, 1/ acetic anhydride
UN risk phrases, R	R34,R37,R14	
US safety phrases, S	S26,R36/37/39,S45,S8	
UN/NA class	-	3265
USE & PERFORMANCE		
Manufacturer	Wacker Chemie AG	
Outstanding properties	used in mineral-filled polymers improved the mechanical and electrical properties, as well increased moisture resistance. Used as a co-monomer in polymers produced binders which improved wet scrub resistance and increased scratch resistance due to crosslinking and strong adhesion to the substrate	
Recommended for polymers	polyacrylates, PA, polyether, epoxy resin, melamine	
Recommended for products	silicone sealants	
Recommended applications	used in acetoxy crosslinking silicone sealants, in which it functions as a crosslinker and/or adhesion promoter	

Geniosil GF 69

PARAMETER	UNIT	VALUE
GENERAL INFORMATION		
Name		Geniosil GF 69
CAS #	-	26115-70-8
EC number	-	247-465-8
Composition	tris-[3-(trimethoxysilyl)propyl]-isocyanurate	
Empirical formula	C21H45N3O12Si3	
Formula		
Molecular mass	daltons	615.90
RTECS number	-	XZ2025000
Chemical class	silane	
Active matter	wt%	>95
Functional organic group	cyanurate	
Purity	wt%	>97
PHYSICAL PROPERTIES		
State	-	liquid
Odor	-	characteristic
Color	clear, light yellowish	
Boiling point	°C	250 at 0.13kpa
Density at 20°C	kg/m³	1170
Kinematic viscosity at 25°C	cSt	325-350
Refractive index at 20°C	-	1.461
HEALTH & SAFETY		
Flash point	°C	102
Flash point method	-	CC
Hazardous combustion products	Carbon oxides, SiOx, formaldehyde.	
Agency rating, listed	EINECS Europe, DSL Canada, ECL Korea, TSCA USA	
Animal testing, acute toxicity, Rat oral LD50	mg/kg	>2000
Animal testing, acute toxicity, Rat dermal LD50	mg/kg	>2000
Effect of exposure, inhalation (human)	May cause allergy or asthma symptoms or breathing difficulties if inhaled.	
Exposure, personal protection	Safety glasses, protective clothing based on chemical resistance data, chemical-resistant gloves, general and local exhaust ventilation.	

Geniosil GF 69

PARAMETER	UNIT	VALUE
First aid, eye		Rinse cautiously with water for several minutes. Remove contact lenses, if present and easy to do. Continue rinsing. If eye irritation persists: Get medical advice/attention.
First aid, inhalation		Keep the patient calm. Protect against loss of body heat. Seek medical advice and clearly identify substance.
First aid, skin		Remove contaminated or soaked clothing. Wash off with plenty of water or water and soap immediately for 10-15 minutes. In serious cases, use emergency shower immediately. Seek medical advice and clearly identify substance.
ACGIH, TLV	ppm	200/methanol
NIOSH, REL	ppm	ST250/methanol
OSHA, PEL	ppm	200/methanol
UN risk phrases, R	R23/25,R36/38	
US safety phrases, S	S26,S25	
UN/NA class	-	2810
USE & PERFORMANCE		
Manufacturer	Wacker Chemie AG	
Outstanding properties	improves both adhesion and crosslinking characteristics and influenced the mechanical properties	

GENIOSIL GF 80

PARAMETER	UNIT	VALUE
GENERAL INFORMATION		
Name		GENIOSIL GF 80
CAS #	-	2530-83-8
EC number	-	219-784-2
General description		amino/vinyl/ methoxysilane.
Composition		>97% (3-(2,3-epoxypropoxy)propyl) trimethoxysilane
Common synonym		γ-glycidoxypropyltrimethoxysilane
Empirical formula		C9H20O5Si
Formula		
Molecular mass	daltons	236.34
Chemical class		silane
Functional organic group		glycidoxy
Epoxy content	%	17.5
Purity	%	>97
PHYSICAL PROPERTIES		
State	-	liquid
Odor	-	ester-like
Color	-	clear, colorless
Boiling point	°C	248
Melting point	°C	-70
Density at 20°C	kg/m³	1,070
Hydrolysis half-life	18 h/pH 6.5 at 24.5°C	
Refractive index at 25°C	-	1.427
Solubility (diluents)	solvent	
Solubility in water at 25°C	hydrolytic decomposition releasing methanol	
Vapor pressure at 20°C	kPa	0.1
Vapor pressure at 50°C	kPa	0.5
Viscosity at 25°C	mPas	3.65
HEALTH & SAFETY		
Carcinogenicity		animal tests have not revealed any carcinogenic effects
Mutagenicity		animal tests have shown no indications of possibility of damage to embryo and impairment of fertility
DOT class		not regulated

GENIOSIL GF 80

PARAMETER	UNIT	VALUE
TDG class	not regulated	
ICAO/IATA class	not regulated	
IMDG class	not regulated	
Autoignition temperature	°C	400/liquid
Flash point	°C	122
Flash point method	-	PMCC
Explosive LEL	wt%	0.7
Explosive UEL	wt%	13.6
Hazardous combustion products	Carbon oxides, NOx, SiOx, formaldehyde.	
Agency rating, listed	EINECS Europe, ECL Korea, ENCS Japan, AICS Australia, IECSC China, DSL Canada, PICCS Philippines, TSCA USA, NZIoC-New Zealand, TCSI Taiwan, REACH-EU	
Hazardous products of hydrolysis	methanol	
Animal testing, acute toxicity, Rat oral LD50	mg/kg	8025
Animal testing, acute toxicity, Rabbit dermal LD50	mg/kg	4250
Animal testing, acute toxicity, Rat inhalation, LC50	mg/m³	>5300/4H/spray
Effect of exposure, eye (human)	Causes serious eye damage, eye irritation	
Effect of exposure, swallowing (human)	Harmful if swallowed.	
Exposure, personal protection	Safety glasses, protective clothing based on chemical resistance data, chemical-resistant gloves, general and local exhaust ventilation.	
First aid, eye	Rinse cautiously with water for several minutes. Remove contact lenses, if present and easy to do. Continue rinsing. If eye irritation persists: Get medical advice/attention.	
First aid, inhalation	Keep the patient calm. Protect against loss of body heat. Seek medical advice and clearly identify substance.	
First aid, skin	Remove contaminated or soaked clothing. Wash off with plenty of water or water and soap immediately for 10-15 minutes. In serious cases, use emergency shower immediately. Seek medical advice and clearly identify substance.	
NIOSH, REL	mg/m³	ST325/methanol

GENIOSIL GF 80

PARAMETER	UNIT	VALUE
OSHA, PEL	mg/m³	260/methanol
ACGIH, TLV	ppm	200/methanol
NIOSH, REL	ppm	ST250/methanol
OSHA, PEL	ppm	200/methanol
ECOLOGICAL PROPERTIES		
Aquatic toxicity, *Green algae*, 96-h EC50	mg/l	119/7d
Aquatic toxicity, *Daphnia magna*, 48-h LC50	mg/l	324
Aquatic toxicity, *Rainbow trout*, 96-h LC50	mg/l	237
Biodegradation probability	37%/28d/not readily biodegradable	
USE & PERFORMANCE		
Manufacturer	Wacker Chemie AG	
Outstanding properties	improved filler dispersibility reduced its sedimentation tendency and greatly lowered the resin's viscosity. In addition, it leads to higher filler loading and a marked increase in water (vapor) resistance, as well as resistance to acids and bases. As a component of adhesives and sealants, GENIOSIL® GF 80 improves both adhesion to the substrate and mechanical properties such as flexural strength, tensile strength and, modulus of elasticity.	
Recommended for polymers	polyacrylates, epoxy resins, urethane, melamine resins, EPDM, and polysulfides	
Recommended for products	adhesives and sealants, coatings and paints, primers	
Recommended applications	coupling agent in mineral-filled plastics, used in the treatment of inorganic fillers (e.g., glass, mineral and glass wools, ATH, kaolin, mica, metallic oxides) for various polymers and used as an additive or primer in coatings, paints, adhesives, and sealants.	
Conditions to avoid	contact with moisture must be avoided during processing to prevent undesired hydrolysis with moisture	

Geniosil GF 82

PARAMETER	UNIT	VALUE
GENERAL INFORMATION		
Name	Geniosil GF 82	
CAS #	-	2602-34-8
EC number	-	220-011-6
Composition	(3-glycidyloxypropyl)-triethoxysilane	
Common synonym	3-(2,3-epoxypropyloxy)propyltriethoxysilane	
Empirical formula	C12H26O5Si	
Formula		

PARAMETER	UNIT	VALUE
Molecular mass	daltons	278.4
Chemical class	silane	
Functional organic group	glycidoxy	
Epoxy content	%	15
Purity	wt%	>96
PHYSICAL PROPERTIES		
State	-	liquid
Color	-	clear, colorless
Color, Platinum-cobalt scale	-	<30
Boiling point	°C	143 at 13 hPa
Melting point	°C	<-50
Density at 25°C	kg/m³	1,010
Refractive index at 25°C	-	1.425
Viscosity at 25°C	mPas	3
HEALTH & SAFETY		
Autoignition temperature	°C	225
Flash point	°C	>100
Hazardous combustion products	Carbon oxides, NOx, SiOx, formaldehyde.	
Agency rating, listed	EINECS Europe, ECL Korea, ENCS Japan, AICS Australia, IECSC China, DSL Canada, PICCS Philippines, TSCA USA, NZIoC-New Zealand, TCSI Taiwan, REACH-EU	
Hazardous products of hydrolysis	methanol	
Effect of exposure, eye (human)	Causes serious eye irritation.	
Effect of exposure, inhalation (human)	Irritating to respiratory system.	

Geniosil GF 82

PARAMETER	UNIT	VALUE
Effect of exposure, skin (human)	Causes skin irritation.	
Effect of exposure, swallowing (human)	Harmful if swallowed.	
Exposure, personal protection	Safety glasses, protective clothing based on chemical resistance data, chemical-resistant gloves, general and local exhaust ventilation.	
First aid, eye	Rinse cautiously with water for several minutes. Remove contact lenses, if present and easy to do. Continue rinsing. If eye irritation persists: Get medical advice/attention.	
First aid, inhalation	Keep the patient calm. Protect against loss of body heat. Seek medical advice and clearly identify substance.	
First aid, skin	Remove contaminated or soaked clothing. Wash off with plenty of water or water and soap immediately for 10-15 minutes. In serious cases, use emergency shower immediately. Seek medical advice and clearly identify substance.	
UN risk phrases, R	-	R36/37/38,R20
US safety phrases, S	-	S26
ECOLOGICAL PROPERTIES		
Aquatic toxicity, *Rainbow trout*, 96-h LC50	mg/l	37
USE & PERFORMANCE		
Manufacturer	Wacker Chemie AG	
Outstanding properties	improved mechanical properties such as flexural strength, tensile strength, and modulus of elasticity	
Recommended for polymers	epoxy resins, urethane, melamine resins, EPDM, and polysulfides	
Recommended for products	electronic potting compounds, as an additive or primer in coatings, paints, adhesives, and sealants, and as a component of inorganic polysiloxane-based coatings	
Conditions to avoid	contact with moisture must be avoided during processing to prevent undesired hydrolysis	
Concentrations used	wt%	0.5-2.5 (primers)

Geniosil GF 93

PARAMETER	UNIT	VALUE
GENERAL INFORMATION		
Name		Geniosil GF 93
CAS #	-	919-30-2
EC number	-	213-048-4
Composition	>99% 3-aminopropyltriethoxysilane	
Common synonym	3-(triethoxysilyl)-propylamin	
Empirical formula	(C2H5O)3SiC3H6NH2	
Formula		
Molecular mass	daltons	261.34
RTECS number	-	TX2100000
Amine number	mmol/g	4.5
Purity	wt%	>97
PHYSICAL PROPERTIES		
State	-	liquid
Odor	-	amine-like
Color	-	colorless
Boiling point	°C	217
Melting point	°C	-70
Density at 25⁴C	kg/m³	940
Hydrolysis half-life	8.5 h/pH 7 at 24.5°C	
pH	10-11 at 25°C (10 g/l H_2O)	
Refractive index at 25°C	-	1.42
Solubility (diluents)	organic solvents such as ethers and hydrocarbons	
Solubility in water at 25°C	hydrolytic decomposition releasing methanol	
Viscosity at 25°C	mPas	1.6
HEALTH & SAFETY		
Carcinogenicity	IARC, OSHA, NTP: no ingredient of this product present at levels greater than or equal to 0.1% is identified as probable, possible or confirmed human carcinogen	
Teratogenicity	negative	
ICAO/IATA class	Dangerous Goods, Corrosive liquid, basic, organic, n.o.s. (Aminopropyltriethoxysilane)8,II	

Geniosil GF 93

PARAMETER	UNIT	VALUE
IMDG class		Dangerous Goods, Corrosive liquid, basic, organic, n.o.s. (Aminopropyltriethoxysilane)8,II
Autoignition temperature	°C	300
Flash point	°C	93
Flash point method	-	PMCC
Explosive LEL	wt%	0.8
Explosive UEL	wt%	4.5
Hazardous combustion products		Carbon oxides, SiOx, formaldehyde.
Hazardous ingredients, labelling		Aminopropyltriethoxysilane
Animal testing acute toxicity, Rat inhalation, LC50	mg/m^3	>16ppm/6/female, >5ppm/6H/male
Effect of exposure, eye (human)		Causes serious eye damage.
Effect of exposure, skin (human)		Causes severe skin burns. May cause an allergic skin reaction.
Effect of exposure, swallowing (human)		Harmful if swallowed.
Exposure, personal protection		Safety glasses, protective clothing based on chemical resistance data, chemical-resistant gloves, general and local exhaust ventilation.
First aid, eye		Rinse cautiously with water for several minutes. Remove contact lenses, if present and easy to do. Continue rinsing. If eye irritation persists: Get medical advice/attention.
First aid, inhalation		Keep the patient calm. Protect against loss of body heat. Seek medical advice and clearly identify substance.
First aid, skin		Remove contaminated or soaked clothing. Wash off with plenty of water or water and soap immediately for 10-15 minutes. In serious cases, use emergency shower immediately. Seek medical advice and clearly identify substance.
NIOSH, REL	mg/m^3	ST325/methanol
OSHA, PEL	mg/m^3	260/methanol
ACGIH, TLV	ppm	200/methanol
NIOSH, REL	ppm	ST250/methanol
OSHA, PEL	ppm	200/methanol
UN/NA class	-	3267

Geniosil GF 93

PARAMETER	UNIT	VALUE
ECOLOGICAL PROPERTIES		
Aquatic toxicity, *Daphnia magna*, 48-h LC50	mg/l	331
Aquatic toxicity, *Zebra fish*, 96-h LC50	mg/l	934
Bioconcentration factor	BCF	3.4
Biodegradation probability	67%/28d/not readily biodegradable	
USE & PERFORMANCE		
Manufacturer	Wacker Chemie AG	
Outstanding properties	improved the dispersibility of the filler and the mechanical properties of the composites, such as flexural strength, tensile strength, and modulus, reduced the filler's sedimentation tendency in the uncured polymer, increased water (vapor) and corrosion resistance.	
Recommended for polymers	silane-terminated polyethers or polyurethanes and polysiloxanes	
Recommended for products	construction adhesives and sealants	
Recommended applications	adhesion promoter in sealants, adhesives and coatings, as a surface modifier for fillers and for production of silane-modified polymers as a binder in sealants and adhesives.	
Concentrations used	1-2% of the total formulation	
Conditions to avoid	contact with moisture must be avoided during processing to prevent undesired hydrolysis with moisture	

Geniosil GF 94

PARAMETER	UNIT	VALUE
GENERAL INFORMATION		
Name	Geniosil GF 94	
CAS #	-	5089-72-5
EC number	-	225-806-1
Composition	3-(2-aminomethylamino)methyldimethoxysilane	
Common synonym	aminoethylaminopropyltriethoxysilane	
Empirical formula	(C2H5O)3SiC3H6NHC2H4NH2	
Formula		
Molecular mass	daltons	264.40
Chemical class	silane	
Functional organic group	primary amine	
PHYSICAL PROPERTIES		
State	-	liquid
Odor	-	slight
Color	-	colorless
Boiling point	°C	308.74
Density at 20°C	kg/m³	953
Refractive index at 20°C	-	1.438
Solubility in water at 25°C	slower hydrolytic decomposition	
Vapor pressure at 25°C	kPa	0.000133
HEALTH & SAFETY		
HMIS classification	Flammability	1
	Health	3
	Reactivity	1
Flash point	°C	140
Hazardous combustion products	Carbon oxides, SiOx, formaldehyde.	
Effect of exposure, eye (human)	Causes eye irritation.	
Effect of exposure, inhalation (human)	Causes respiratory system irritation.	
Effect of exposure, skin (human)	Causes severe skin burns damage.	
Exposure, personal protection	Safety glasses, protective clothing based on chemical resistance data, chemical-resistant gloves, general and local exhaust ventilation.	

Geniosil GF 94

PARAMETER	UNIT	VALUE
First aid, eye		Rinse cautiously with water for several minutes. Remove contact lenses, if present and easy to do. Continue rinsing. If eye irritation persists: Get medical advice/attention.
First aid, inhalation		Keep the patient calm. Protect against loss of body heat. Seek medical advice and clearly identify substance.
First aid, skin		Remove contaminated or soaked clothing. Wash off with plenty of water or water and soap immediately for 10-15 minutes. In serious cases, use emergency shower immediately. Seek medical advice and clearly identify substance.
UN risk phrases, R		R36/37/38
US safety phrases, S		S26,R36/37/39
USE & PERFORMANCE		
Manufacturer		Wacker Chemie AG
Recommended for polymers		epoxy, phenolic aldehyde, melamine and hot-fusible resins such as PS, PA
Recommended applications		coupling of thermosetting resins

Geniosil GF 95

PARAMETER	UNIT	VALUE
GENERAL INFORMATION		
Name		Geniosil GF 95
CAS #		3069-29-2, (impurity: 175394-72-6, 618914-52-6, 1079259-16-7, 107-15-5)
EC number	-	217-164-6
Composition		<100% N-(2-aminoethyl)-3-aminopropylmethyldimethoxysilane + (impurity: <0.5% N,N'-bis(3-(dimethoxy(methyl)silyl)propyl)-1,2ethanediamine, <0.5% 1-(2-aminoethyl)-2-methoxy-2-methyl-1-aza-2silacyclopentane, <0.2%1,3-di(3-(2-aminoethyl)amino)propyl-1,3-dimethoxy-1,3dimethyl-disiloxan, 0.5% 1,2-diaminoethane)
Empirical formula		C8H22N2O2Si
Formula		
Molecular mass	daltons	206.4
RTECS number	-	KV7400000
Chemical class	silane	
Functional organic group	primary amine	
PHYSICAL PROPERTIES		
State	-	liquid
Odor	-	slight
Color	-	colorless
Boiling point	°C	110 at 0.3 kPa
Melting point	°C	-50
Density at 20°C	kg/m³	930
Hydrolysis half-life	15/h/pH 7 at 20-25°C	
pH	10.5 at 25°C (10 g/l H_2O)	
Solubility in water at 25°C	hydrolytic decomposition releasing methanol	
HEALTH & SAFETY		
Carcinogenicity		IARC, OSHA, NTP: no ingredient of this product present at levels greater than or equal to 0.1% is identified as probable, possible or confirmed human carcinogen
Mutagenicity	negative	
Teratogenicity	negative	
DOT class	not regulated	

Geniosil GF 95

PARAMETER	UNIT	VALUE
TDG class	not regulated	
ICAO/IATA class	not regulated	
IMDG class	not regulated	
Autoignition temperature	°C	290
Hazardous combustion products	Carbon oxides, SiOx, formaldehyde.	
Agency rating, listed	EINECS Europe, ECL Korea, ENCS Japan, AICS Australia, IECSC China, DSL Canada, PICCS Philippines, TSCA USA, NZIoC-New Zealand, TCSI Taiwan, REACH-EU	
Hazardous ingredients, labelling	3-(trimethoxysilyl) propylamine, Methyl(3-trimethoxysilyl)propyl ether	
Animal testing, acute toxicity, Rat oral LD50	mg/kg	200 - 2000
Animal testing, acute toxicity, Rabbit dermal LD50	mg/kg	155200
Animal testing, acute toxicity, Rat inhalation, LC50	mg/m^3	5200/4H
Effect of exposure, eye (human)	Causes serious eye damage.	
Effect of exposure, inhalation (human)	May cause allergy or asthma symptoms or breathing difficulties if inhaled.	
Effect of exposure, skin (human)	Causes skin irritation. May cause an allergic skin reaction.	
Effect of exposure, swallowing (human)	Harmful if swallowed.	
Effect of repeated or overexposure (human)	Target organs: respiratory system.	
Exposure, personal protection	Safety glasses, protective clothing based on chemical resistance data, chemical-resistant gloves, general and local exhaust ventilation.	
First aid, eye	Rinse cautiously with water for several minutes. Remove contact lenses, if present and easy to do. Continue rinsing. If eye irritation persists: Get medical advice/attention.	
First aid, inhalation	Keep the patient calm. Protect against loss of body heat. Seek medical advice and clearly identify substance.	
First aid, skin	Remove contaminated or soaked clothing. Wash off with plenty of water or water and soap immediately for 10-15 minutes. In serious cases, use emergency shower immediately. Seek medical advice and clearly identify substance.	
NIOSH, REL	mg/m^3	ST325/methanol

Geniosil GF 95

PARAMETER	UNIT	VALUE
OSHA, PEL	mg/m^3	260/methanol
ACGIH, TLV	ppm	200/methanol
NIOSH, REL	ppm	ST250/methanol
OSHA, PEL	ppm	200/methanol
UN risk phrases, R	R41,R43,R34	
US safety phrases, S	S26,S39,S45,S36/37/39	
UN/NA class	-	3267
ECOLOGICAL PROPERTIES		
Aquatic toxicity, *Daphnia magna*, 48-h LC50	mg/l	>100
Aquatic toxicity, *Zebra fish*, 96-h LC50	mg/l	597
Biodegradation probability	39%/28d/not readily biodegradable	
USE & PERFORMANCE		
Manufacturer	Wacker Chemie AG	
Outstanding properties	increased filler dispersibility and improved numerous mechanical properties (e.g., flexural strength, tensile strength, modulus of elasticity) of the composites.	
Recommended for polymers	polyacrylates, butyl rubber, neoprene, phenolic resin, epoxies, PA, PU, melamine	
Recommended for products	sealants, seals, gaskets	
Recommended applications	adhesion promoter in glass fiber-reinforced or filler-modified plastics.	

Geniosil GF 96

PARAMETER	UNIT	VALUE
GENERAL INFORMATION		
Name		Geniosil GF 96
CAS #		13822-56-5, (impurity: 157923-76-7)
EC number	-	237-511-5
Composition		>98% 3-(trimethoxysilyl) propylamine + (impurity <1% 2,2-dimethoxy-1-aza-2-silacyclopentane)
Common synonym		3-trimethoxysilylpropan-1-amine
Empirical formula		C6H17NO3Si
Formula		
Molecular mass	daltons	179.3
RTECS number	-	V7400000
Chemical class	silane	
Amine number	mmol/g	5.5
Functional organic group	primary amine	
Purity	wt%	>95
PHYSICAL PROPERTIES		
State	-	liquid
Odor	-	amine-like
Color	-	clear, colorless
Boiling point	°C	210
Melting point	°C	<-60
Density at 25°C	kg/m³	1,014
Hydrolysis half-life	4h/pH 7 at 20°C	
Kinematic viscosity at 25°C	cSt	1.6
Refractive index at 25°C	-	1.424
Solubility in water at 25°C	hydrolytic decomposition releasing methanol	
Vapor pressure at 20°C	kPa	0.017
Vapor pressure at 50°C	kPa	0.17
Vapor pressure at 100°C	kPa	3.4
HEALTH & SAFETY		
Carcinogenicity		IARC, OSHA, NTP: no ingredient of this product present at levels greater than or equal to 0.1% is identified as probable, possible or confirmed human carcinogen
Mutagenicity		negative

Geniosil GF 96

PARAMETER	UNIT	VALUE
DOT class		not regulated in containers up to 119 Gal./450 L each
ICAO/IATA class		not regulated
IMDG class		not regulated
Autoignition temperature	°C	300
Flash point	°C	79
Flash point method	-	PMCC
Hazardous combustion products		Carbon oxides, SiOx, formaldehyde.
Agency rating listed		EINECS Europe, ECL Korea, ENCS Japan, AICS Australia, IECSC China, DSL Canada, PICCS Philippines, TSCA USA, NZIoC-New Zealand, TCSI Taiwan, REACH-EU
Hazardous ingredients, labelling		3-(Trimethoxysilyl) propylamine, Methyl(3-trimethoxysilyl)propyl ether
Hazardous products of hydrolysis		methanol
Effect of exposure, eye (human)		Causes serious eye damage.
Effect of exposure, skin (human)		Causes skin irritation.
Effect of exposure, swallowing (human)		Harmful if swallowed.
Exposure, personal protection		Safety glasses, protective clothing based on chemical resistance data, chemical-resistant gloves, general and local exhaust ventilation.
First aid, eye		Rinse cautiously with water for several minutes. Remove contact lenses, if present and easy to do. Continue rinsing. If eye irritation persists: Get medical advice/attention.
First aid, inhalation		Keep the patient calm. Protect against loss of body heat. Seek medical advice and clearly identify substance.
First aid, skin		Remove contaminated or soaked clothing. Wash off with plenty of water or water and soap immediately for 10-15 minutes. In serious cases, use emergency shower immediately. Seek medical advice and clearly identify substance.
OSHA, PEL	mg/m^3	260/methanol
ACGIH, TLV	ppm	200/methanol
OSHA, PEL	ppm	200/methanol
UN risk phrases, R		R36/38,R34
US safety phrases, S		S26,S39,S45,S36/37/39

Geniosil GF 96

PARAMETER	UNIT	VALUE
ECOLOGICAL PROPERTIES		
Aquatic toxicity, *Green algae*, 96-h EC50	mg/l	>1000/72H
Aquatic toxicity, *Daphnia magna*, 96-h LC50	mg/l	331
Aquatic toxicity, *Zebra fish*, 96-h LC50	mg/l	>934
Biodegradation probability	67%/28d/not readily biodegradable	
USE & PERFORMANCE		
Manufacturer	Wacker Chemie AG	
Outstanding properties	improved dispersibility of the filler and the mechanical properties - such as flexural strength, tensile strength and modulus - of the composites, also reduces the filler's sedimentation tendency in the uncured polymer. Geniosil GF 96 also greatly increases water (vapor) and corrosion resistance.	
Recommended for polymers	epoxy resins, polyamides, polyacrylates, polyurethanes, ethyl/vinyl acetate polymers, and phenolic resins	
Recommended for products	sealants, seals, gaskets	
Recommended applications	adhesion promoter in formulations and primers, as well as a surface modifier in fillers (e.g., glass, mineral wool, mica, metal oxides) and pigments used in various plastics,	

Geniosil GF 98

PARAMETER	UNIT	VALUE
GENERAL INFORMATION		
Name	Geniosil GF 98	
CAS #	-	23843-64-3
EC number	-	245-904-8
Composition	1-[3-(trimethoxysilyl)propyl]urea	
Common synonym	3-ureidopropyl)trimethoxysilane	
Empirical formula	C7H18N2O4Si	
Formula		

Molecular mass	daltons	222.3
Chemical class	silane	
Active matter	wt%	>95
Functional organic group	primary amine	
PHYSICAL PROPERTIES		
State	-	liquid
Odor	-	aromatic
Color	clear, colorless to light yellowish	
Boiling point	°C	>300
Melting point	°C	-5
Density at 25°C	kg/m³	1,115
Refractive index at 25°C	-	1.46
Solubility (diluents)	alcohols, esters, ketones; mixing with ketones results in imine formation, while mixing with alcohols other than methanol leads to an autocatalytic exchange of alkoxy groups until the system reaches thermodynamic equilibrium	
Solubility in water at 25°C	hydrolytic decomposition releasing methanol	
Vapor pressure at 20°C	kPa	0.133
HEALTH & SAFETY		
Carcinogenicity	IARC, OSHA, NTP: no ingredient of this product present at levels greater than or equal to 0.1% is identified as probable, possible or confirmed human carcinogen	
ICAO/IATA class	not regulated	
IMDG class	not regulated	
Autoignition temperature	°C	300

Geniosil GF 98

PARAMETER	UNIT	VALUE
Flash point	°C	99
Flash point method	-	PMCC
Hazardous combustion products	Carbon oxides, SiOx, formaldehyde.	
Agency rating, listed	EINECS Europe, ECL Korea, ENCS Japan, AICS Australia, IECSC China, DSL Canada, PICCS Philippines, TSCA USA, NZIoC-New Zealand, TCSI Taiwan, REACH-EU	
Animal testing, acute toxicity, Rat oral LD50	mg/kg	>5000
Animal testing, acute toxicity, Rabbit dermal LD50	mg/kg	no skin irritation
Animal testing, acute toxicity, Rat dermal LD50	mg/kg	>2000
Animal testing, acute toxicity, Rat inhalation, LC50	mg/m^3	<5000/4H
Effect of exposure, swallowing (human)	Not expected in industrial use. May be harmful if swallowed.	
Exposure, personal protection	Safety glasses, protective clothing based on chemical resistance data, chemical-resistant gloves, general and local exhaust ventilation.	
First aid, eye	Rinse cautiously with water for several minutes. Remove contact lenses, if present and easy to do. Continue rinsing. If eye irritation persists: Get medical advice/attention.	
First aid, inhalation	Keep the patient calm. Protect against loss of body heat. Seek medical advice and clearly identify substance.	
First aid, skin	Remove contaminated or soaked clothing. Wash off with plenty of water or water and soap immediately for 10-15 minutes. In serious cases, use emergency shower immediately. Seek medical advice and clearly identify substance.	
USE & PERFORMANCE		
Manufacturer	Wacker Chemie AG	
Outstanding properties	enhanced dispersibility in organic binders and plastics, and thereby improve mechanical properties (flexural strength, tensile strength, modulus, etc.)	
Recommended for polymers	polyacrylates, butyl rubber, neoprene, phenolic resin, epoxies, PA, PU, melamine	
Recommended for products	sealants, adhesives and coatings	

Geniosil GF 98

PARAMETER	UNIT	VALUE
Recommended applications		coupling agent for modifying fillers and pigments to enhance dispersibility in organic binders and plastics
Concentrations used		1-2% added to the formulation

Geniosil XL 10

PARAMETER	UNIT	VALUE
GENERAL INFORMATION		
Name	Geniosil XL 10	
CAS #	-	2768-02-7
EC number	-	220-449-8
Composition	>98% vinyltrimethoxysilane + 0.3% methanol	
Empirical formula	C5H12O3Si	
Formula		

$$H_3CO-\underset{\underset{OCH_3}{|}}{\overset{\overset{OCH_3}{|}}{Si}}-CH_2$$

PARAMETER	UNIT	VALUE
Molecular mass	daltons	148.2
Chemical class	silane	
Functional organic group	vinyl	
Dimer content	%	<0.3
Methanol content	%	<0.3
Purity	wt%	>99
PHYSICAL PROPERTIES		
State	-	liquid
Odor	-	slight
Color	-	colorless
Boiling point	°C	122
Melting point	°C	-97
Density at 25°C	kg/m³	970
Hydrolyzable chloride as HCl	mg/kg	<10
Refractive index at 25°C	-	1.391
Solubility in water at 25°C	hydrolytic decomposition releasing methanol	
Specific gravity at 25°C	-	0.97
Vapor pressure at 20°C	kPa	1.19
Viscosity at 25°C	mPas	0.6
HEALTH & SAFETY		
TDG class	Dangerous Goods Flammable liquid, n.o.s..(Trimethoxyvinylsilane) 3,III	
ICAO/IATA class	Dangerous Goods Flammable liquid, n.o.s..(Trimethoxyvinylsilane) 3,III	
IMDG class	Dangerous Goods Flammable liquid, n.o.s..(Trimethoxyvinylsilane) 3,III	
Autoignition temperature	°C	240
Flash point	°C	25

Geniosil XL 10

PARAMETER	UNIT	VALUE
Flash point method	-	PMCC
Explosive LEL	wt%	1.4
Explosive UEL	wt%	19.9
Hazardous combustion products	Carbon oxides, NOx, SiOx, formaldehyde.	
Agency rating, listed	EINECS Europe, ECL Korea, ENCS Japan, AICS Australia, IECSC China, DSL Canada, PICCS Philippines, TSCA USA, NZIoC-New Zealand, TCSI Taiwan, REACH-EU	
Hazardous ingredients, labelling	Trimethoxyvinylsilane	
Hazardous products of hydrolysis	methanol	
Animal testing, acute toxicity, Rat oral LD50	mg/kg	7120-7236
Animal testing, acute toxicity, Rabbit dermal LD50	mg/kg	>3200
Animal testing, acute toxicity, Rat dermal LD50	mg/kg	>2000
Animal testing, acute toxicity, Rat inhalation, LC50	mg/m^3	168000/4H
Effect of exposure, eye (human)	Avoid contact with eyes.	
Effect of exposure, inhalation (human)	Harmful if inhaled. Avoid breathing vapors/spray	
Effect of exposure, skin (human)	Avoid contact with skin.	
Effect of exposure, swallowing (human)	Harmful if swallowed.	
Exposure, personal protection	Safety glasses, protective clothing based on chemical resistance data, chemical-resistant gloves, general and local exhaust ventilation.	
First aid, eye	Rinse cautiously with water for several minutes. Remove contact lenses, if present and easy to do. Continue rinsing. If eye irritation persists: Get medical advice/attention.	
First aid, inhalation	Keep the patient calm. Protect against loss of body heat. Seek medical advice and clearly identify substance.	
First aid, skin	Remove contaminated or soaked clothing. Wash off with plenty of water or water and soap immediately for 10-15 minutes. In serious cases, use emergency shower immediately. Seek medical advice and clearly identify substance.	
NIOSH, REL	mg/m^3	ST325/methanol
OSHA, PEL	mg/m^3	260/methanol

Geniosil XL 10

PARAMETER	UNIT	VALUE
ACGIH, TLV	ppm	200/methanol
NIOSH, REL	ppm	ST250/methanol
OSHA, PEL	ppm	200/methanol
UN/NA class	-	1993
ECOLOGICAL PROPERTIES		
Aquatic toxicity, *Daphnia magna*, 48-h LC50	mg/l	169
Aquatic toxicity, *Rainbow trout*, 96-h LC50	mg/l	191
Biodegradation probability	51%/28d/not readily biodegradable	
USE & PERFORMANCE		
Manufacturer	Wacker Chemie AG	
Outstanding properties	improved resistance to heat and weathering, as well improved electrical properties of products from silane-crosslinked polyethylene (PE-Xb)	
Recommended for polymers	crosslinked polyethylene	
Recommended for products	pipes and cable	
Recommended applications	used as co-monomers in the production of silane-modified binders for surface coatings and used as an additive in silane-crosslinked polyethylene (PE-Xb) formulations as water scavenger during the production of silane-crosslinking adhesive and sealant formulations	
Conditions to avoid	contact with moisture must be avoided during processing to prevent undesired hydrolysis with moisture must be avoided to prevent undesired hydrolysis	

Geniosil XL 12

PARAMETER	UNIT	VALUE
GENERAL INFORMATION		
Name		Geniosil XL 12
CAS #	-	16753-62-1
EC number	-	240-816-6
Composition		60-100% dimethoxymethyl vinylsilane + methanol (varies)
Common synonym		vinyldimethoxymethylsilane
Empirical formula		C5H12O2Si
Formula		

PARAMETER	UNIT	VALUE
Molecular mass	daltons	132.20
Functional organic group	vinyl	
Purity	wt%	>97
PHYSICAL PROPERTIES		
State	-	liquid
Odor	-	characteristic
Color	-	colorless
Boiling point	°C	104
Melting point	°C	<-70
Density at 20°C	kg/m³	880
Kinematic viscosity at 25°C	cSt	0.7
Refractive index at 20°C	-	1.395
Solubility in water at 25°C	hydrolytic decomposition releasing methanol	
Viscosity at 25°C	mPas	0.6
HEALTH & SAFETY		
HMIS classification	Flammability	3
	Health	1
	Reactivity	1
Carcinogenicity	OSHA, WHMIS: no ingredient of this product present at levels greater than or equal to 0.1% is identified as probable, possible or confirmed human carcinogen	
DOT class	Dangerous Goods, Flammable liquid, n.o.s. (dimethoxymethylvinylsilane) 3,II	
TDG class	Dangerous Goods, Flammable liquid, n.o.s. (dimethoxymethylvinylsilane) 3,II	

Geniosil XL 12

PARAMETER	UNIT	VALUE
ICAO/IATA class		Dangerous Goods, Flammable liquid, n.o.s. (dimethoxymethylvinylsilane) 3,II
IMDG class		Dangerous Goods, Flammable liquid, n.o.s. (dimethoxymethylvinylsilane) 3,II
Autoignition temperature	°C	245/liquid
Flash point	°C	15
Flash point method	-	PMCC
Hazardous combustion products		Carbon oxides, SiOx, formaldehyde.
Agency rating, listed		EINECS Europe, TCSI Taiwan, TSCA USA,
Hazardous ingredients, labelling		Flammable liquid, n.o.s. (dimethoxymethylvinylsilane)
Hazardous products of hydrolysis		methanol
Animal testing, acute toxicity, Rat oral LD50	mg/kg	>5000
Effect of exposure, eye (human)		May cause slight eye irritation
Effect of exposure, inhalation (human)		Inhalation is not expected due to low vapor pressure.
Effect of exposure, skin (human)		No acute toxic effects are known. Irritation can not be excluded
Effect of exposure, swallowing (human)		Not expected in industrial use. May be harmful if swallowed.
Exposure, personal protection		Safety glasses, protective clothing based on chemical resistance data, chemical-resistant gloves, general and local exhaust ventilation.
First aid, eye		Rinse cautiously with water for several minutes. Remove contact lenses, if present and easy to do. Continue rinsing. If eye irritation persists: Get medical advice/attention.
First aid, inhalation		Keep the patient calm. Protect against loss of body heat. Seek medical advice and clearly identify substance.
First aid, skin		Remove contaminated or soaked clothing. Wash off with plenty of water or water and soap immediately for 10-15 minutes. In serious cases, use emergency shower immediately. Seek medical advice and clearly identify substance.
OSHA, PEL	mg/m³	260/methanol
ACGIH, TLV	ppm	200/methanol
OSHA, PEL	ppm	200/methanol
UN/NA class	-	1993

Geniosil XL 12

PARAMETER	UNIT	VALUE
ECOLOGICAL PROPERTIES		
Bioaccumulative potential		not bioaccumulative
USE & PERFORMANCE		
Manufacturer		Wacker Chemie AG
Outstanding properties		improved electrical resistance to heat and weathering, improved wet scrub resistance, and higher abrasion resistance.
Recommended for polymers		polyacrylates, PE
Recommended for products		cable, wire, pipe hose, fittings
Recommended applications		used in the production of pipes and cables made of silane-crosslinked polyethylene (PE-Xb), as a water scavenger during the production of silane-crosslinking adhesive and sealant formulations

Geniosil XL 32

PARAMETER	UNIT	VALUE
GENERAL INFORMATION		
Name		Geniosil XL 32
CAS #	-	121177-93-3
Composition		(methacryloxymethyl)methyldimethoxysilane
Empirical formula		C8H16O4Si
Formula		

PARAMETER	UNIT	VALUE
Molecular mass	daltons	204.28
Chemical class		silane
Functional organic group		methacryl
PHYSICAL PROPERTIES		
State	-	liquid
Odor	-	slight
Color	-	colorless
Boiling point	°C	205
Melting point	°C	<-100
Density at 20°C	kg/m³	1,020
Kinematic viscosity at 25°C	cSt	1.53
Solubility in water at 25°C		hydrolytic decomposition releasing methanol
Vapor pressure at 20°C	kPa	0.025
HEALTH & SAFETY		
DOT class		Dangerous Goods, Flammable liquid, n.o.s. (alkoxysilanes)3,ll, DOT regulated as a Combustible Liquid when packaged in bulk containers (>119 Gallons).
ICAO/IATA class		not regulated
IMDG class		not regulated
Autoignition temperature	°C	300
Flash point	°C	82
Flash point method	-	CC
Hazardous combustion products		Carbon oxides, SiOx, formaldehyde.
Hazardous ingredients, labelling		Combustible Liquid, n.o.s. (alkoxysilanes)
Hazardous products of hydrolysis		methanol
Animal testing, acute toxicity, Rat oral LD50	mg/kg	>2000

Geniosil XL 32

PARAMETER	UNIT	VALUE
Animal testing, acute toxicity, Rabbit dermal LD50	mg/kg	no skin irritation
Animal testing acute toxicity, Rat dermal LD50	mg/kg	>2000
Animal testing, acute toxicity, Rat inhalation, LC50	mg/m³	530/4H (highest possible concentration no mortality in animal test)
Effect of exposure, swallowing (human)	Not expected in industrial use. May be harmful if swallowed.	
Exposure, personal protection	Safety glasses, protective clothing based on chemical resistance data, chemical-resistant gloves, general and local exhaust ventilation.	
First aid, eye	Rinse cautiously with water for several minutes. Remove contact lenses, if present and easy to do. Continue rinsing. If eye irritation persists: Get medical advice/attention.	
First aid, inhalation	Keep the patient calm. Protect against loss of body heat. Seek medical advice and clearly identify substance.	
First aid, skin	Remove contaminated or soaked clothing. Wash off with plenty of water or water and soap immediately for 10-15 minutes. In serious cases, use emergency shower immediately. Seek medical advice and clearly identify substance.	
US safety phrases, S	S23,S24/25	
UN/NA class	-	1993
ECOLOGICAL PROPERTIES		
Aquatic toxicity, *Daphnia magna*, 48-h LC50	mg/l	>200
Aquatic toxicity, *Zebra fish*, 96-h LC50	mg/l	>200
Biodegradation probability	74%/28d/readily biodegradable	
USE & PERFORMANCE		
Manufacturer	Wacker Chemie AG	
Outstanding properties	improved dispersibility. Offers improved mechanical properties.	
Recommended for polymers	polyacrylates, TPE, TPV	
Recommended for products	cable, wire, pipe hose, fittings	
Recommended applications	adhesion promoter	

Geniosil XL 33

PARAMETER	UNIT	VALUE
GENERAL INFORMATION		
Name	Geniosil XL 33	
CAS #	-	54586-78-6
Composition	methacryloxymethyl-trimethoxysilane	
Common synonym	(methacryloxymethyl)triethoxysilane	
Acronym		MMTMS
Formula		
Molecular mass	daltons	220.30
Chemical class	silane	
Functional organic group	methacryl	
PHYSICAL PROPERTIES		
State	-	liquid
Odor	-	characteristically sweet
Color	-	clear, colorless
Boiling point	°C	217
Density at 20°C	kg/m³	1070
HEALTH & SAFETY		
Autoignition temperature	°C	295
Flash point	°C	92
Flash point method	-	CC
Hazardous combustion products	Carbon oxides, SiOx, formaldehyde.	
Animal testing, acute toxicity, Rat oral LD50	mg/kg	>2000
Effect of exposure, eye (human)	Causes eye irritation.	
Effect of exposure, inhalation (human)	Causes respiratory system irritation.	
Effect of exposure, skin (human)	Causes severe skin burns damage.	
Exposure, personal protection	Safety glasses, protective clothing based on chemical resistance data, chemical-resistant gloves, general and local exhaust ventilation.	
First aid, eye	Rinse cautiously with water for several minutes. Remove contact lenses, if present and easy to do. Continue rinsing. If eye irritation persists: Get medical advice/attention.	

Geniosil XL 33

PARAMETER	UNIT	VALUE
First aid, inha ation		Keep the patient calm. Protect against loss of body heat. Seek medical advice and clearly identify substance.
First aid, skin		Remove contaminated or soaked clothing. Wash off with plenty of water or water and soap immediately for 10-15 minutes. In serious cases, use emergency shower immediately. Seek medical advice and clearly identify substance.
UN risk phrases, R		R38,R36/38,R36/37/38,R37/38,R36,R36/37,R37
USE & PERFORMANCE		
Manufacturer		Wacker Chemie AG
Recommended applications		adhesion promoter in paints and industrial coatings, used in coatings for packaging, paper and film, industrial such as metal, plastic or wood and architectural both plaster and paints

Geniosil XL 65

PARAMETER	UNIT	VALUE
GENERAL INFORMATION		
Name	Geniosil XL 65	
CAS #	-	23432-65-7
Composition	N-dimethoxy(methyl)silylmethyl-O-methyl-carbamate +(impurity <3% methanol, <1% (Isocyanatomethyl)(dimethoxy) methylsilane, <0.1% N,N-dimethylformamide)	
Empirical formula	(CH3O)2CH3SiCH2NH-CO-OCH3	
Formula		
Molecular mass	daltons	193.30
Chemical class	silane	
Functional organic group	carbamato	
PHYSICAL PROPERTIES		
State	-	liquid
Odor	-	characteristic
Color	colorless to light yellowish	
Boiling point	°C	211.1
Melting point	°C	<-100
Density at 20°C	kg/m³	1,102.5
Kinematic viscosity at 25°C	cSt	3.25
Solubility in water at 25°C	reacts rapidly in aqueous media generating heat and releasing methanol	
Vapor pressure at 20°C	kPa	0.0008
Vapor pressure at 25°C	kPa	0.0011
HEALTH & SAFETY		
Teratogenicity	May damage the unborn child	
DOT class	not regulated	
TDG class	not regulated	
ICAO/IATA class	not regulated	
IMDG class	not regulated	
Autoignition temperature	°C	250
Flash point	°C	112
Flash point method	-	CC
Hazardous combustion products	Carbon oxides, SiOx, formaldehyde.	
Agency rating, listed	EINECS Europe, DSL Canada, ECL Korea, TSCA USA	

Geniosil XL 65

PARAMETER	UNIT	VALUE
Hazardous ingredients, labelling		N,N-dimethylformamide, (Isocyanato-methyl)(dimethoxy)methylsilane
Animal testing, acute toxicity, Rat oral LD50	mg/kg	>2000
Animal testing, acute toxicity, Rat dermal LD50	mg/kg	>2000
Effect of exposure, inhalation (human)		May cause allergy or asthma symptoms or breathing difficulties if inhaled.
Exposure, personal protection		Safety glasses, protective clothing based on chemical resistance data, chemical-resistant gloves, general and local exhaust ventilation.
First aid, eye		Rinse cautiously with water for several minutes. Remove contact lenses, if present and easy to do. Continue rinsing. If eye irritation persists: Get medical advice/attention.
First aid, inhalation		Keep the patient calm. Protect against loss of body heat. Seek medical advice and clearly identify substance.
First aid, skin		Remove contaminated or soaked clothing. Wash off with plenty of water or water and soap immediately for 10-15 minutes. In serious cases, use emergency shower immediately. Seek medical advice and clearly identify substance.
ACGIH, TLV	ppm	200/methanol
NIOSH, REL	ppm	ST250/methanol
OSHA, PEL	ppm	200/methanol
ECOLOGICAL PROPERTIES		
Aquatic toxicity, *Green algae*, 96-h EC50	mg/l	117/72H
Aquatic toxicity, *Daphnia magna*, 48-h LC50	mg/l	>100
Aquatic toxicity, *Rainbow trout*, 96-h LC50	mg/l	>100
Biodegradation probability		19%/28d/not readily biodegradable
USE & PERFORMANCE		
Manufacturer		Wacker Chemie AG
Outstanding properties		highly reactivity toward moisture, a direct consequence of the structural proximity of the nitrogen atom to the silicon atom (alpha-effect). Improved the storage stability

Geniosil XL 65

PARAMETER	UNIT	VALUE
Recommended applications		used as a water scavenger in the production of moisture-curing adhesive and sealant formulations. It is important particularly in systems using highly reactive silane-modified polymers based on α-organofunctional silanes. A bifunctional silane functions additionally as a chain extender and is thus ideally suited for highly flexible products.

NXT

PARAMETER	UNIT	VALUE
GENERAL INFORMATION		
Name	NXT	
CAS #	-	220727-26-4
EC number	-	436-690-9
Composition	90-100% - S-(3-(triethoxysilyl)propyl) octanethioate	
Formula		
Chemical class	silane	
PHYSICAL PROPERTIES		
State	-	liquid
Color	pale yellow to amber	
Initial boiling point	°C	>400
Melting point	°C	<-70
Density at 25°C	kg/m^3	980
pH	-	9
Solubility in water at 23°C	mg/l	<280
Vapor density	-	>1
Vapor pressure	hPa	<0.0133
Volatility	ethanol (VOC) emission essentially reduced during mixing and use	
HEALTH & SAFETY		
HMIS classification	Flammability	1
	Health	2
	Physical hazard	1
Flash point	°C	150
Flash point method	-	PMCC
Hazardous products of hydrolysis	ethanol	
Animal testing, acute toxicity, Rat oral LD50	mg/kg	>2000
Animal testing, acute toxicity, Rabbit dermal LD50	mg/kg	>2000
Effect of repeated or overexposure (human)	Repeated exposure to ethanol may aggravate liver injury produced from other causes.	
Exposure, personal protection	Safety glasses, protective clothing based on chemical resistance data, chemical-resistant gloves, general and local exhaust ventilation.	

NXT

PARAMETER	UNIT	VALUE
First aid, eye		Rinse immediately with plenty of water, also under the eyelids, for at least 15 minutes. Get medical attention.
First aid, inhalation		Move the exposed person to fresh air at once. If respiratory problems, artificial respiration/oxygen. Call a physician or poison control center immediately.
First aid, skin		Wash off promptly and flush contaminated skin with water. Promptly remove clothing if soaked through and flush skin with water. Wash contaminated clothing before reuse. Get medical attention.
NIOSH, REL	mg/m^3	1900/ethanol
OSHA, PEL	mg/m^3	1900/ethanol
ACGIH, TLV	ppm	1000/ethanol
NIOSH, REL	ppm	1000/ethanol
OSHA, PEL	ppm	1000/ethanol
ECOLOGICAL PROPERTIES		
Aquatic toxicity, *Zebra fish*, 96-h LC50	mg/l	>100
Aquatic toxicity, *Daphnia magna*, 48-h LC50	mg/l	>100
USE & PERFORMANCE		
Manufacturer		Momentive
Outstanding properties		NXT silane provides options for enhanced tire performance and overall systems cost-efficiencies for tire manufacturers. NXT silane, a thiocarboxylate functional silane, has been shown to enable reduced rolling resistance without loss of wet traction while increasing overall production efficiency for tire manufacturers as compared to standard sulfur silanes.
Recommended for polymers		synthetic rubber (SBR, NBR)
Recommended for products		tires
Recommended applications		superior coupling agent typically can be used with high surface area silica and functionalized polymers for easy processing of high-performance compounds.

NXT Z 45S

PARAMETER	UNIT	VALUE
GENERAL INFORMATION		
Name		NXT Z 45S
CAS #	-	922519-17-3
Composition		S-(3-(triethoxysilyl)propyl)octanethioate
Chemical class		silane
PHYSICAL PROPERTIES		
State	-	liquid
Color	-	pale yellow
Odor	-	sulfur
Melting point	°C	<-41
Initial boiling point	°C	241.9
Density at 25°C	kg/m^3	1060.4
Solubility in water at 25°C	hydrolytic decomposition releasing ethanol	
Vapor density	-	>1
Vapor pressure at 20°C	kPa	0.2
VOC	g/l	287
HEALTH & SAFETY		
Flash point	°C	>100
Autoignition temperature	°C	315
Effect of repeated or overexposure (human)	Repeated exposure to ethanol may aggravate liver injury produced from other causes.	
Exposure, personal protection	Safety glasses, protective clothing based on chemical resistance data, chemical-resistant gloves, general and local exhaust ventilation.	
First aid, eye	Rinse immediately with plenty of water and seek medical advice. Get medical attention.	
First aid, inhalation	Move the exposed person to fresh air at once. Get medical attention.	
First aid, skin	Take off immediately all contaminated clothing. Wash with soap and water. Get medical attention.	
Animal testing, acute toxicity, Rat oral LD50	mg/kg	>2000
Animal testing, acute toxicity, Rabbit dermal LD50	mg/kg	>2000
NIOSH, REL	mg/m^3	1900/ethanol

NXT Z 45S

PARAMETER	UNIT	VALUE
OSHA, PEL	mg/m^3	1900/ethanol
ACGIH, TLV	ppm	1000/ethanol
NIOSH, REL	ppm	1000/ethanol
OSHA, PEL	ppm	1000/ethanol
ECOLOGICAL PROPERTIES		
Aquatic toxicity, *Daphnia magna*, 48-h LC50	mg/l	23
USE & PERFORMANCE		
Manufacturer	Momentive Performance Materials	
Outstanding properties	reduced ethanol (VOC) emissions during tire manufacturing and use. Lower system cost due to fewer non-productive mixing steps, faster processing, and VOC abatement cost avoidance. High temperature mixing without viscosity increases or premature vulcanization. Enhanced vulcanization speed. Substantial reduction in small strain non-linearity, G'. Very low maximum in tan over the 0 to 25% strain range. Excellent dynamic properties at low temperatures (+5° to -20°C). Excellent storage stability of uncured rubber due to low silica reagglomeration. NXT silanes can have a positive environmental impact by enabling improved rolling resistance, which can help reduce fuel consumption and CO$_2$ emissions.	
Recommended for polymers	synthetic rubber (SBR, NBR)	
Recommended for products	tires	
Recommended applications	coupling agent for silica-reinforced tire tread compounds. NXT Z 45S silane is providing significant reductions in ethanol evolved during tire manufacturing and use and can help eliminate the need to mix silica compounds in multiple steps. It can also provide a faster cure rate and some improvements in compound performance properties. Compared to silanes that bear three ethoxy groups per silicon atom, NXT Z 45S silane can reduce ethanol emissions by more than 66%. NXT Z 45S silane may lead to overall systems-cost efficiencies for tire manufacturers.	

Si 69™

PARAMETER	UNIT	VALUE
GENERAL INFORMATION		
Name	Si 69®	
CAS #	-	40272-72-3
EC #	-	254-896-5
General description	bifunctional, sulfur-containing organosilane for rubber applications in combination with white fillers containing silanol groups.	
Acronym	-	TESPTS
Common synonym	3,3'-tetrathiobis(propyl-triethoxysilane)	
Formula		
Molecular mass	daltons	532
Sulfur content	wt%	22.5
Average sulfur chain length, HPLC	-	3.7
PHYSICAL PROPERTIES		
State	-	liquid
Color	-	yellowish
Density at 20°C	kg/m^3	1100
USE & PERFORMANCE		
Manufacturer	Evonik	
Outstanding properties	Si 69® reacts with silanol groups of white fillers during mixing and with the polymer during vulcanization under the formation of covalent chemical bonds. Si 69® imparts greater tensile strength, higher moduli, reduced compression set, increased abrasion resistance, and optimized dynamic properties. The product is also available as a dry blend (1:1) with carbon black, named X 50-S®.	
Recommended for polymers	rubber	
Recommended applications	low rolling resistance tires	
Guidelines for use	Aqueous acids and bases accelerate the transformation of organosilanes to polymer siloxanes. Therefore, organosilanes should be stored in a dry place, away from acids and bases.	

Si 75™

PARAMETER	UNIT	VALUE
GENERAL INFORMATION		
Name	Si 75®	
CAS #	-	56706-10-6
EC #	-	260-350-7
General description	bifunctional, sulfur-containing organosilane for rubber applications in combination with white fillers containing silanol groups.	
Acronym	-	TESPTS
Common synonym	3,3'-dithiobis(propyl-triethoxysilane)	
Formula	$H_5C_2O-\underset{\underset{OC_2H_5}{|}}{\overset{\overset{OC_2H_5}{|}}{Si}}-(CH_2)_3-S_2-(CH_2)_3-\underset{\underset{OC_2H_5}{|}}{\overset{\overset{OC_2H_5}{|}}{Si}}-OC_2H_5$	
Molecular mass	daltons	448
Sulfur content	wt%	22.5
Average sulfur chain length, HPLC	-	3.7
PHYSICAL PROPERTIES		
State	-	liquid
Color	-	yellowish
Odor	-	characteristic
Density at 20°C	kg/m^3	1000
Melting point	°C	-20
Boiling point	°C	269
Vapor pressure at 20°C	kPa	0.009
Viscosity	mPas	8
HEALTH & SAFETY		
Autoignition temperature	°C	230
Flash point	°C	129
ECOLOGICAL PROPERTIES		
Partition coefficient, log K_{oc}	-	5.2
USE & PERFORMANCE		
Manufacturer	Evonik	
Outstanding properties	Si 75® reacts with silanol groups of white fillers during mixing and with the polymer during vulcanization under the formation of covalent chemical bonds. Si 75® imparts greater tensile strength, higher moduli, reduced compression set, increased abrasion resistance, and optimized dynamic properties.	

Si 75™

PARAMETER	UNIT	VALUE
Recommended for polymers	rubber	
Recommended applications	low rolling resistance tires	
Guidelines for use	Aqueous acids and bases accelerate the transformation of organosilanes to polymer siloxanes. Therefore, organosilanes should be stored in a dry place, away from acids and bases.	

Si 264™

PARAMETER	UNIT	VALUE
GENERAL INFORMATION		
Name	Si 264™	
CAS #	-	34708-08-2
EC number	-	252-161-3
General description	bifunctional, sulfur-containing organosilane for rubber applications in combination with white fillers containing silanol groups.	
Common synonym	3-thiocyanatopropyltriethoxysilane	
Empirical formula	C10H21NO3SSi	
Formula		

$$H_5C_2O-\underset{\underset{OC_2H_5}{|}}{\overset{\overset{OC_2H_5}{|}}{Si}}-(CH_2)_3-SCN$$

PARAMETER	UNIT	VALUE
Molecular mass	daltons	263.43
Sulfur content	wt%	12.5
Residue on ignition	wt%	23
Purity	wt%	96
PHYSICAL PROPERTIES		
State	-	liquid
Color	colorless to yellowish	
Boiling point	°C	95
Density at 20°C	kg/m³	1000
Refractive index at 20°C	-	1.446
HEALTH & SAFETY		
Flash point	°C	112
USE & PERFORMANCE		
Manufacturer	Evonik	
Outstanding properties	Si 264™ reacts with silanol groups of white fillers during mixing and with the polymer during vulcanization under the formation of covalent chemical bonds. This imparts greater tensile strength, higher moduli, reduced compression set, increased abrasion resistance, and optimized dynamic properties.	
Recommended for polymers	rubber	
Recommended applications	mechanical rubber goods, shoe soles	

Si 264™

PARAMETER	UNIT	VALUE
Guidelines for use		compared to polysulfidic silanes Si 264™ leads to lower compression set values and better abrasion resistance. Aqueous acids and bases accelerate the transformation of organosilanes to polymer siloxanes. Therefore, organosilanes should be stored in a dry place, away from acids and bases.

Si 266™

PARAMETER	UNIT	VALUE
GENERAL INFORMATION		
Name	Si 266®	
General description	bifunctional, sulfur-containing organosilane for rubber applications in combination with white fillers containing silanol groups.	
Common synonym	bis(triethoxysilpropyl)polysulfide	
Formula	$$H_5C_2O-\underset{\underset{OC_2H_5}{\vert}}{\overset{\overset{OC_2H_5}{\vert}}{Si}}-(CH_2)_3-S_x-(CH_2)_3-\underset{\underset{OC_2H_5}{\vert}}{\overset{\overset{OC_2H_5}{\vert}}{Si}}-OC_2H_5$$	
Molecular mass	daltons	480
Sulfur content	wt%	14.4
Average sulfur chain length, HPLC	-	2.15
PHYSICAL PROPERTIES		
State	-	liquid
Color	-	yellowish
Density at 20°C	kg/m³	1030
USE & PERFORMANCE		
Manufacturer	Evonik	
Outstanding properties	Si 266® reacts with silanol groups of white fillers during mixing and with the polymer during vulcanization under the formation of covalent chemical bonds. Si 266® imparts greater tensile strength, higher moduli, reduced compression set, increased abrasion resistance, and optimized dynamic properties. The product is also available as a dry blend (1:1) with carbon black, named X 266-S®.	
Recommended for polymers	rubber	
Recommended applications	low rolling resistance tires	
Guidelines for use	Aqueous acids and bases accelerate the transformation of organosilanes to polymer siloxanes. Therefore, organosilanes should be stored in a dry place, away from acids and bases.	

Silcat 17 Industrial Silane

PARAMETER	UNIT	VALUE
GENERAL INFORMATION		
Name		Silcat 17 Industrial Silane
CAS #	-	2768-02-7
Common synonym		vinyltrimethoxysilane
Composition		50-<100 vinyltrimethoxysilane, 3-<5 dibutyltin dilaurate
Formula		
Molecular mass	daltons	148.23
PHYSICAL PROPERTIES		
State	-	liquid
Odor	-	ester-like
Color	-	clear
Initial boiling point	°C	>122
Melting point	°C	<-70
Density at 20°C	kg/m³	968
Vapor pressure at 20°C	kPa	<0.133
Viscosity at 25°C	cSt	1.4
HEALTH & SAFETY		
HMIS classification	Flammability	3
	Health	3
	Physical hazard	1
Flash point	°C	25
Flash point method	-	TCC
Flammability limit - upper	%	19.9
Flammability limit - lower	%	1.4
Animal testing, acute toxicity, Rat oral LD50	mg/kg	>7300
Animal testing, acute toxicity, Rabbit dermal LD50	mg/kg	>3400
Animal testing, acute toxicity, Rat inhalation LC50	mg/l	16.79
First aid, eye		Rinse immediately with plenty of water, also under the eyelids, for at least 15 minutes. Get medical attention.
First aid, inhalation		Move the exposed person to fresh air at once

Silcat 17 Industrial Silane

PARAMETER	UNIT	VALUE
First aid, skin	Wash off promptly and flush contaminated skin with water. Promptly remove clothing if soaked through and flush skin with water. Wash contaminated clothing before reuse. Get medical attention.	
UN #	-	1933
ECOLOGICAL PROPERTIES		
Aquatic toxicity, *Green algae*, 72-h EC50	mg/l	>100
Aquatic toxicity, *Daphnia magna*, 48-h EC50	mg/l	>100
USE & PERFORMANCE		
Manufacturer	Momentive	
Outstanding properties	modified vinyltrimethoxysilane containing a crosslinking tin catalyst. In order to be used as one-step (Monosil) crosslinking system, the addition of a suitable peroxide is required.	
Recommended for polymers	polyethylene	

Silcat RHE

PARAMETER	UNIT	VALUE
GENERAL INFORMATION		
Name		Silcat RHE
CAS #	-	2768-02-7
Common synonym		vinyltrimethoxysilane
Composition		50-<100 vinyltrimethoxysilane, 3-<5 dibutyltin dilaurate
Formula		

$$H_3CO-\underset{\underset{OCH_3}{|}}{\overset{\overset{OCH_3}{|}}{Si}} \diagup\!\!\!\diagup$$

PARAMETER	UNIT	VALUE
Molecular mass	daltons	148.23
PHYSICAL PROPERTIES		
State	-	liquid
Odor	-	ester-like
Color	-	light straw
Initial boiling point	°C	>122
Melting point	°C	<-70
Density at 20°C	kg/m³	962
Vapor pressure at 20°C	hPa	<1.33
Viscosity at 25°C	mPas	2.2
HEALTH & SAFETY		
HMIS classification	Flammability	3
	Health	4
	Physical hazard	1
Flash point	°C	23
Flash point method	-	TCC
Flammability limit - upper	%	19.9
Flammability limit - lower	%	1.4
Animal testing, acute toxicity, Rat oral LD50	mg/kg	>7300
Animal testing, acute toxicity, Rabbit dermal LD50	mg/kg	>3,460-4,000
Animal testing, acute toxicity, Rat inhalation LC50	mg/l	16.79
First aid, eye		Rinse immediately with plenty of water, also under the eyelids, for at least 15 minutes. Get medical attention.
First aid, inhalation		Move the exposed person to fresh air at once. If respiratory problems, artificial respiration/oxygen. Get medical attention.

Silcat RHE

PARAMETER	UNIT	VALUE
First aid, skin		Wash off promptly and flush contaminated skin with water. Promptly remove clothing if soaked through and flush skin with water. Wash contaminated clothing before reuse. Get medical attention.
UN #	-	1933
ECOLOGICAL PROPERTIES		
Aquatic toxicity, *Green algae*, 72-h EC50	mg/l	>100
Aquatic toxicity, *Daphnia magna*, 48-h EC50	mg/l	>100
USE & PERFORMANCE		
Manufacturer		Momentive
Outstanding properties		crosslinking system (silane, peroxide and catalyst) for the manufacture of crosslinked LDPE & LLDPE polyethylene LV & MV cables using the Monosil(1) one-step process. It provides excellent performance on equipment designed for Monosil technology. A high onset temperature of the silane crosslinking agent improves process stability and minimizes pregrafted/crosslinked particles in the insulation layer
Recommended for polymers		LDPE & LLDPE
Recommended for products		LV & MV cables

Silox 23

PARAMETER	UNIT	VALUE
GENERAL INFORMATION		
Name		Silox 23
CAS #	-	2768-02-7
Common synonym		vinyltrimethoxysilane
Composition		50-<100% vinyltrimethoxysilane, 1-3% vinyl silsesquioxane
Formula		

$$H_3CO-\underset{\underset{OCH_3}{|}}{\overset{\overset{OCH_3}{|}}{Si}}\diagup\diagup$$

PARAMETER	UNIT	VALUE
Molecular mass	daltons	148.23
PHYSICAL PROPERTIES		
State	-	liquid
Odor	-	ester-like
Color	-	colorless
Initial boiling point	°C	>100
Density at 20°C	kg/m^3	970
Vapor pressure at 20°C	kPa	0.7
HEALTH & SAFETY		
HMIS classification	Flammability	3
	Health	1
	Physical hazard	1
Flash point	°C	23
Flash point method	-	TCC
Animal testing, acute toxicity, Rat oral LD50	mg/kg	>7300
Animal testing, acute toxicity, Rabbit dermal LD50	mg/kg	>3,460-4,000
Animal testing, acute toxicity, Rat inhalation LC50	mg/l	16.79
First aid, eye		Rinse immediately with plenty of water, also under the eyelids, for at least 15 minutes. Get medical attention.
First aid, inhalation		Move the exposed person to fresh air at once.
First aid, skin		Wash off promptly and flush contaminated skin with water. Promptly remove clothing if soaked through and flush skin with water. Wash contaminated clothing before reuse. Get medical attention.
UN #	-	1933

Silox 23

PARAMETER	UNIT	VALUE
ECOLOGICAL PROPERTIES		
Aquatic toxicity, *Green algae*, 72-h EC50	mg/l	>100
Aquatic toxicity, *Daphnia magna*, 48-h EC50	mg/l	>100
USE & PERFORMANCE		
Manufacturer	Momentive	
Outstanding properties	excellent candidate to consider for crosslinking high- and medium-density polyethylene (PEXb) for use in potable water pipes. This product is a vinyl silane and peroxide combination that can provide high grafting efficiency in a Sioplas process. The extrapolated service life of continuous and intermittent hot water pipe of 134 and 536 years. Maximum service temperature of 110°C	
Recommended for polymers	polyethylene	
Recommended for products	water pipes	

Silquest A-137

PARAMETER	UNIT	VALUE
GENERAL INFORMATION		
Name	Silquest A-137	
CAS #	-	2943-75-1
Composition	50-<100% octyltriethoxysilane	
Formula		
Molecular mass	daltons	276.49
Active material content	%	95
PHYSICAL PROPERTIES		
State	-	liquid
Color	water-white to straw-colored	
Color, Pt/Co	-	100
Initial boiling point	°C	250
Melting point	°C	<-74
Density at 25°C	kg/m³	876
Vapor pressure at 20°C	kPa	<0.133
VOC	wt%	<3
HEALTH & SAFETY		
HMIS classification	Flammability	2
	Health	2
	Physical hazard	1
Flash point	°C	82
Flash point method	-	PMCC
Animal testing, acute toxicity, Rat oral LD50	mg/kg	>=5110
Animal testing, acute toxicity, Rat dermal LD50	mg/kg	6730 (male)
Animal testing, acute toxicity, Rat inhalation LC50	mg/l	>=22
First aid, eye	Rinse immediately with plenty of water, also under the eyelids, for at least 15 minutes. Get medical attention.	
First aid, inhalation	Move the exposed person to fresh air at once. If respiratory problems, artificial respiration/oxygen. Call a physician or poison control center immediately	

Silquest A-137

PARAMETER	UNIT	VALUE
First aid, skin	Wash off promptly and flush contaminated skin with water. Promptly remove clothing if soaked through and flush skin with water. Wash contaminated clothing before reuse. Get medical attention.	
UN #	-	1993, 3082
USE & PERFORMANCE		
Manufacturer	Momentive	
Outstanding features	when exposed to moisture, is reactive with the minerals contained in concrete, masonry, and other substrates. The monomer can penetrate deeply into a substrate and then react to form a protective surface surrounding each microscopic particle. This protective surface is extremely durable and leads to significant improvements in weather resistance. The silane is easily dissolved in organic solvents to form clear, colorless, stable solutions.	
Recommended for products	penetrating sealers, highway bridge decks	

Silquest A-151NT

PARAMETER	UNIT	VALUE
GENERAL INFORMATION		
Name	Silquest A-151NT	
CAS #	-	78-08-0
EC number	-	201-081-7
Composition	95-100% vinyltriethoxysilane, 1-5% related silane <1% ethanol	
Empirical formula	CH2=CHSi(OCH2CH3)3	
Formula		
Molecular mass	daltons	222.32
Functional organic group	vinyl/triethoxy	
PHYSICAL PROPERTIES		
State	-	liquid
Odor	-	ester-like
Color	clear, colorless to light straw	
Boiling point	°C	160.5
Melting point	°C	<0
Density at 25°C	kg/m³	905
Kinematic viscosity at 25°C	cSt	0.7
Refractive index at 25°C	-	1.397
Solubility in water at 25°C	hydrolytic decomposition releasing ethanol	
Vapor density	-	>1.0
Vapor pressure at 20°C	kPa	0.893
HEALTH & SAFETY		
DOT class	FLAMMABLE LIQUID, N.O.S. (Vinyltriethoxysilane) 3, III	
ICAO/IATA class	FLAMMABLE LIQUID, N.O.S. (Vinyltriethoxysilane) 3, III	
IMDG class	FLAMMABLE LIQUID, N.O.S. (Vinyltrimethoxysilane) 3, III	
Flash point	°C	44
Flash point method	-	TCC
Hazardous ingredients, labelling	FLAMMABLE LIQUID, N.O.S. (Vinyltriethoxysilane)	
Hazardous products of hydrolysis	ethanol	
Effect of exposure, skin (human)	Causes skin irritation.	
Effect of exposure, swallowing (human)	Harmful if swallowed.	

Silquest A-151NT

PARAMETER	UNIT	VALUE
Effect of repeated or overexposure (human)		Repeated exposure to ethanol may aggravate liver injury produced from other causes.
Exposure, personal protection		Safety glasses, protective clothing based on chemical resistance data, chemical-resistant gloves, general and local exhaust ventilation.
First aid, eye		Rinse immediately with plenty of water, also under the eyelids, for at least 15 minutes. Get medical attention.
First aid, inhalation		Move the exposed person to fresh air at once. If respiratory problems, artificial respiration/oxygen. Call a physician or poison control center immediately.
First aid, skin		Wash off promptly and flush contaminated skin with water. Promptly remove clothing if soaked through and flush skin with water. Wash contaminated clothing before reuse. Get medical attention.
NIOSH, REL	mg/m^3	1900/ethanol
OSHA, PEL	mg/m^3	1900/ethanol
ACGIH, TLV	ppm	1000/ethanol
NIOSH, REL	ppm	1000/ethanol
OSHA, PEL	ppm	1000/ethanol
UN risk phrases, R	-	R37/38,R41
US safety phrases, S	-	S26,S36/37/39
UN/NA class	-	1993
USE & PERFORMANCE		
Manufacturer		Momentive
Outstanding properties		highly resistant to exposure to moisture, chemicals, and UV. Siloxane crosslinks tend to not generate color and are resistant to environmental factors, such as acid. Bonds to inorganic substrates to provide excellent wet and dry adhesion. Excellent adhesion to a wide array of substrates can be obtained by the addition of adhesion-promoting silanes, such as Silquest A-1110 silane or Silquest A-1120 silane
Recommended for polymers		polyolefins

Silquest A-151NT

PARAMETER	UNIT	VALUE
Recommended applications		adhesion promoter. Silquest A-151NT silane offers vinyl and silane functionality, making them suitable for crosslinking organic polymers.

Silquest A-171

PARAMETER	UNIT	VALUE
GENERAL INFORMATION		
Name	Silquest A-171	
CAS #	-	2768-02-7
EC number	-	220-449-8
Composition	97.5-100 % vinyltrimethoxysilane, <2% methanol, <1% ethyltrimethoxysilane	
Empirical formula	CH2=CHSi(OCH3)3	
Formula		
Molecular mass	daltons	148.20
Chemical class	silane	
Functional organic group	vinyl/trimethoxy	
PHYSICAL PROPERTIES		
State	-	liquid
Odor	-	ester-like
Color	colorless to light straw	
Boiling point	°C	122
Melting/freezing point	°C	<-70
Density at 25°C	kg/m³	967
Refractive index at 25°C	-	1.3905
Solubility in water at 20°C	g/l	9400
Vapor pressure at 20°C	kPa	1.197
Vapor density	-	heavier than air
HEALTH & SAFETY		
HMIS classification	Flammability	3
	Health	1
	Physical hazard	1
DOT class	FLAMMABLE LIQUID, N.O.S. (Vinyltriethoxysilane) 3, III	
ICAO/IATA class	FLAMMABLE LIQUID, N.O.S. (Vinyltrimethoxysilane) 3, III	
IMDG class	FLAMMABLE LIQUID, N.O.S. (Vinyltrimethoxysilane) 3, III	
Flash point	°C	28
Explosive LEL	wt%	1.4
Explosive UEL	wt%	19.9
Hazardous ingredients, labelling	FLAMMABLE LIQUID, N.O.S. (Vinyltrimethoxysilane)	

Silquest A-171

PARAMETER	UNIT	VALUE
Hazardous products of hydrolysis	ethanol	
Effect of expcsure, inhalation (human)	Irritating to respiratory system.	
Effect of exposure, swallowing (human)	Harmful if swallowed.	
Effect of repeated or overexposure (human)	May cause damage to organs through prolonged or repeated exposure to vapor (kidneys, liver)	
Exposure, personal protection	Safety glasses, protective clothing based on chemical resistance data, chemical-resistant gloves, general and local exhaust ventilation.	
First aid, eye	Rinse immediately with plenty of water, also under the eyelids, for at least 15 minutes. Get medical attention.	
First aid, inhalation	Move the exposed person to fresh air at once. If respiratory problems, artificial respiration/oxygen. Call a physician or poison control center immediately.	
First aid, skin	Wash off promptly and flush contaminated skin with water. Promptly remove clothing if soaked through and flush skin with water. Continue to rinse for at least 15 minutes. Get medical attention. Wash contaminated clothing before reuse. After contact with skin, remove product mechanically	
Animal testing, acute toxicity, Rat oral LD50	mg/kg	>7000
Animal testing, acute toxicity, Rabbit dermal LD50	mg/kg	3000-4000
ACGIH, TLV	ppm	5/Vinyltrimethoxysilane, 200/methanol
UN/NA class	-	1993
USE & PERFORMANCE		
Manufacturer	Momentive	
Outstanding properties	Silquest* A-171 silane results in Si-O-Si crosslink sites that are highly resistant to exposure to moisture, chemicals, and UV. Siloxane crosslinks without changing color, and they are resistant to environmental factors, such as acid. Bonds to inorganic substrates to provide excellent wet and dry adhesion. Excellent adhesion to a wide array of substrates can be obtained by the addition of adhesion promoting silane, such as Silquest A-1110 silane or Silquest A-1120 silane.	

Silquest A-171

PARAMETER	UNIT	VALUE
Recommended for polymers		PE, polyester, PU, SBR
Recommended applications		Silquest A-171 silane is a monomeric vinyl functional silane in vinyl, vinyl acrylic, and acrylic resins. The vinyl silanes can be added as monomers during emulsion polymerization to form silane-modified latexes. The silanes in such latexes function as crosslinkers, forming very stable Si-O-Si linkages. Vinyl silanes can also be grafted to select unsaturated polymers such as PE, polyester, and styrene-butadiene copolymers, via free radical chemistry. Once grafted to the resin, the resin exhibits silane functionality through which the resin can be crosslinked via an ambient moisture cure mechanism.

Silquest A-172NT

PARAMETER	UNIT	VALUE
GENERAL INFORMATION		
Name		Silquest A-172NT
CAS #	-	1067-53-4
Composition		50-100% vinyl tris-(2-methoxyethoxy) silane
Common synonym		2,5,7,10-tetraoxa-6-silaundecane, 6-ethenyl-6-(2methoxyethoxy)-
Empirical formula		$CH_2=CHSi(OCH_2CH_2OCH_3)_3$
Formula		
Molecular mass	daltons	280.40
Chemical class		silane
Functional organic group		vinyl/trimethoxy
PHYSICAL PROPERTIES		
State	-	liquid
Odor	-	ester-like
Color	-	light yellow
Initial boiling point	°C	285
Melting/freezing point	°C	-130
Evaporation rate (butyl acetate=1)	-	<1
Refractive index at 25°C	-	1.427
Solubility (diluents)		methanol, ethanol, toluene, benzene, xylene, acetone
Solubility in water at 20°C		71 g/l reactive, hydrolytic decomposition releasing methanol
Specific gravity at 25°C	-	1.039
Vapor pressure at 20°C	kPa	<0.665
HEALTH & SAFETY		
HMIS classification	Flammability	2
	Health	2
	Reactivity	1
Mutagenicity		negative in a mammalian cell gene mutation test
Teratogenicity		May damage the unborn child. May damage fertility. Developmental effects: no known significant effects or critical hazards.

Silquest A-172NT

PARAMETER	UNIT	VALUE
DOT class	Combustible liquid, n.o.s. III (It is regulated for transport in the US in container >119 gallons).	
Autoignition temperature	°C	210
Flash point	°C	92
Flash point method	-	TCC
Hazardous combustion products	Carbon monoxide, carbon dioxide, nitrogen oxides	
Agency rating, listed	TSCA USA, DSL Canada, AICS Australia, MITI Japan, EINECS Europe, ECL Korea, PICCS Philippines, Taiwan CSNN	
Hazardous ingredients, labelling	Combustible liquid, n.o.s.	
Hazardous products of hydrolysis	methanol	
Animal testing, acute toxicity, Rat oral LD50	mg/kg	>2000
Animal testing, acute toxicity, Rabbit dermal LD50	mg/kg	Non-irritant to skin
Animal testing, acute toxicity, Rat dermal LD50	mg/kg	>2000
Effect of exposure, eye (human)	No known significant effects or critical hazards	
Effect of exposure, inhalation (human)	May be harmful if inhaled.	
Effect of exposure, skin (human)	No known significant effects or critical hazards	
Effect of exposure, swallowing (human)	May be harmful if swallowed	
Effect of repeated or overexposure (human)	May cause damage to organs through prolonged or repeated exposure to methanol (kidneys, liver)	
Exposure, personal protection	Safety glasses, protective clothing based on chemical resistance data, chemical-resistant gloves, general and local exhaust ventilation.	
First aid, eye	In case of contact, immediately flush eyes with plenty of water for at least 15 minutes. Get medical attention.	
First aid, inhalation	Move the exposed person to fresh air at once. Get medical attention.	
First aid, skin	Take off immediately all contaminated clothing. Wash area with soap and water. Wash contaminated clothing before reuse.	
NIOSH, REL	mg/m³	ST325/methanol
OSHA, PEL	mg/m³	260/methanol

Silquest A-172NT

PARAMETER	UNIT	VALUE
ACGIH, TLV	ppm	200/methanol
NIOSH, REL	ppm	200/methanol
OSHA, PEL	ppm	200/methanol
ECOLOGICAL PROPERTIES		
Partition coefficient, log P_{oc}	-	0.26 (pH=7)
USE & PERFORMANCE		
Manufacturer	Momentive	
Outstanding properties	improved electrical properties and strength of mineral-filled ethylene/propylene rubber and crosslinked polyethylene and other polymer or resin systems, particularly after wet-conditioning. Enhanced the strength performance of cured, filled, or reinforced polyester and other resin composites, both initially and after wet-conditioning. Reduced water absorption in cured polyester molding compounds, particularly diallyl phthalate compounds, thus improving the wet electrical and mechanical properties. Improved the bond of glass filament to polyester and other resins in fiberglass reinforced applications, rendering the cured composites more resistant to wet environmental conditions. Increased the adhesion of printing inks or pastes and coatings to glass, ceramics, or metal, providing economic advantages over costlier ceramic-fusing techniques. Improved the adhesion of silicone rubber applied to polyester or glass surfaces, especially important in high-temperature applications	
Recommended applications	coupling agent that promotes adhesion among unsaturated, polyester-type resins or crosslinked polyethylene resins or elastomers and inorganic substrates, including fiberglass, silica, silicates, and many metal oxides	

Silquest A-174NT

PARAMETER	UNIT	VALUE
GENERAL INFORMATION		
Name	Silquest A-174NT	
CAS #	-	2530-85-0
Composition	>90% γ-methacryloxypropyltrimethoxy-silane, <1% methanol	
Formula		
Molecular mass	daltons	248.10
Chemical class	silane	
Functional organic group	methacryloxy	
PHYSICAL PROPERTIES		
State	-	liquid
Odor	-	ester-like
Color	-	clear, pale yellow
Initial boiling point	°C	255
Melting point	°C	<-48
Density at 25°C	kg/m³	1,045
Evaporation rate (butyl acetate=1)	-	<1
Refractive index at 25°C	-	1.429
Solubility in water at 20°C	slow hydrolytic decomposition releasing methanol	
Vapor density	-	>1
Vapor pressure at 20°C	kPa	<0.133
VOC	g/l	1,041.94
HEALTH & SAFETY		
NFPA classification	Flammability	1
	Health	2
	Reactivity	1
HMIS classification	Flammability	1
	Health	2
	Reactivity	1
Carcinogenicity	IARC, OSHA, NTP: no ingredient of this product present at levels greater than or equal to 0.1% is identified as probable, possible or confirmed human carcinogen	
Mutagenicity	not determined	
Teratogenicity	no data available	

Silquest A-174NT

PARAMETER	UNIT	VALUE
Flash point	°C	108
Flash point method	-	TCC
Hazardous combustion products	Carbon monoxide, carbon dioxide, nitrogen oxides	
Agency rating, listed	TSCA USA, DSL Canada, AICS Australia, MITI Japan, EINECS Europe, ECL Korea, PICCS Philippines, Taiwan CSNN	
Hazardous products of hydrolysis	methanol	
Animal testing, acute toxicity, Rat oral LD50	mg/kg	>2000
Animal testing, acute toxicity, Rabbit dermal LD50	mg/kg	slight irritation
Animal testing, acute toxicity, Rat dermal LD50	mg/kg	>2000
Effect of exposure, eye (human)	May cause mild irritation.	
Effect of exposure, inhalation (human)	Respiratory tract irritation. Narcotic effects. Causes damage to central nervous system (CNS)	
Effect of exposure, skin (human)	May cause allergic skin reaction. Skin contact may aggravate an existing dermatitis.	
Effect of repeated or overexposure (human)	May aggravate an existing kidney disease and existing liver disease through prolong or repeated exposure to methanol	
Exposure, personal protection	Safety glasses, protective clothing based on chemical resistance data, chemical-resistant gloves, general and local exhaust ventilation.	
First aid, eye	Rinse immediately with plenty of water, also under the eyelids, for at least 15 minutes. Get medical attention.	
First aid, inhalation	Move the exposed person to fresh air at once. If respiratory problems, artificial respiration/oxygen. Call a physician or poison control center immediately.	
First aid, skin	Wash off promptly and flush contaminated skin with water. Promptly remove clothing if soaked through and flush skin with water. Continue to rinse for at least 15 minutes. Get medical attention. Wash contaminated clothing before reuse. After contact with skin, remove product mechanically.	
NIOSH, REL	mg/m³	ST325/methanol

Silquest A-174NT

PARAMETER	UNIT	VALUE
ACGIH, TLV	ppm	200/methanol
NIOSH, REL	ppm	200/methanol
OSHA, PEL	ppm	200/methanol
USE & PERFORMANCE		
Manufacturer	Momentive	
Outstanding properties	excellent wet and dry adhesion. Improved wet electrical performance of many mineral-filled thermoplastic materials such as crosslinked polyethylene and polyvinylchloride	
Recommended applications	used in adhesives and sealants based on resin systems such as polyesters, polyurethanes, acrylics, and many thermoplastics such as PVC, polyolefins, and polyurethanes.	

Silquest A-178

PARAMETER	UNIT	VALUE
GENERAL INFORMATION		
Name		Silquest A-178
Composition		<100% methacrylamidosilane
Formula		$$\underset{\underset{\displaystyle O}{\|\|}}{H_2C\!=\!C}CNHCH_2CH_2CH_2Si(OCH_2CH_3)_{3\text{-}a}(OCH_3)_a \quad \overset{CH_3}{}$$
Empirical formula		CH2=CH3CONHCH2CH2CH2Si(OCH2CH3)3-a(OCH3)a
Molecular mass	daltons	280.40
Chemical class	-	silane
Active matter	wt%	100
Functional organic group	methacryloxy	
PHYSICAL PROPERTIES		
State	-	liquid
Odor	-	ester-like
Color	-	clear, colorless
Initial boiling point	°C	322
Melting/freezing point	°C	-60
Solubility (diluents)	methanol, ethanol, acetone, toluene, methyl Cellosolve solvent, reacts with alcoholic solvents	
Solubility in water at 20°C	mg/l	26100 (reactive)
Specific gravity at 25°C	-	1.0192
Vapor pressure at 30°C	kPa	0.00388
VOC	g/l	167
HEALTH & SAFETY		
HMIS classification	Flammability	1
	Health	0
	Physical hazard	1
ICAO/IATA class	not regulated	
IMDG class	not regulated	
Autoignition temperature	°C	265
Flash point	°C	101.7
Flash point method	-	PMCC
Hazardous combustion products	Carbon monoxide, carbon dioxide, nitrogen oxides	
Effect of exposure, swallowing (human)	Harmful if swallowed.	

Silquest A-178

PARAMETER	UNIT	VALUE
Exposure, personal protection	Safety glasses, protective clothing based on chemical resistance data, chemical-resistant gloves, general and local exhaust ventilation.	
First aid, eye	Rinse immediately with plenty of water, also under the eyelids, for at least 15 minutes. Get medical attention.	
First aid, inhalation	Move the exposed person to fresh air at once. If respiratory problems, artificial respiration/oxygen. Call a physician or poison control center immediately.	
First aid, skin	Wash off promptly and flush contaminated skin with water. Promptly remove clothing if soaked through and flush skin with water. Continue to rinse for at least 15 minutes. Get medical attention. Wash contaminated clothing before reuse. After contact with skin, remove product mechanically.	
Animal testing, acute toxicity, Rat oral LD50	mg/kg	>2000
Animal testing, acute toxicity, Rat dermal LD50	mg/kg	>2000
ECOLOGICAL PROPERTIES		
Aquatic toxicity, *Daphnia magna*, 48-h LC50	mg/l	>100
Aquatic toxicity, *Rainbow trout*, 96-h LC50	mg/l	>100
USE & PERFORMANCE		
Manufacturer	Momentive	
Outstanding properties	improved strand integrity for better fiber processing and composite fabrication. Improved hygrothermal aging properties of glass fiber and particulate filler-reinforced composites. Provides glass fiber protection. Contains no flammable or combustible solvent. Offers low VOC emissions.	
Recommended for polymers	unsaturated polyester, vinyl ester, acrylic, polybutylene, and polyolefins	

Silquest A-178

PARAMETER	UNIT	VALUE
Recommended applications		coupling agent for glass fiber and particulate filler reinforced composites. May be used to promote adhesion between a wide range of resins and substrates and reinforcements. Reactive with a large number of resin systems, such unsaturated polyester, vinyl ester, acrylic, polybutylene and polyolefins. Compatible with many typical glass fiber size and coating ingredients, such as film formers, anti-static agents, surfactants, lubricants and other coupling agents

Silquest A-186

PARAMETER	UNIT	VALUE
GENERAL INFORMATION		
Name		Silquest A-186
CAS #	-	3388-04-3
Composition		60-100% beta-(3,4-epoxycyclohexyl) ethyltrimethoxysilane
Formula		
PHYSICAL PROPERTIES		
State	-	liquid
Color	-	clear, pale
Initial boiling point	°C	310
Melting point	°C	<0
Density at 25°C	kg/m³	1065
Refractive index at 25°C	-	1.448
Vapor pressure at 20°C	hPa	1.33
HEALTH & SAFETY		
HMIS classification	Flammability	1
	Health	2
	Physical hazard	1
Flash point	°C	113
Flash point method	-	TCC
First aid, eye		Rinse immediately with plenty of water, also under the eyelids, for at least 15 minutes. Get medical attention.
First aid, inhalation		Move the exposed person to fresh air at once. If respiratory problems, artificial respiration/oxygen. Call a physician or poison control center immediately.
First aid, skin		Wash off promptly and flush contaminated skin with water. Promptly remove clothing if soaked through and flush skin with water. Wash contaminated clothing before reuse. Get medical attention.
USE & PERFORMANCE		
Manufacturer		Momentive

Silquest A-186

PARAMETER	UNIT	VALUE
Outstanding properties		epoxy ring reacts with many organic functionalities. Epoxy functionality offers non-yellowing adhesion in many resin systems. Excellent wet and dry adhesion. Very fast hydrolysis rate.
Recommended for polymers		urethanes, epoxy, polysulfides, silicones, and acrylics
Recommended for products		caulks, coatings, sealants and adhesives, electronic encapsulants, and packaging materials
Concentrations used	wt%	0.5-1.5

Silquest A-187

PARAMETER	UNIT	VALUE
GENERAL INFORMATION		
Name		Silquest A-187
CAS #	-	2530-83-8
Composition		50-100% gamma-glycidoxypropyltrime-thoxysilane, <1% allyl glycidyl ether
Formula		
Molecular mass	daltons	236.1
Chemical class	-	silane
Active matter	wt%	100
Functional organic group	epoxy	
PHYSICAL PROPERTIES		
State	-	liquid
Odor	-	ester-like
Color	-	clear, pale yellow
Initial boiling point	°C	233
Melting point	°C	<-70
Density at 25°C	kg/m³	1,069
Kinematic viscosity at 20°C	cSt	2.0
Refractive index at 25°C	-	1.427
Solubility (diluents)		alcohol, acetone and most aliphatic esters
Solubility in water at 20°C		36.5 g/l, hydrolytic decomposition releasing methanol
Vapor pressure at 20°C	hPa	0.011
HEALTH & SAFETY		
HMIS classification	Flammability	1
	Health	2
	Reactivity	1
Carcinogenicity	not determined	
Mutagenicity	not determined	
Teratogenicity	not determined	
DOT class	not regulated	
TDG class	not regulated	
ICAO/IATA class	not regulated	
IMDG class	not regulated	
Autoignition temperature	°C	400

Silquest A-187

PARAMETER	UNIT	VALUE
Flash point	°C	110
Flash point method	-	PMCC
Hazardous combustion products		Carbon monoxide (CO), Carbon dioxide (CO2), Nitrogen Oxides
Agency rating listed		TSCA USA, DSL Canada, AICS Australia, MITI Japan, ECL Korea, PICCS Philippines, Taiwan CSNN, IECSC China
Hazardous products of hydrolysis		methanol
Effect of exposure, eye (human)		Causes serious eye damage
Effect of exposure, skin (human)		Causes minor skin irritation.
Effect of exposure, swallowing (human)		Harmful if swallowed. May cause burns to mouth, throat and stomach.
Effect of repeated or overexposure (human)		Long-term repeated overexposure to methanol vapor concentrations of 3000 ppm or greater may allow a cumulative effect to occur with resulting nausea, vomiting, headache, ringing in the ears, insomnia, trembling, unsteady gait, vertigo, clouded and double vision. Liver and/or kidney injury may occur. Prolonged overexposure at levels of 800-1000 ppm may result in severe eye damage in some persons.
Exposure, personal protection		Safety glasses, protective clothing based on chemical resistance data, chemical-resistant gloves, general and local exhaust ventilation.
First aid, eye		Immediately rinse with water for several minutes. Remove contact lenses, if present and easy to do. Continue rinsing. Call a physician or poison control center immediately.
First aid, inhalation		Move the exposed person to fresh air at once. If respiratory problems, artificial respiration/oxygen. Call a physician or poison control center immediately.
First aid, skin		Wash off promptly and flush contaminated skin with water. Promptly remove clothing if soaked through and flush skin with water. Wash contaminated clothing before reuse. Get medical attention.
NIOSH, REL	mg/m^3	22/ allyl glycidyl ether
OSHA, PEL	mg/m^3	22/ allyl glycidyl ether

Silquest A-187

PARAMETER	UNIT	VALUE
ACGIH, TLV	ppm	1/allyl glycidyl ether
NIOSH, REL	ppm	5/allyl glycidyl ether
OSHA, PEL	ppm	5/ allyl glycidyl ether, 200/methanol
ECOLOGICAL PROPERTIES		
Aquatic toxicity, *Daphnia magna*, 48-h LC50	mg/l	324
Partition coefficient, log K_{oc}	-	0.5
USE & PERFORMANCE		
Manufacturer	Momentive	
Outstanding properties	excellent wet and dry adhesion. Improves flexibility of systems vs. other adhesion promoters. Very fast hydrolysis rate	
Recommended for polymers	acrylic, epoxy, PU, polysulfide, silane-terminated polyurethane	
Recommended applications	adhesion promoters in SPUR+, urethane, epoxy, polysulfide, silicone, and acrylic caulks, coatings, sealants, and adhesives	
Concentrations used	wt%	0.5-1.5

Silquest A-189

PARAMETER	UNIT	VALUE
GENERAL INFORMATION		
Name		Silquest A-189
CAS #		4420-74-0, 2530-87-2
EC number	-	224-588-5
General description		50-100% gamma-mercaptopropyltrime-thoxysilane, 5-10% Silane, (3-chloropro-pyl)trimethoxy, <1% methanol
Empirical formula		HSCH2CH2CH2Si(OCH3)3
Formula		
Molecular mass	daltons	196.4
Chemical class	silane	
Functional organic group	mercapto	
PHYSICAL PROPERTIES		
State	-	liquid
Odor	-	ester-like
Color	-	clear, light straw
Initial boiling point	°C	212
Melting/freezing point	°C	<0
Density at 25°C	kg/m³	1,050
Evaporation rate (butyl acetate=1)	-	>1
Refractive index at 25°C	-	1.44
Solubility (diluents)		methanol, ethanol, isopropanol, mineral spirits, acetone, benzene, toluene, xylene
Solubility in water at 20°C		soluble in water with hydrolytic decomposition releasing methanol
Vapor pressure	hPa	<1.33
VOC	g/l	999.86
HEALTH & SAFETY		
HMIS classification	Flammability	2
	Health	2
	Reactivity	1
Carcinogenicity		No known significant effects or critical hazards.
Mutagenicity		no known significant effects or critical hazards
Teratogenicity		Suspected of damaging the unborn child

Silquest A-189

PARAMETER	UNIT	VALUE
DOT class	Combustible liquid, n.o.s. (γ-Mercaptopropyltrimethoxysilane) III	
ICAO/IATA class	ENVIRONMENTALLY HAZARDOUS SUBSTANCE, LIQUID, N.O.S. (γ-Mercaptopropyltrimethoxysilane) 9, III	
IMDG class	ENVIRONMENTALLY HAZARDOUS SUBSTANCE, LIQUID, N.O.S. (γ-Mercaptopropyltrimethoxysilane) 9, III	
Flash point	°C	88
Flash point method	-	TCC
Hazardous combustion products	Carbon monoxide, carbon dioxide, sulfur oxides	
Agency rating, listed	TSCA USA, DSL Canada, AICS Australia, MITI Japan, ECL Korea, PICCS Philippines, Taiwan CSNN, IECSC China	
Hazardous ingredients, labelling	ENVIRONMENTALLY HAZARDOUS SUBSTANCE, LIQUID, N.O.S. (γ-Mercaptopropyltrimethoxysilane)	
Animal testing, acute toxicity, Rat oral LD50	mg/kg	850/female
Animal testing, acute toxicity, Rat dermal LD50	mg/kg	1922/female
Effect of exposure, eye (human)	No known significant effects or critical hazards	
Effect of exposure, inhalation (human)	Respiratory tract irritation (CNS, optic nerves)/methanol	
Effect of exposure, skin (human)	May cause an allergic skin reaction.	
Effect of exposure, swallowing (human)	Harmful if swallowed	
Exposure, personal protection	Safety glasses, protective clothing based on chemical resistance data, chemical-resistant gloves, general and local exhaust ventilation.	
First aid, eye	Rinse immediately with plenty of water, also under the eyelids, for at least 15 minutes. Get medical attention.	
First aid, inhalation	Move the exposed person to fresh air at once. If respiratory problems, artificial respiration/oxygen. Call a physician or poison control center immediately.	
First aid, skin	Wash off promptly and flush contaminated skin with water. Promptly remove clothing if soaked through and flush skin with water. Wash contaminated clothing before reuse. Get medical attention.	
NIOSH, REL	mg/m^3	ST325/methanol

Silquest A-189

PARAMETER	UNIT	VALUE
OSHA, PEL	mg/m^3	260/methanol
ACGIH, TLV	ppm	200/methanol
NIOSH, REL	ppm	200/methanol
OSHA, PEL	ppm	200/methanol
UN/NA class	-	1995/DOT, 3082/ IMDG, ICAO)

ECOLOGICAL PROPERTIES		
Aquatic toxicity, *Green algae*, 96-h EC50	mg/l	40/96H
Aquatic toxicity, *Bluegill sunfish*, 96-h LC50	mg/l	12.3

USE & PERFORMANCE	
Manufacturer	Momentive
Outstanding properties	improved mineral-filled elastomer properties, including modulus, tensile and tear strength, heat buildup, abrasion resistance, resilience, compression set, and cure time
Recommended for polymers	sulfur-cured and metal oxide-cured elastomers
Recommended for products	shoe soles, rubber rollers and wheels, white sidewalls, wire and cable insulation
Recommended applications	coupling agent for fillers. Silquest A-189 silane chemically bonds reinforcing minerals such as silica, clay, mica, and talc to the polymer matrix. This silane is especially useful in sulfur-cured and metal oxide-cured elastomers. Silquest A-189 silane is substituted for commonly used polysulfide-type silanes. Mineral-reinforced articles can be produced with lower silane loadings.

Silquest A-1100

PARAMETER	UNIT	VALUE
GENERAL INFORMATION		
Name	Silquest A-1100	
CAS #	-	919-30-2
EC number	-	213-048-4
Composition	50-100% γ-aminopropyltriethoxysilane , <1% ethanol	
Common synonym	1-propanamine, 3-(triethoxysilyl)-	
Empirical formula	H2NCH2CH2CH2Si(OCH2CH3)3	
Formula		
Molecular mass	daltons	179
RTECS number	-	TX2100000
Chemical class	silane	
Functional organic group	primary amine/triethoxy	
PHYSICAL PROPERTIES		
State	-	liquid
Odor	-	amine-like
Color	-	light straw
Boiling point	°C	210
Melting point	°C	-60
Density at 25°C	kg/m³	1014
Evaporation rate (butyl acetate=1)	-	<1
Refractive index at 25°C	-	1.442
Solubility (diluents)	alcohol, aromatic and aliphatic hydrocarbons	
Solubility in water at 20°C	g/l	180
Vapor density	-	>1
Vapor pressure at 20°C	hPa	0.18
Volatility	g/l	947
HEALTH & SAFETY		
NFPA classification	Flammability	2
	Health	1
	Reactivity	3
HMIS classification	Flammability	2
	Health	3
	Physical hazards	1
Carcinogenicity	not determined	

Silquest A-1100

PARAMETER	UNIT	VALUE
Mutagenicity	no data available	
DOT class	CORROSIVE LIQUID, BASIC, ORGANIC, N.O.S.(3-aminopropyltriethoxysilane) 8, II	
ICAO/IATA class	CORROSIVE LIQUID, BASIC, ORGANIC, N.O.S.(3aminopropyltriethoxysilane) 8, II	
IMDG class	CORROSIVE LIQUID, BASIC, ORGANIC, N.O.S.(3aminopropyltriethoxysilane) 8, II	
Autoignition temperature	°C	295
Flash point	°C	82
Flash point method	-	PMCC
Hazardous combustion products	Carbon monoxide, carbon dioxide, nitrogen oxides	
Agency rating, listed	TSCA USA, DSL Canada, AICS Australia, EINECS Europa, MITI Japan, ECL Korea, PICCS Philippines	
Hazardous ingredients, labelling	CORROSIVE LIQUID, BASIC, ORGANIC, N.O.S.(3-aminopropyltriethoxysilane)	
Animal testing, acute toxicity, Rat oral LD50	mg/kg	3000
Animal testing, acute toxicity, Rat dermal LD50	mg/kg	11000
Effect of exposure, eye (human)	Causes serious eye damage.	
Effect of exposure, inhalation (human)	Irritating to respiratory system.	
Effect of exposure, skin (human)	Causes severe burns.	
Effect of exposure, swallowing (human)	Harmful if swallowed	
Exposure, personal protection	Safety glasses, protective clothing based on chemical resistance data, chemical-resistant gloves, general and local exhaust ventilation.	
First aid, eye	Rinse immediately with plenty of water, also under the eyelids, for at least 15 minutes. Get medical attention.	
First aid, inhalation	Move the exposed person to fresh air at once. If respiratory problems, artificial respiration/oxygen. Call a physician or poison control center immediately.	
First aid, skin	Wash off promptly and flush contaminated skin with water. Promptly remove clothing if soaked through and flush skin with water. Wash contaminated clothing before reuse. Get medical attention.	
OSHA, PEL	mg/m^3	1900/ethanol

Silquest A-1100

PARAMETER	UNIT	VALUE
ACGIH, TLV	ppm	1000/ethanol
NIOSH, REL	ppm	1000/ethanol
OSHA, PEL	ppm	1000/ethanol
UN/NA class	-	3267
ECOLOGICAL PROPERTIES		
Partition coefficient, log K_{oc}	-	0.2
USE & PERFORMANCE		
Manufacturer	Momentive	
Outstanding properties	improved the flexural, compressive, and interlaminar shear strengths before and after exposure to humidity and greatly improved wet electrical properties. As a phenolic resin binder additive, it can impart moisture resistance and allow recovery after compression.	
Recommended for polymers	EP, PU	
Recommended for products	adhesives & sealants exceptional adhesion to PVC, ABS, polystyrene, and polyamide, and improvement to wet adhesion to glass and various metals.	
Recommended applications	adhesion promoter and coupling agent, used over a broad range of applications to provide superior bonds between inorganic substrates and organic polymers and for glass fiber.	

Silquest A-1102

PARAMETER	UNIT	VALUE
GENERAL INFORMATION		
Name		Silquest A-1102
CAS #		919-30-2, 13497-18-2, 64-17-5, 7664-41-7, 18784-74-2, 108-88-3
EC number	-	213-048-4
Composition		50-100% γ-aminopropyltriethoxysilane, 5-<10% bis(3-(triethoxysilyl)propyl) amine, 1-<5% ethanol, 1-<3% ammonia anhydrous, 1-<5% propane-1-amine, 3-(triethoxysilyl)-N,N-bis-3-(triethoxysilyl)-propyl-, 0.1-<1% toluene
Common syncnym		1-propanamine, 3-(triethoxysilyl)-
Empirical formula		H2NCH2CH2CH2Si(OCH2CH3)3
Formula		
Chemical class		silane
Functional organic group		amino/triethoxy
PHYSICAL PROPERTIES		
State	-	liquid
Odor	-	amine-like
Color	-	pale yellow
Initial boiling point	°C	>217
Melting point	°C	<0
Density at 25°C	kg/m³	950
Evaporation rate (butyl acetate=1)	-	<1
Solubility (diluents)		aromatic and aliphatic hydrocarbons
Solubility in water at 20°C	-	soluble
Vapor density	-	1.0
Vapor pressure at 20°C	hPa	<1.33
VOC	g/l	947.22
HEALTH & SAFETY		
HMIS classification	Flammability	1
	Health	3
	Reactivity	2
Carcinogenicity		not determined
Mutagenicity		no data available
Teratogenicity		no data available

Silquest A-1102

PARAMETER	UNIT	VALUE
DOT class		CORROSIVE LIQUID, BASIC, ORGANIC, N.O.S.(3-aminopropyltriethoxysilane) 8, II
ICAO/IATA class		CORROSIVE LIQUID, BASIC, ORGANIC, N.O.S.(3-aminopropyltriethoxysilane) 8, II
IMDG class		CORROSIVE LIQUID, BASIC, ORGANIC, N.O.S.(3-aminopropyltriethoxysilane) 8, II
Autoignition temperature	°C	270
Flash point	°C	93
Flash point method	-	PMCC
Hazardous combustion products		Carbon monoxide, carbon dioxide, nitrogen oxides
Agency rating, listed		TSCA USA, DSL Canada, AICS Australia, EINECS Europa, MITI Japan, ECL Korea, PICCS Philippines
Hazardous ingredients, labelling		Environmentally hazardous substance, liquid, n.o.s.
Hazardous products of hydrolysis		methanol
Animal testing, acute toxicity, Rat oral LD50	mg/kg	700 (male)
Animal testing, acute toxicity, Rabbit dermal LD50	mg/kg	12000 (female)
Effect of exposure, eye (human)		Causes severe eye irritation.
Effect of exposure, inhalation (human)		Irritating to respiratory system.
Effect of exposure, skin (human)		Causes severe burns.
Effect of exposure, swallowing (human)		Harmful if swallowed.
Exposure, personal protection		Safety glasses, protective clothing based on chemical resistance data, chemical-resistant gloves, general and local exhaust ventilation.
First aid, eye		Rinse immediately with plenty of water, also under the eyelids, for at least 15 minutes. Obtain medical attention without delay, preferably from an ophthalmologist.
First aid, inhalation		Move the exposed person to fresh air at once.
First aid, skin		Wash off promptly and flush contaminated skin with water. Promptly remove clothing if soaked through and flush skin with water. Wash contaminated clothing before reuse.

Silquest A-1102

PARAMETER	UNIT	VALUE
OSHA, PEL	mg/m^3	1900/ethanol
ACGIH, TLV	ppm	1000/ethanol
NIOSH, REL	ppm	1000/ethanol
OSHA, PEL	ppm	1000/ethanol
UN/NA class	-	3267
ECOLOGICAL PROPERTIES		
Aquatic toxicity, *Daphnia magna*, 48-h LC50	mg/l	331
Biodegradation probability	67%/28d	
USE & PERFORMANCE		
Manufacturer	Momentive	
Outstanding properties	improved the flexural, compressive, and interlaminar shear strengths before and after exposure to humidity and greatly improves wet electrical properties. As a phenolic resin binder additive, it can impart moisture resistance and allow recovery after compression.	
Recommended for polymers	PA, PC, PBT, EP	
Recommended applications	adhesion promoter and coupling agent, used over a broad range of applications to provide superior bonds between inorganic substrates and organic polymers (amino reactive resins) and in glass fiber	

Silquest A-1106

PARAMETER	UNIT	VALUE
GENERAL INFORMATION		
Name	Silquest A-1106	
CAS #	7732-18-5, 68400-07-7, 58160-99-9	
EC number	-	261-145-5
Composition	60-90% water, 10-30% aminopropyl-silsesquioxanes, <5% ethanol, <5% aminopropylsilanol	
Empirical formula	(H2NCH2CH2CH2SiO1.5)n	
Molecular mass	daltons	oligomer
Chemical class	silane	
Functional organic group	amino/silanol	
PHYSICAL PROPERTIES		
State	-	liquid
Odor	-	amine-like
Color	-	clear, straw
Initial boiling point	°C	>100
Melting point	°C	-1.0
Density at 25°C	kg/m³	1076
Kinematic viscosity at 25°C	cSt	4.0
Vapor pressure at 20°C	hPa	<27
VOC	g/l	9
HEALTH & SAFETY		
DOT class	not regulated	
TDG class	not regulated	
ICAO/IATA class	not regulated	
IMDG class	not regulated	
Flash point	°C	100
Flash point method	-	PMCC
Hazardous combustion products	Carbon monoxide (CO), Carbon dioxide (CO2), Nitrogen Oxides	
Agency rating, listed	TSCA USA, DSL Canada, AICS Australia, EINECS Europa, MITI Japan, ECL Korea, PICCS Philippines	
Hazardous ingredients, labelling	Environmentally hazardous substance, liquid, n.o.s.	
Animal testing, acute toxicity, Rat oral LD50	mg/kg	>2000
Animal testing, acute toxicity, Rabbit dermal LD50	mg/kg	>2000
Effect of exposure, eye (human)	May cause eye irritation.	

Silquest A-1106

PARAMETER	UNIT	VALUE
Effect of exposure, skin (human)	May cause skin irritation.	
Effect of exposure, swallowing (human)	Harmful if swallowed.	
Exposure, personal protection	Safety glasses, protective clothing based on chemical resistance data, chemical-resistant gloves, general and local exhaust ventilation.	
First aid, eye	In the event of contact with the eyes, rinse thoroughly with clean water. Get medical attention if any discomfort continues.	
First aid, inhalation	Move to fresh air. Get medical attention if symptoms persist	
First aid, skin	Wash area with soap and water. Get medical attention if symptoms occur.	
ACGIH, TLV	ppm	1000/ethanol
NIOSH, REL	ppm	1000/ethanol
OSHA, PEL	ppm	1000/ethanol
USE & PERFORMANCE		
Manufacturer	Momentive	
Outstanding properties	improved adhesion and coupling between organic resins and inorganic surfaces, easily diluted with water for safer and easier handling. Reduced VOC and HAPS emissions without sacrificing performance. Promotes bonding to many inorganic materials.	
Recommended for polymers	melamine, phenolic	
Recommended for products	waterborne sealants, coatings, adhesives, and primers	
Recommended applications	adhesion promoter used in aqueous systems where the alcohol of hydrolysis causes instability of the formulation or where a high flash point additive is required. Adheres or couples with inorganic materials, such as siliceous fillers, glass, and metal substrates.	

Silquest A-1110

PARAMETER	UNIT	VALUE
GENERAL INFORMATION		
Name	Silquest A-1110	
CAS #	13822-56-5, 67-56-1	
EC number	-	237-511-5
Composition	>98% γ-aminopropyltrimethoxysilane, <2% methanol	
Common synonym	3-aminopropyltrimethoxysilane	
Empirical formula	H2NCH2CH2CH2Si(OCH3)3	
Formula		
Molecular mass	daltons	179
Chemical class	silane	
Functional organic group	primary amine/trimethoxy	
PHYSICAL PROPERTIES		
State	-	liquid
Odor	-	amine-like
Color	-	light straw
Boiling point	°C	210
Melting point	°C	<-60
Kinematic viscosity at 25°C	cSt	2.0
Solubility in water at 25°C	hydrolytic decomposition releasing methanol	
Specific gravity at 25°C	-	1.014
HEALTH & SAFETY		
TDG class	not regulated	
Flash point	°C	82
Flash point method	-	TCC
Hazardous combustion products	Carbon monoxide (CO), Carbon dioxide (CO2), Nitrogen Oxides	
Agency rating, listed	TSCA USA, DSL Canada, AICS Australia, EINECS Europa, MITI Japan, ECL Korea, PICCS Philippines	
Hazardous ingredients, labelling	Environmentally hazardous substance, liquid, n.o.s.	
Hazardous products of hydrolysis	methanol	
Effect of exposure, eye (human)	Causes severe eye irritation.	
Effect of exposure, skin (human)	May cause skin irritation.	
Effect of exposure, swallowing (human)	Harmful if swallowed.	

Silquest A-1110

PARAMETER	UNIT	VALUE
Exposure, personal protection	Safety glasses, protective clothing based on chemical resistance data, chemical-resistant gloves, general and local exhaust ventilation.	
First aid, eye	Immediately flush eyes with plenty of water, occasionally lifting the upper and lower eyelids. Check for and remove any contact lenses. Continue to rinse for at least 10 minutes. Get medical attention if irritation occurs.	
First aid, inhalation	Remove to fresh air. If not breathing, give artificial respiration. If breathing is difficult, give oxygen. If irritation persists, obtain medical advice.	
First aid, skin	Wash off with soap and water. Remove and wash contaminated clothing before re-use. Get medical attention if irritation persists.	
NIOSH, REL	mg/m^3	ST325/methanol
OSHA, PEL	mg/m^3	260/methanol
ACGIH, TLV	ppm	200/methanol
NIOSH, REL	ppm	200/methanol
UN/NA class	-	1993
USE & PERFORMANCE		
Manufacturer	Momentive	
Outstanding properties	excellent wet and dry adhesion to inorganic substrates. Provide exceptional adhesion to many plastics (PVC, ABS), improved adhesion to glass, aluminum.	
Recommended for polymers	ABS, PA, PS, PVC, 2K PU, silicone	
Recommended for products	adhesives & sealants	
Recommended applications	adhesion promoter in silane-modified urethanes (such as SPUR + prepolymer), filled silicone, epoxy, and sealants. Silquest A-1110 silane has been utilized as an adhesion promoter in two-part urethane system, demonstrating an excellent bond to glass, metal, and many difficult plastics, such as PVC, polystyrene, and nylon. It has also been utilized on the polyol side of a two-part urethane adhesive or sealant to enhance wet adhesion.	

Silquest A-1120

PARAMETER	UNIT	VALUE
GENERAL INFORMATION		
Name		Silquest A-1120
CAS #		1760-24-3, 67-56-1, 107-15-3
EC number	-	212-164-2
General description		bifunctional
Composition		50-100% N-(3-(trimethoxysilyl)propyl) ethylenediamine, <5% methanol, <1% 1,2-ethylenediamine
Empirical formula		H2NCH2CH2NHCH2CH2CH2Si(OCH3)3
Formula		
Molecular mass	daltons	222.40
Chemical class	silane	
Mixture	-	yes
Functional organic group		diamino/trimethoxy
PHYSICAL PROPERTIES		
State	-	liquid
Odor	-	amine-like
Color		colorless to light yellow
Initial boiling point	°C	259
Melting point	°C	<-36
Density at 25°C	kg/m^3	1,030
Kinematic viscosity at 25°C	cSt	6.0
Refractive index at 25°C	-	1.448
Vapor pressure at 20°C	hPa	<1.333
VOC	g/l	198.9
HEALTH & SAFETY		
HMIS classification	Flammability	1
	Health	3
	Reactivity	2
Carcinogenicity		IARC, OSHA, NTP: no ingredient of this product present at levels greater than or equal to 0.1% is identified as probable, possible or confirmed human carcinogen
Mutagenicity		no evidence
Teratogenicity		no evidence
DOT class		not regulated

Silquest A-1120

PARAMETER	UNIT	VALUE
TDG class	not regulated	
ICAO/IATA class	not regulated	
IMDG class	not regulated	
Flash point	°C	138
Flash point method	-	PMCC
Hazardous combustion products	Carbon monoxide (CO), Carbon dioxide (CO2), Nitrogen Oxides	
Agency rating, listed	TSCA USA, DSL Canada, AICS Australia, EINECS Europa, MITI Japan, ECL Korea, PICCS Philippines	
Animal testing, acute toxicity, Rat oral LD50	mg/kg	8000
Animal testing, acute toxicity, Rat dermal LD50	mg/kg	16000
Animal testing, acute toxicity, Rat inhalation, LC50	mg/m^3	1490-2440 N-(3-(trimethoxysilyl)propyl)ethylenediamine
Effect of exposure, eye (human)	Causes serious eye damage.	
Effect of exposure, inhalation (human)	Harmful if inhaled. Causes damage to central nervous system (CNS), optic nerve, respiratory tract irritation. May cause allergy or asthma symptoms or breathing difficulties if inhaled.	
Effect of exposure, skin (human)	May cause an allergic skin reaction.	
Effect of exposure, swallowing (human)	May cause burns to mouth, throat and stomach.	
Effect of repeated or overexposure (human)	May cause damage to organs through prolonged or repeated exposure to vapor (kidneys, liver)	
Exposure, personal protection	Safety glasses, protective clothing based on chemical resistance data, chemical-resistant gloves, general and local exhaust ventilation.	
First aid, eye	Immediately rinse with water for several minutes. Remove contact lenses, if present and easy to do. Continue rinsing. Call a physician or poison control center immediately.	
First aid, inhalation	Move the exposed person to fresh air at once. If respiratory problems, artificial respiration/oxygen. Call a physician or poison control center immediately.	

Silquest A-1120

PARAMETER	UNIT	VALUE
First aid, skin		Wash off promptly and flush contaminated skin with water. Promptly remove clothing if soaked through and flush skin with water. Continue to rinse for at least 15 minutes. Get medical attention. Wash contaminated clothing before reuse. After contact with skin, remove product mechanically.
NIOSH, REL	mg/m^3	ST325/methanol, 25/1,2-ethylenediamine
OSHA, PEL	mg/m^3	260/methanol, 25/1,2-ethylenediamine
ACGIH, TLV	ppm	200/methanol, 10/1,2-ethylenediamine
NIOSH, REL	ppm	200/methanol, 10/1,2-ethylenediamine
OSHA, PEL	ppm	10/1,2-ethylenediamine
UN/NA class	-	1993
ECOLOGICAL PROPERTIES		
Aquatic toxicity, *Green algae*, 96-h EC50	mg/l	8.8
Aquatic toxicity, *Daphnia magna*, 48-h EC50	mg/l	87.4
Partition coefficient, log K$_{oc}$	-	-0.77 methanol
USE & PERFORMANCE		
Manufacturer		Momentive
Outstanding properties		excellent adhesion to inorganic substrates such as metal, glass, etc. Superior adhesion to plastics when employed in silylated polyurethane resin (SPUR) technology-based adhesives or sealants
Recommended for polymers		polysulfide, PVC plastisol, 2K PU, EP, silicones, 1K silylated PU

Silquest A-1120

PARAMETER	UNIT	VALUE
Recommended applications		adhesion promoter for polysulfide, polyvinyl chloride plastisol, silicone two-part urethanes and epoxy adhesives and sealants; adhesion promoter in one-part silylated urethane adhesives and sealants based on the Momentive Performance Materials SPUR prepolymer technology. Used as additives in phenolic and epoxy molding compounds and additives to latex coatings, adhesives and sealants
Concentrations used	wt%	0.5-1.5
Guidelines for use		use of Silquest A-1120 silane can eliminate the need for primers normally required to achieve adhesion to surfaces

Silquest A-1128

PARAMETER	UNIT	VALUE
GENERAL INFORMATION		
Name		Silquest A-1128
CAS #		42965-91-3, 67-56-1
EC number	-	256-023-3
Composition		benzylamino-silane
Formula		

PARAMETER	UNIT	VALUE
PHYSICAL PROPERTIES		
State	-	liquid
Odor	-	alcohol
Color	-	clear, brown
Boiling point	°C	>65
Melting/freezing point	°C	<-75
Evaporation rate	-	5.9
Specific gravity at 25°C	-	0.942
Vapor pressure at 20°C	hPa	64
HEALTH & SAFETY		
Flash point	°C	9
Flash point method	-	PMCC
Flammability limit - upper	%	36
Flammability limit - lower	%	6
Hazardous combustion products		Carbon monoxide, carbon dioxide, nitrogen oxides
Hazardous ingredients, labelling		Environmentally hazardous substance, liquid, n.o.s.
Effect of exposure, eye (human)		Causes severe eye irritation.
Effect of exposure, inhalation (human)		Cause respiratory tract irritation.
Effect of exposure, skin (human)		Causes skin irritation.
Exposure, personal protection		Safety glasses, protective clothing based on chemical resistance data, chemical-resistant gloves, general and local exhaust ventilation.
First aid, eye		Rinse immediately with plenty of water, also under the eyelids, for at least 15 minutes. Get medical attention.
First aid, inhalation		Move the exposed person to fresh air at once.

Silquest A-1128

PARAMETER	UNIT	VALUE
First aid, skin		Wash area with soap and water. Get medical attention. Wash contaminated clothing before reuse. Discard contaminated shoes and clothing.
UN #		1230
USE & PERFORMANCE		
Manufacturer		Momentive
Outstanding properties		improved adhesion under hot and humid conditions.
Recommended applications		coupling agent used over a broad range of applications to provide superior bonds between inorganic substrates and organic polymers. Improved substrate adhesion in adhesives, sealants, and coatings, especially under hot and humid conditions.

Silquest A-1130

PARAMETER	UNIT	VALUE
GENERAL INFORMATION		
Name		Silquest A-1130
CAS #		35141-30-1, 103526-27-8, 162339-40-4, 67-56-1, 111-40-0, 97763-30-9
EC number	-	252-390-9
Composition		50%-100% polyaminofunctional silane, 5-<10% 1,2-ethanediamine, N1-(2-aminoethyl)-N2-[3(trimethoxysilyl)propyl]-, 5-<10% 1,2-ethanediamine, N1-(2-aminoethyl)-N2-[3-(trimethoxysilyl) propyl]-, homopolymer, 1-<3% diethylenetriamine, 1-<3% methanol, 1-<3% 1,2-ethanediamine, N1-(2-aminoethyl)-N2-[3-(trimethoxysilyl)propyl]-, hydrochloride (1:1)
Formula		
Molecular mass	daltons	265.40
Chemical class	silane	
Functional organic group	amino/trimethoxy	
PHYSICAL PROPERTIES		
State	-	liquid
Odor	-	amine-like
Color	-	pale yellow
Initial boiling point	°C	>250
Melting point	°C	<0
Density at 25°C	kg/m³	1,030
Evaporation rate (butyl acetate=1)	-	<1
Solubility (diluents)	toluene, ethyl ether	
Solubility in water at 20°C	-	soluble
Vapor density	-	>1
Vapor pressure at 20°C	hPa	<1.33
VOC	g/l	1026.98
HEALTH & SAFETY		
HMIS classification	Flammability	1
	Health	2
	Reactivity	2
Carcinogenicity	no data available	

Silquest A-1130

PARAMETER	UNIT	VALUE
Mutagenicity	no data available	
Teratogenicity	no data available	
DOT class	ENVIRONMENTALLY HAZARD-OUS SUBSTANCE, LIQUID, N.O.S.(Polyaminofunctional silane) 9, III	
ICAO/IATA class	Environmentally hazardous substance liquid, N.O.S. (Polyaminofunctional silane) 9, III	
IMDG class	Environmentally hazardous substance liquid, N.O.S. (Polyaminofunctional silane) 9, III	
Flash point	°C	125
Flash point method	-	TCC
Hazardous combustion products	Carbon monoxide, carbon dioxide, nitrogen oxides	
Agency rating, listed	TSCA USA, DSL Canada, AICS Australia, MITI Japan, ECL Korea, IECSC China, CSNN Taiwan	
Hazardous ingredients, labelling	ENVIRONMENTALLY HAZARDOUS SUBSTANCE, LIQUID, N.O.S. (Polyaminofunctional silane)	
Hazardous products of hydrolysis	methanol	
Animal testing, acute toxicity, Rat oral LD50	mg/kg	2295
Animal testing, acute toxicity, Rat dermal LD50	mg/kg	>2000
Effect of exposure, eye (human)	Causes severe eye irritation.	
Effect of exposure, inhalation (human)	Cause respiratory tract irritation. May causes damage to organs: (central nervous system (CNS), optic nerve.	
Effect of exposure, skin (human)	Causes skin irritation. May cause an allergic skin reaction	
Effect of repeated or overexposure (human)	May cause damage to organs through prolonged or repeated exposure (gastrointestinal tract, kidneys, liver, respiratory tract, skin).	
Exposure, personal protection	Safety glasses, protective clothing based on chemical resistance data, chemical-resistant gloves, general and local exhaust ventilation.	
First aid, eye	Immediately flush with plenty of water for up to 15 minutes. Remove any contact lenses and open eyelids widely. If irritation persists: Continue flushing during transport to hospital. Take along these instructions.	

Silquest A-1130

PARAMETER	UNIT	VALUE
First aid, inhalation		Move the exposed person to fresh air at once. After inhalation of aerosol/mist seek medical advice immediately. Get medical attention if symptoms persist.
First aid, skin		Get medical attention if symptoms persist. Wash off promptly and flush contaminated skin with water. Promptly remove clothing if soaked through and flush skin with water.
NIOSH, REL	mg/m³	ST325/methanol
OSHA, PEL	mg/m³	260/methanol
ACGIH, TLV	ppm	200/methanol
NIOSH, REL	ppm	200/methanol
OSHA, PEL	ppm	200/methanol
UN/NA class	-	3082
ECOLOGICAL PROPERTIES		
Aquatic toxicity, *Daphnia magna*, 48-h LC50	mg/l	81
Partition coefficient, log K_{oc}	-	-0.77 methanol
USE & PERFORMANCE		
Manufacturer	Momentive	
Outstanding properties	improved chemical resistance, weatherability, rheology and durability.	
Recommended for polymers	acrylic and acrylic copolymer	
Recommended applications	coupling agent used over a broad range of applications to provide superior bonds between inorganic substrates and organic polymers.	

Silquest A-1160

PARAMETER	UNIT	VALUE
GENERAL INFORMATION		
Name		Silquest A-1160
CAS #		116912-64-2, 67-56-1
EC number	-	245-876-7
Composition		50% γ-ureidopropyltrialkoxysilane, <50% methanol
Empirical formula		H2NCONHCH2CH2CH2Si(OCH3) x(OCH2CH3)3-x
Formula		
Molecular mass	daltons	264.40
Chemical class	silane	
Functional organic group	ureido/trialkoxy	
PHYSICAL PROPERTIES		
State	-	liquid
Odor	-	alcohol
Color	-	colorless
Initial boiling point	°C	>65
Evaporation rate	-	5.9
Kinematic viscosity at 25°C	cSt	2.2
Specific gravity at 25°C	-	0.92
Vapor pressure	hPa	<130
VOC	g/l	559
HEALTH & SAFETY		
Flash point	°C	14
Flammability limit - upper	%	36
Flammability limit - lower	%	6
Hazardous combustion products		Carbon monoxide, carbon dioxide, nitrogen oxides
Animal testing, acute toxicity, Rat oral LD50	mg/kg	>2457 (male)
Animal testing, acute toxicity, Rat dermal LD50	mg/kg	>2457 (male)
USE & PERFORMANCE		
Manufacturer		Momentive

Silquest A-1160

PARAMETER	UNIT	VALUE
Outstanding properties		exhibits excellent urethane adhesive and sealant adhesion to oily metals. Exhibits chemical and corrosion resistance. Does not generate color with time and aging, improved scrub resistance and moisture-initiated crosslinking of resins.
Recommended for polymers		epoxy, phenolic, furan, and melamine resin, PA, PBT
Recommended applications		coupling agent in a wide range of applications including coatings

Silquest A-1170

PARAMETER	UNIT	VALUE
GENERAL INFORMATION		
Name		Silquest A-1170
CAS #		82985-35-1, 13822-56-5, 82984-64-3, 67-56-1
EC number	-	280-084-5
Composition		90-100% bis(trimethoxysilylpropyl) amine, 1-5% γ-minopropyltrimethoxysilane, 1-5% tris(trimethoxysilylpropyl) amine, 0.1-0.5% methanol,
Common synonym		Bis-(γ-trimethoxysilylpropyl)amine
Empirical formula		C12H31NO6Si2
Formula		
Molecular mass	daltons	341.50
RTECS number	-	TX2101000
Chemical class	silane	
Active matter	wt%	100
Purity	wt%	94
Functional organic group	diamino/di-(trimethoxy)	
PHYSICAL PROPERTIES		
State	-	liquid
Odor	-	ester-like
Color	-	clear, pale
Boiling point	°C	152 at 0.533 kPa
Density	kg/m³	1042.3
HEALTH & SAFETY		
TDG class		Environmentally hazardous substance, liquid, n.o.s. (Bis(trimethoxysilylpropyl) amine) 9, III
ICAO/IATA class		Environmentally hazardous substance, liquid, n.o.s. (Bis(trimethoxysilylpropyl) amine) 9, III
IMDG class		Environmentally hazardous substance, liquid, n.o.s. (Bis(trimethoxysilylpropyl) amine) 9, III
Flash point	°C	112.7
Flash point method	-	PMCC
Hazardous combustion products		Carbon monoxide, carbon dioxide, nitrogen oxides

Silquest A-1170

PARAMETER	UNIT	VALUE
Agency rating, listed	TSCA USA, DSL Canada, AICS Australia, EINECS Europa, MITI Japan, ECL Korea, PICCS Philippines	
Hazardous ingredients, labelling	Environmentally hazardous substance, liquid, n.o.s.	
Animal testing, acute toxicity, Rat dermal LD50	mg/kg	>2000
Animal testing, acute toxicity, Rat inhalation, LC50	mg/m³	>7.35/4H/aerosols
Effect of exposure, eye (human)	Causes serious eye damage.	
Effect of exposure, inhalation (human)	Irritating to respiratory system. May cause dizziness and drowsiness	
Effect of exposure, skin (human)	Causes severe burns.	
Effect of exposure, swallowing (human)	Harmful if swallowed. May cause eye damage and blindness if swallowed	
Effect of repeated or overexposure (human)	May cause damage to organs through prolonged or repeated exposure (heart, kidneys, liver)	
Exposure, personal protection	Safety glasses, protective clothing based on chemical resistance data, chemical-resistant gloves, general and local exhaust ventilation.	
First aid, eye	Immediately flush eyes with plenty of water, occasionally lifting the upper and lower eyelids. Check for and remove any contact lenses. Continue to rinse for at least 10 minutes. Get medical attention if irritation occurs.	
First aid, inhalation	Remove to fresh air. If not breathing, give artificial respiration. If breathing is difficult, give oxygen. If irritation persists, obtain medical advice.	
First aid, skin	Wash off with soap and water. Remove and wash contaminated clothing before re-use. Get medical attention if irritation persists.	
NIOSH, REL	mg/m³	ST325/methanol
OSHA, PEL	mg/m³	260/methanol
ACGIH, TLV	ppm	200/methanol
NIOSH, REL	ppm	200/methanol
UN/NA class	-	3082

Silquest A-1170

PARAMETER	UNIT	VALUE
USE & PERFORMANCE		
Manufacturer		Momentive
Outstanding properties		improved mechanical properties such as tensile and flexural strengths. Improved shelf stability in various resin systems. Rapid cure or adhesion build due to fast reaction with moisture. Bis-silyl functionality - generally affords greater bonding or adhesion promotion to inorganic substrates. Secondary amino functionality - can typically improve shelf stability in various resin systems.
Recommended for polymers		EP, PU, melamine, polyimide, phenolic, and furan thermosetting resins, and many thermoplastics, such as PA and polyesters
Recommended for products		coatings, adhesives, and sealants
Recommended applications		adhesion promoter between inorganic substrates and organic polymers

Silquest A-1387

PARAMETER	UNIT	VALUE
GENERAL INFORMATION		
Name		Silquest A-1387
Composition		polyazamide, 50% solution in methanol
Chemical class		silane
Mixture	-	yes
Functional organic group		amino
PHYSICAL PROPERTIES		
State	-	liquid
Color	-	clear, amber
Initial boiling point	°C	>65
Melting/freezing point	°C	<0
Density at 25°C	kg/m³	969
Evaporation rate (n-butylacetate=1)	-	5.9
Vapor pressure	hPa	64
HEALTH & SAFETY		
Flash point	°C	8
Flash point method	-	TCC
Flammability limit - upper	%	36
Flammability limit - lower	%	6
Animal testing, acute toxicity, Rat oral LD50	mg/kg	16600
Animal testing, acute toxicity, Rabbit dermal LD50	mg/kg	>15500
Effect of exposure, inhalation (human)		Respiratory tract irritation (CNS, optic nerves)/methanol
Effect of exposure, swallowing (human)		Harmful if swallowed.
Effect of repeated or overexposure (human)		May cause damage to organs through prolonged or repeated exposure to methanol (kidneys, liver)
Exposure, personal protection		Safety glasses, protective clothing based on chemical resistance data, chemical-resistant gloves, general and local exhaust ventilation.
First aid, eye		Rinse immediately with plenty of water, also under the eyelids, for at least 15 minutes. Get medical attention.
First aid, inhalation		Move the exposed person to fresh air at once. If respiratory problems, artificial respiration/oxygen.

Silquest A-1387

PARAMETER	UNIT	VALUE
First aid, skin		Wash area with soap and water. Get medical attention. Wash contaminated clothing before reuse. Discard contaminated shoes and clothing.
USE & PERFORMANCE		
Manufacturer		Momentive
Recommended applications		adhesion promoter

Silquest A-1524

PARAMETER	UNIT	VALUE
GENERAL INFORMATION		
Name		Silquest A-1524
CAS #	-	23843-64-3
EC number	-	245-904-8
General description		bifunctional
Composition		100% γ-ureidopropyltrimethoxysilane
Common synonym		1-[3-(trimethoxysilyl)propyl]urea
Empirical formula		H2NCONHCH2CH2CH2Si(OCH3)3
Formula		
Molecular mass	daltons	222.40
Chemical class	silane	
Active matter	wt%	100
Functional organic group	ureido/trimethoxy	
PHYSICAL PROPERTIES		
State	-	viscous liquid
Odor	-	amine-like
Color	-	pale yellow
Initial boiling point	°C	217-250
Melting/freezing point	°C	<-5
Density	kg/m³	1150
Refractive index at 25°C	-	1.459
Solubility (diluents)		acetone, toluene, methanol, ethanol (reacts with alcoholic solvents)
Solubility in water at 20°C		hydrolytic decomposition, water solution 1.00 wt% stable for 72H
Vapor pressure at 20°C	hPa	<1.33
Vapor density	-	>1
VOV	g/l	172
HEALTH & SAFETY		
Flash point	°C	99
Flash point method	-	TCC
Hazardous combustion products		Carbon monoxide, carbon dioxide, nitrogen oxides
Hazardous products of hydrolysis		methanol
USE & PERFORMANCE		
Manufacturer		Momentive

Silquest A-1524

PARAMETER	UNIT	VALUE
Outstanding properties		excellent urethane adhesive and sealant adhesion to oily metals, contains no flammable or combustible solvent, low VOC emissions, good water solubility, longer pot life (stability), does not generate color with time and aging. Rapid attainment of bond strength.
Recommended for polymers		phenolic, urea-melamine, EP resins, PA, PU
Recommended for products		insulation, woven glass fabrics, filler treatment, foundry sand, coatings, sealants and caulks, primers
Recommended applications		adhesion promoter between a wide range of resins and substrates, fillers, or reinforcements. Contains no alcoholic solvent diluent, making it suitable for use in reactive polymer systems such as isocyanate-terminated polyurethane polymers. Recommended when inorganic surfaces such as fiberglass, particulate fillers, or metals are combined or over-coated with phenolic, urea-melamine, epoxy resins, polyamide and/or polyurethane polymers. The silane improves adhesion between the inorganic filler or fiber.
Concentrations used	wt%	0.-1

Silquest A-1871

PARAMETER	UNIT	VALUE
GENERAL INFORMATION		
Name	Silquest A-1871	
CAS #	-	2602-34-8
Composition	>97% [3-(2,3-epoxypropoxy)propyl] triethoxysilane	
Common synonym	γ-glycidoxypropyltriethoxysilane	
Formula		

PARAMETER	UNIT	VALUE
Molecular mass	daltons	278.42
Chemical class	silane	
Active matter	wt%	>97.00
Functional organic group	epoxy and alkoxy	
PHYSICAL PROPERTIES		
State	-	liquid
Odor	-	ester-like
Color	-	clear, pale
Boiling point	°C	304
Melting point	°C	<0.0
Evaporation rate (butyl acetate=1)	-	>1
Refractive index at 25°C	-	1.426
Solubility (diluents)	methyl acetate, acetone, alcohol	
Solubility in water at 20°C	hydrolytic decomposition releasing ethanol	
Specific gravity at 25°C	-	1.0034
Vapor density	-	>1
HEALTH & SAFETY		
Flash point	°C	118
Flash point method	-	TCC
Hazardous products of hydrolysis	ethanol	
Effect of repeated or overexposure (human)	Repeated exposure to ethanol may aggravate liver injury produced from other causes.	
Exposure, personal protection	Safety glasses, protective clothing based on chemical resistance data, chemical-resistant gloves, general and local exhaust ventilation.	

Silquest A-1871

PARAMETER	UNIT	VALUE
First aid, eye		Immediately flush with plenty of water for at least 15 minutes. If easy to do, remove contact lenses. Call a physician or poison control center immediately. In case of irritation from airborne exposure, move to fresh air. Get medical attention if symptoms persist
First aid, inhalation		Remove to fresh air. If not breathing, give artificial respiration. If breathing is difficult, give oxygen. If irritation persists, obtain medical advice.
First aid, skin		Immediately flush with plenty of water for at least 15 minutes while removing contaminated clothing and shoes. Wash contaminated clothing before reuse. Immediately call a POISON CENTER or physician
USE & PERFORMANCE		
Manufacturer		Momentive
Outstanding properties		improved wet and dry adhesion without the potential for yellowing. Provides outstanding chemical resistance and mechanical properties to coatings, adhesives or sealants. Improves shelf stability over aminosilane alternatives. Enhances electrical properties of epoxy based electronic encapsulants and packaging materials.
Recommended for polymers		acrylic & acrylic copolymers, epoxies, PU, polysulfide, styrene copolymers
Recommended for products		electronic encapsulants and packaging materials, building and construction sealants
Recommended applications		used as an adhesion promoter in polysulfide, urethane, epoxy, and acrylic caulks, sealants. Improves substrate adhesion in many waterborne acrylic and vinyl-acrylic caulks, adhesives, and coatings. Enhances adhesion to glass and metal substrates in polyurethane resins. Improves wet and dry adhesion, tensile strength and water resistance of quartz-filled epoxy encapsulants, sand-filled epoxy concrete repair materials, and metal-filled epoxy resins for mold die tools.

Silquest A-1891

PARAMETER	UNIT	VALUE
GENERAL INFORMATION		
Name	Silquest A-1891	
CAS #	-	14814-09-6
EC number	-	238-883-1
General description	>98% gamma-mercaptopropyl-triethoxysilane	
Empirical formula	HSCH2CH2CH2Si(OCH2CH3)3	
Formula		
Molecular mass	daltons	238.40
RTECS number	-	TZ7760000
Chemical class	silane	
Active matter	wt%	>98
Functional organic group	mercapto	
PHYSICAL PROPERTIES		
State	-	liquid
Odor	-	ester-like
Color	-	clear, colorless
Initial boiling point	°C	240
Melting/freezing point	°C	<0
Density at 25°C	kg/m^3	990
Refractive index at 25°C	-	1.433-1438
Solubility (diluents)	soluble in a variety of organic solvents	
Solubility in water at 20°C	hydrolytic decomposition	
Vapor pressure	hPa	<1.33
VOC	g/l	987
HEALTH & SAFETY		
Autoignition temperature	°C	200
Flash point	°C	88
Flash point method	-	TCC
Hazardous products of hydrolysis	ethanol	
Effect of exposure, eye (human)	May cause eye irritation.	
Effect of exposure, inhalation (human)	May cause respiratory tract irritation.	
Effect of exposure, skin (human)	Causes skin irritation	
Effect of repeated or overexposure (human)	Repeated exposure to ethanol may aggravate liver injury produced from other causes.	

Silquest A-1891

PARAMETER	UNIT	VALUE
Exposure, personal protection		Safety glasses, protective clothing based on chemical resistance data, chemical-resistant gloves, general and local exhaust ventilation.
First aid, eye		Rinse immediately with plenty of water, also under the eyelids, for at least 15 minutes. Get medical attention.
First aid, inhalation		Move the exposed person to fresh air at once. If respiratory problems, artificial respiration/oxygen. Call a physician or poison control center immediately.
First aid, skin		Wash off promptly and flush contaminated skin with water. Promptly remove clothing if soaked through and flush skin with water. Wash contaminated clothing before reuse. Get medical attention.
Animal testing, acute toxicity, Rat oral LD50	mg/kg	6108 (male)
NIOSH, REL	mg/m^3	1900/ethanol
OSHA, PEL	mg/m^3	1900/ethanol
ACGIH, TLV	ppm	1000/ethanol
NIOSH, REL	ppm	1000/ethanol
OSHA, PEL	ppm	1000/ethanol
UN risk phrases, R	R36/R37/R38	
US safety phrases, S	S26,S36	
ECOLOGICAL PROPERTIES		
Aquatic toxicity, *Daphnia magna*, 48-h LC50	mg/l	6.7
Partition coefficient, log K_{oc}	-	methanol -0.77
USE & PERFORMANCE		
Manufacturer	Momentive	
Outstanding properties	improved in mineral-filled elastomer properties, including modulus, tensile and tear strength, heat buildup, abrasion resistance, resilience, compression set and cure time. Increased adhesion to resins and other polymers	
Recommended for polymers	synthetic rubber (SBR, NBR)	
Recommended for products	tire	
Recommended applications	coupling agent for fillers, especially for fumed silica treatment. Acts as a coupling agent for surface treatment for fillers (glass fiber, kaolin silica, aluminum, etc.)	

Silquest A-2120

PARAMETER	UNIT	VALUE
GENERAL INFORMATION		
Name	Silquest A-2120	
CAS #	3069-29-2, 67-56-1	
EC number	-	221-336-6
General description	bifunctional	
Composition	>99% N-[3-(dimethoxymethylsilyl)propyl] ethylenediamine, <0.2% methanol	
Empirical formula	H2NCH2CH2NHCH2CH2CH2SiCH3(OCH3)2	
Formula		
Molecular mass	daltons	206.40
Chemical class	silane	
Functional organic group	amino/diamino/dimethoxy	
PHYSICAL PROPERTIES		
State	-	liquid
Odor	-	amine-like
Color	colorless to light yellow	
Initial boiling point	°C	85
Melting/freezing point	°C	<0
Density at 25°C	kg/m³	980
Vapor pressure at 20°C	hPa	1.333
HEALTH & SAFETY		
HMIS classification	Flammability	1
	Health	3
	Reactivity	2
DOT class	not regulated	
TDG class	not regulated	
ICAO/IATA class	not regulated	
IMDG class	not regulated	
Flash point	°C	93.4
Flash point method	-	PMCC
Hazardous combustion products	Carbon monoxide, carbon dioxide, nitrogen oxides	
Agency rating, listed	TSCA USA, DSL Canada, AICS Australia, EINECS Europa, MITI Japan, ECL Korea, PICCS Philippines	
Effect of exposure, eye (human)	Causes serious eye damage.	

Silquest A-2120

PARAMETER	UNIT	VALUE
Effect of exposure, inhalation (human)	Harmful if inhaled. Causes damage to central nervous system (CNS), optic nerve, respiratory tract irritation. May cause allergy or asthma symptoms or breathing difficulties if inhaled.	
Effect of expcsure, swallowing (human)	May cause burns to mouth, throat and stomach.	
Effect of repeated or overexposure (human)	May cause damage to organs through prolonged or repeated exposure to vapor (kidneys, liver)	
Exposure, personal protection	Safety glasses, protective clothing based on chemical resistance data, chemical-resistant gloves, general and local exhaust ventilation.	
First aid, eye	Rinse immediately with plenty of water, also under the eyelids, for at least 15 minutes. Get medical attention.	
First aid, inhalation	Move the exposed person to fresh air at once. If respiratory problems, artificial respiration/oxygen. Call a physician or poison control center immediately.	
First aid, skin	Wash off promptly and flush contaminated skin with water. Promptly remove clothing if soaked through and flush skin with water. Get medical attention. Wash contaminated clothing before reuse.	
Animal testing, acute toxicity, Rat oral LD50	mg/kg	>2000
Animal testing, acute toxicity, Rabbit dermal LD50	mg/kg	>2000
NIOSH, REL	mg/m^3	ST325/methanol
ACGIH, TLV	ppm	200/methanol
NIOSH, REL	ppm	200/methanol
ECOLOGICAL PROPERTIES		
Aquatic toxicity, *Daphnia magna*, 48-h LC50	mg/l	87.4
USE & PERFORMANCE		
Manufacturer	Momentive	
Outstanding properties	excellent adhesion to inorganic substrates such as metal, glass, etc. Superior adhesion to plastics when employed in silylated polyurethane resin (SPUR) technology-based adhesives or sealants	
Recommended for polymers	polysulfide, PVC plastisol, 2K PU, EP, silicones, 1K silylated PU	

Silquest A-2120

PARAMETER	UNIT	VALUE
Recommended applications		adhesion promoter in polysulfide, polyvinyl chloride plastisol, silicone two-part urethanes, and epoxy adhesives and sealants and adhesion promoters in one-part silylated urethane adhesives and sealants based on the Momentive Performance Materials SPUR prepolymer technology. Used as additives in phenolic and epoxy molding compounds and additives to latex coatings, adhesives, and sealants.
Concentrations used	wt%	0.50-1.50
Guidelines for use		use of Silquest A-2120 silane can eliminate the need for primers normally required to achieve adhesion to surfaces.

Silquest A-2387 silylated polyazamide

PARAMETER	UNIT	VALUE
GENERAL INFORMATION		
Name		Silquest A-2387 silylated polyazamide
Composition		silylated polyazamide silane in ethanol
Active mater al content	%	50 in ethanol
Functionality	-	amino and amido
PHYSICAL PROPERTIES		
State	-	liquid
Odor	-	alcohol
Color	-	clear, amber
Color, Pt/Co	-	100
Initial boiling point	°C	78.1
Density at 25°C	kg/m³	949
pH	-	11-12
Solvents	water, alcohol	
Vapor pressure at 20°C	hPa	<127.68
HEALTH & SAFETY		
HMIS classification	Flammability	3
	Health	2
	Physical hazard	1
Flash point	°C	14
Flash point method	-	PMCC
Animal testing, acute toxicity, Rat inhalation LC50	mg/l	38.3
First aid, eye		Rinse immediately with plenty of water, also under the eyelids, for at least 15 minutes. Get medical attention.
First aid, inhalation		Move the exposed person to fresh air at once. If respiratory problems, artificial respiration/oxygen. Call a physician or poison control center immediately
First aid, skin		Wash off promptly and flush contaminated skin with water. Promptly remove clothing if soaked through and flush skin with water. Wash contaminated clothing before reuse. Get medical attention.
UN #	-	1170
USE & PERFORMANCE		
Manufacturer		Momentive

Silquest A-2387 silylated polyazamide

PARAMETER	UNIT	VALUE
Outstanding features		excellent reactivity and compatibility with a wide range of thermosetting and thermoplastic resins. Due to its distinct molecular structure, Silquest A-2387 silane can perform several functions essential to the performance of glass fiber and other reinforcing materials and can also help improve fiberglass processing via reduced fuzz and enhanced chopping efficiency.
Recommended for polymers		epoxy, phenolic, melamine, and furan thermosets, as well as polyamide, polyester, polypropylene, and polyurethane thermoplastic polymers
Recommended for products		composites

Silquest A-Link 15

PARAMETER	UNIT	VALUE
GENERAL INFORMATION		
Name	Silquest A-Link 15 Silane	
CAS #	-	227085-51-0
Composition	50-100% n-ethyl-gamma-aminoisobutyl trimethoxysilane	
Common synonym	N-ethyl-3-trimethoxysilyl-2-methyl-propanamine	
Formula		
Chemical class	silane	
Functional organic group	diamino/trimethoxy	
PHYSICAL PROPERTIES		
State	-	liquid
Odor	-	amine-like
Color	-	colorless
Boiling point	°C	216.6
Melting point	°C	<-70
Density at 25°C	kg/m³	954
Evaporation rate (butyl acetate=1)	-	<1
Solubility in water at 20°C	hydrolytic decomposition releasing methanol	
Vapor density	-	>1
Vapor pressure at 20°C	hPa	14.4
VOC	g/l	984
HEALTH & SAFETY		
HMIS classification	Flammability	2
	Health	3
	Reactivity	1
Carcinogenicity	not determined	
Mutagenicity	negative	
Teratogenicity	no data available	
DOT class	Combustible liquid, n.o.s.(1-Propanamine, N-ethyl-2-methyl-3(trimethoxysilyl)-) III	
ICAO/IATA class	not regulated	
IMDG class	not regulated	
Flash point	°C	74.8

Silquest A-Link 15

PARAMETER	UNIT	VALUE
Flash point method	-	PMCC
Hazardous combustion products	Carbon monoxide, carbon dioxide, nitrogen oxides	
Agency rating, listed	TSCA USA, DSL Canada, AICS Australia, MITI Japan, ECL Korea, IECSC China, CSNN Taiwan	
Hazardous ingredients, labelling	Combustible liquid, n.o.s.(1-Propanamine, N-ethyl-2-methyl-3(trimethoxysilyl)-)	
Hazardous products of hydrolysis	methanol	
Animal testing, acute toxicity, Rat oral LD50	mg/kg	>2000
Animal testing, acute toxicity, Rabbit dermal LD50	mg/kg	slightly irritating to the skin.
Animal testing, acute toxicity, Rat dermal LD50	mg/kg	>2000
Effect of exposure, eye (human)	Causes severe eye irritation.	
Effect of exposure, swallowing (human)	Harmful if swallowed	
Exposure, personal protection	Safety glasses, protective clothing based on chemical resistance data, chemical-resistant gloves, general and local exhaust ventilation.	
First aid, eye	Immediately flush with plenty of water for at least 15 minutes. If easy to do, remove contact lenses. Obtain medical attention without delay, preferably from an ophthalmologist.	
First aid, inhalation	Move the exposed person to fresh air at once. If respiratory problems, artificial respiration/oxygen. Call a physician or poison control center immediately.	
First aid, skin	Wash area with soap and water. Get medical attention. Wash contaminated clothing before reuse. Discard contaminated shoes and clothing.	
UN/NA class	-	1993
ECOLOGICAL PROPERTIES		
Aquatic toxicity, *Green algae*, 96-h EC50	mg/l	100/72H
Aquatic toxicity, *Zebra fish*, 96-h LC50	mg/l	100

Silquest A-Link 15

PARAMETER	UNIT	VALUE
USE & PERFORMANCE		
Manufacturer		Momentive
Outstanding properties		excellent thermal, chemical, and UV stable performance, excellent wet adhesion to glass, metal and other inorganic substrates. Excellent durability and joint movement. Good tensile strength and elasticity. Provides a controlled reaction with -NCO, -COOH, and epoxy-functional materials, providing a moisture cure silane crosslink mechanism.
Recommended for polymers		PVC, polystyrene, PA, ABS, acrylics
Recommended applications		adhesion promoter in epoxy, silicone, hybrid, polyurethane, adhesives, and sealants. Excellent primerless adhesion to glass, aluminum, and other metals. Unusually good adhesion to plastics such as PVC, polystyrene, Nylon, ABS, acrylics, when an amino silane such as Silquest A-1110 silane is utilized as an adhesion promoter in conjunction with the silylated resin. It may be used as a crosslinker for an isocyanate functional prepolymer system.

Silquest A-Link 25

PARAMETER	UNIT	VALUE
GENERAL INFORMATION		
Name	Silquest A-Link 25	
CAS #	-	24801-88-5
EC number	-	246-467-5
Composition	3-isocyanatopropyltriethoxysilane	
Common synonym	gamma-Isocyanatopropyltriethoxysilane	
Empirical formula	OCNCH2CH2CH2Si(OCH2CH3)3	
Formula		
Molecular mass	daltons	247
RTECS number	-	VV6691000
Chemical class	silane	
Active matter	wt%	>95
Functional organic group	isocyanate	
PHYSICAL PROPERTIES		
State	-	liquid
Odor	-	ester-like
Color	-	colorless
Initial boiling point	°C	238
Melting point	°C	<0
Solubility in water at 20°C	hydrolytic decomposition releasing methanol	
Specific gravity at 25°C	-	0.999
Vapor density	-	>1
Vapor pressure at 20°C	hPa	<1.33
VOC	g/l	0.05
HEALTH & SAFETY		
HMIS classification	Flammability	1
	Health	3
	Reactivity	2
Carcinogenicity	not determined	
Mutagenicity	not determined	
DOT class	TOXIC LIQUID, CORROSIVE, ORGANIC, N.O.S.(3-(triethoxysilyl)propyl isocyanate) 6.1, III	

Silquest A-Link 25

PARAMETER	UNIT	VALUE
ICAO/IATA class	TOXIC LIQUID, CORROSIVE, ORGANIC, N.O.S.(3-(triethoxysilyl)propyl isocyanate) 6.1, III	
IMDG class	TOXIC LIQUID, CORROSIVE, ORGANIC, N.O.S.(3-(triethoxysilyl)propyl isocyanate) 6.1, III	
Flash point	°C	77
Flash point method	-	PMCC
Hazardous combustion products	Carbon monoxide (CO), Carbon dioxide (CO2), Nitrogen Oxides	
Agency rating, listed	TSCA USA, DSL Canada, AICS Australia, MITI Japan, EINECS Europe, ECL Korea, PICCS Philippines, Taiwan CSNN	
Hazardous products of hydrolysis	methanol	
Animal testing, acute toxicity, Rat oral, LD50	mg/kg	700
Animal testing, acute toxicity, Rabbit, dermal LD50	mg/kg	1250
Animal testing, acute toxicity, Rat inhalation, LC50	mg/m³	360/4H
Effect of exposure, eye (human)	Causes severe eye damage.	
Effect of exposure, inhalation (human)	Toxic by inhalation. May cause sensitization by inhalation.	
Effect of exposure, skin (human)	Harmful in contact with skin.	
Effect of exposure, swallowing (human)	Harmful if swallowed.	
Effect of repeated or overexposure (human)	May cause damage to organs through prolonged or repeated exposure to methanol (kidneys, liver)	
Exposure, personal protection	Safety glasses, protective clothing based on chemical resistance data, chemical-resistant gloves, general and local exhaust ventilation.	
First aid, eye	Immediately flush with plenty of water for at least 15 minutes. If easy to do, remove contact lenses. Obtain medical attention without delay, preferably from an ophthalmologist.	
First aid, inhalation	Move the exposed person to fresh air at once. For breathing difficulties, oxygen may be necessary. Call a physician or poison control center immediately.	

Silquest A-Link 25

PARAMETER	UNIT	VALUE
First aid, skin		Immediately remove contaminated clothing. Wash area with soap and water. Wash contaminated clothing before reuse.
UN risk phrases, R		R21/22R26R34,R42
US safety phrases, S		S23,S26,S28,S36/37/39,S45
UN/NA class	-	3390
ECOLOGICAL PROPERTIES		
Aquatic toxicity, *Green algae*, 72-h EC50	mg/l	>1000
Aquatic toxicity, *Daphnia magna*, 21d LC50	mg/l	>100
USE & PERFORMANCE		
Manufacturer		Momentive
Outstanding properties		superior wet adhesion to glass, metal, and other inorganic substrates, excellent thermal, chemical, and UV stability. Slower hydrolysis in the presence of atmospheric moisture for applications requiring greater open time or enhanced self-time. Reactive with -OH, -NH$_2$, and -SH functional polyols and polymers, providing a moisture-cure silane cross-link mechanism
Recommended applications		adhesion promoter for silicone sealants or coatings, improved adhesion to organic substrates on which active hydrogen are present, such as nylon. Adhesion promoter for one-part moisture curable and two-part reactive urethane system.

Silquest A-Link 35

PARAMETER	UNIT	VALUE
GENERAL INFORMATION		
Name	Silquest A-Link 35	
CAS #	-	15396-00-6
EC number	-	239-415-9
Composition	50-100% 3-isocyanatopropyltrime-thoxysilane, 1-10% organofunctional silane, 1-5% organoalkoxysilane	
Empirical formula	OCNCH2CH2CH2Si(OCH3)3	
Formula		
Molecular mass	daltons	205
Chemical class	silane	
Mixture	-	yes
Active matter	wt%	>95
Functional organic group	isocyanate	
PHYSICAL PROPERTIES		
State	-	liquid
Odor	-	faint
Color	-	colorless
Boiling point	°C	>200
Melting point	°C	<0
Density at 25°C	kg/m^3	1,073
Evaporation rate (butyl acetate=1)	-	<1
Solubility in water at 20°C	hydrolytic decomposition releasing methanol	
Vapor density	-	>1
Vapor pressure at 20°C	hPa	1.33
VOC	g/l	2.09
HEALTH & SAFETY		
HMIS classification	Flammability	1
	Health	3
	Reactivity	2
Carcinogenicity	not determined	
Mutagenicity	not determined	
Teratogenicity	no known significant effects or critical hazards	

Silquest A-Link 35

PARAMETER	UNIT	VALUE
DOT class		TOXIC LIQUID, CORROSIVE, ORGANIC, N.O.S.(3-(trimethoxysilyl)propyl isocyanate) 6.1, III
ICAO/IATA class		TOXIC LIQUID, CORROSIVE, ORGANIC, N.O.S.(3-(trimethoxysilyl)propyl isocyanate) 6.1, III
IMDG class		TOXIC LIQUID, CORROSIVE, ORGANIC, N.O.S.(3-(trimethoxysilyl)propyl isocyanate) 6.1, III
Flash point	°C	99
Flash point method	-	PMCC
Hazardous combustion products		Carbon monoxide (CO), Carbon dioxide (CO_2), Nitrogen Oxides
Agency rating, listed		TSCA USA, DSL Canada, AICS Australia, MITI Japan, EINECS Europe, ECL Korea, PICCS Philippines, Taiwan CSNN
Hazardous products of hydrolysis		methanol
Animal testing, acute toxicity, Rat oral LD50	mg/kg	878
Animal testing, acute toxicity, Rat dermal LD50	mg/kg	1190
Animal testing, acute toxicity, Rat inhalation, LC50	mg/m^3	130/vapor
Effect of exposure, eye (human)		Causes severe eye damage.
Effect of exposure, inhalation (human)		Fatal if inhaled. May cause allergy or asthma symptoms or breathing difficulties if inhaled.
Effect of exposure, skin (human)		Harmful in contact with skin. Causes severe skin burns. May cause an allergic skin reaction.
Effect of exposure, swallowing (human)		Harmful if swallowed.
Effect of repeated or overexposure (human)		May cause damage to organs through prolonged or repeated exposure to methanol (kidneys, liver)
Exposure, personal protection		Safety glasses, protective clothing based on chemical resistance data, chemical-resistant gloves, general and local exhaust ventilation.
First aid, eye		Important! Immediately rinse with water for at least 15 minutes. Call a physician or poison control center immediately.

Silquest A-Link 35

PARAMETER	UNIT	VALUE
First aid, inhalation		Move the exposed person to fresh air at once. When breathing is difficult, properly trained personnel may assist affected person by administering 100% oxygen. Consult a physician for specific advice.
First aid, skin		Wash off promptly and flush contaminated skin with water. Promptly remove clothing if soaked through and flush skin with water. Call a physician or poison control center immediately. Wash contaminated clothing before reuse.
UN risk phrases, R		R21/22R26R34,R42
US safety phrases, S		S23,S26,S28,S36/37/39,S45
UN/NA class	-	2927
ECOLOGICAL PROPERTIES		
Aquatic toxicity, *Green algae*, 72-h EC50	mg/l	>1000
Aquatic toxicity, *Daphnia magna*, 48-h EC50	mg/l	>100
USE & PERFORMANCE		
Manufacturer	Momentive	
Outstanding properties		superior wet adhesion to glass, metal, and other inorganic substrates, excellent thermal, chemical, and UV stability. Fast hydrolysis in the presence of atmospheric moisture.
Recommended applications		adhesion promoter for silicone sealants or coatings, improved adhesion to organic substrates on which active hydrogen is present, such as nylon. Adhesion promoter for one-part moisture curable and two-part reactive urethane system.

Silquest A-Link 597

PARAMETER	UNIT	VALUE
GENERAL INFORMATION		
Name		Silquest A-Link 597
CAS #		26115-70-8, 15396-00-6, 67-56-1
EC number	-	247-465-8
Composition		80% tris[(3-methoxysilyl)propyl]isocyanurate, 0.1-1% Isocyanatopropyltrimethoxysilane, 1-<3% methanol
Common synonym		1,3,5-triazine-2,4,6(1H,3H,5H)-trione, 1,3,5-tris[3-(trimethoxysilyl)propyl]-
Empirical formula		C21H45N3O12Si3
Formula		
Molecular mass	daltons	606.00
RTECS number	-	XZ2025000
Mixture	-	yes
PHYSICAL PROPERTIES		
State	-	viscous liquid
Odor	-	ester-like
Color	-	clear, light yellow
Boiling point at 0.05 mm Hg	°C	230
Melting point	°C	<-50
Density at 25°C	kg/m^3	1,173
Evaporation rate (butyl acetate=1)	-	<1
Kinematic viscosity at 25°C	cSt	95
Solubility in water at 20°C	hydrolytic decomposition releasing methanol	
Vapor density	-	>1
Vapor pressure at 20°C	hPa	0.0011
VOC	g/l	97
HEALTH & SAFETY		
HMIS classification	Flammability	1
	Health	2
	Reactivity	1
Carcinogenicity	not determined	
Mutagenicity	not determined	
Teratogenicity	suspected of damaging the unborn child	
DOT class	no data available	

Silquest A-Link 597

PARAMETER	UNIT	VALUE
TDG class	no data available	
Flash point	°C	102/248
Flash point method	-	PMCC/CC
Hazardous combustion products	Carbon monoxide (CO), Carbon dioxide, Nitrogen Oxides	
Agency rating, listed	TSCA USA, DSL Canada, AICS Australia, MITI Japan, EINECS Europe, ECL Korea, PICCS Philippines, Taiwan CSNN	
Hazardous products of hydrolysis	methanol	
Animal testing, acute toxicity, Rat oral LD50	mg/kg	1713
Animal testing, acute toxicity, Rat dermal LD50	mg/kg	19200
Effect of exposure, eye (human)	Non-irritating to the eyes/rabbit	
Effect of exposure, inhalation (human)	Exposure to decomposition products may cause a health hazard. May cause allergy or asthma symptoms or breathing difficulties if inhaled.	
Effect of exposure, skin (human)	May cause an allergic skin reaction.	
Effect of exposure, swallowing (human)	Harmful if swallowed.	
Effect of repeated or overexposure (human)	May cause damage to organs through prolonged or repeated exposure to methanol (kidneys, liver)	
Exposure, personal protection	Safety glasses, protective clothing based on chemical resistance data, chemical-resistant gloves, general and local exhaust ventilation.	
First aid, eye	Immediately flush with plenty of water for at least 15 minutes. If easy to do, remove contact lenses. Get medical attention if symptoms persist.	
First aid, inhalation	If inhaled, remove to fresh air. If not breathing, give artificial respiration. If breathing is difficult, trained personnel should give oxygen. Get medical attention immediately.	
First aid, skin	Immediately flush with plenty of water for at least 15 minutes while removing contaminated clothing and shoes. Get medical attention if symptoms persist. Wash contaminated clothing before reuse. Destroy or thoroughly clean contaminated shoes.	
NIOSH, REL	mg/m^3	ST325/methanol
OSHA, PEL	mg/m^3	260/methanol

Silquest A-Link 597

PARAMETER	UNIT	VALUE
ACGIH, TLV	ppm	200/methanol
NIOSH, REL	ppm	200/methanol
OSHA, PEL	ppm	200/methanol
UN risk phrases, R	R36/R37/R38	
ECOLOGICAL PROPERTIES		
Partition coefficient, log K_{oc}	-	-0.77 methanol
USE & PERFORMANCE		
Manufacturer	Momentive	
Outstanding properties	excellent thermal stability at 200°C, high boiling point, low volatility for better silane retention in hot melt adhesive application conditions, compatible with most hot melt resins. The polar structure provides good solubility in most resins, good wetting of most substrates, imparts thermal resistance and low volatility to the adhesion promoter. Improved adhesion to difficult substrates such as plastics, glass, and metals, including aluminum and steel.	
Recommended for polymers	EVA, PA, polyesters, PU (SPUR+)	
Recommended for products	hot melt adhesives, polyamide and poly-ester hot melt adhesives, sealants	
Recommended applications	adhesion promoter for a variety of sub-strate/matrix resin combinations. It is es-pecially useful in high-performance hot melt adhesives and other applications where exposure to sustained tempera-tures might occur. High boiling point, low volatility for better silane retention in hot melt adhesive application conditions	

Silquest A-Link 599

PARAMETER	UNIT	VALUE
GENERAL INFORMATION		
Name	Silquest A-Link 599	
CAS #	-	220727-26-4
EC number	-	436-690-9
General description	3-octanoylthio-1-propyltriethoxysilane	
Empirical formula	CH3(CH2)6C(=O) SCH2CH2CH2Si(OCH2CH3)3	
Formula		
Molecular mass	daltons	364.60
Chemical class	silane	
Functional organic group	tricarboxylate	
PHYSICAL PROPERTIES		
State	-	liquid
Color	-	clear, pale yellow
Initial boiling point	°C	>400
Melting/freezing point	°C	<-70
Specific gravity at 25°C	-	0.9686
Vapor pressure	hPa	<0.133
HEALTH & SAFETY		
Flash point	°C	110
Flash point method	-	PMCC
Animal testing, acute toxicity, Rat oral LD50	mg/kg	>2000
Exposure, personal protection	Safety glasses, protective clothing based on chemical resistance data, chemical-resistant gloves, general and local exhaust ventilation.	
First aid, eye	In case of contact, immediately flush eyes with plenty of water for at least 15 minutes. Get medical attention if symptoms persist.	
First aid, inhalation	Move the exposed person to fresh air at once.	
First aid, skin	Take off immediately all contaminated clothing. Wash with soap and water. Contact physician if irritation continues. Wash contaminated clothing before reuse.	

Silquest A-Link 599

PARAMETER	UNIT	VALUE
ECOLOGICAL PROPERTIES		
Aquatic toxicity, *Daphnia magna*, 72-h LC50	mg/l	>100
USE & PERFORMANCE		
Manufacturer	Momentive	
Outstanding properties	excellent adhesion with a variety of substrate/resin combinations. Low volatility and low mercapto reactivity until the thiocarboxylate is activated during application of the adhesive or sealant, eliminates the usual mercapto odor during formulation, storage, and initial application of a sealant.	
Recommended for polymers	PU, silyl-modified polymers	
Recommended applications	adhesion promoter for use in a broad range of adhesives and sealants applications	

Silquest G-170

PARAMETER	UNIT	VALUE
GENERAL INFORMATION		
Name		Silquest G-170
Composition		organalkoxysilane
PHYSICAL PROPERTIES		
State	-	liquid
Color	-	clear, light straw
Initial boiling point	°C	216.7
Melting point	°C	<-71
Density at 25°C	kg/m³	1074
Solubility		toluene, xylene, acetone and many alcohols
Vapor pressure at 25°C	Pa	4.6
HEALTH & SAFETY		
HMIS classification	Flammability	1
	Health	1
	Physical hazard	0
Autoignition temperature	°C	>250
Flash point	°C	120
Animal testing, acute toxicity, Rat oral LD50	mg/kg	>2000
Animal testing, acute toxicity, Rat dermal LD50	mg/kg	>2000
First aid, eye		In the event of contact with the eyes, rinse thoroughly with clean water. Get medical attention if any discomfort continues.
First aid, inhalation		Move to fresh air. Get medical attention if symptoms persist
First aid, skin		Wash area with soap and water. Get medical attention if symptoms occur.
ECOLOGICAL PROPERTIES		
Aquatic toxicity, *Daphnia magna*, 48-h LC50	mg/l	>139
Partition coefficient log P_{ow}	-	<0.5-4.52; pH 7.3
USE & PERFORMANCE		
Manufacturer		Momentive
Outstanding properties		proprietary vinyl-functional silane coupling agent, which can substantially reduce the volatile alcohol by-products during the manufacture of wire and cable jacketing
Recommended for polymers		EPDM, polyester, polyolefins, rubbers
Recommended for products		wire and cable jacketing

Silquest G-170

PARAMETER	UNIT	VALUE
Guidelines for use		reacts with water, especially if the pH of the water is adjusted to 3-4.

Silquest PA-1

PARAMETER	UNIT	VALUE
GENERAL INFORMATION		
Name		Silquest PA-1
Composition		organomodified polydimethylsiloxane
PHYSICAL PROPERTIES		
State	-	liquid
Odor	-	aromatic
Color	-	colorless
Initial boiling point	°C	>150
Melting/freezing point	°C	<0
Density at 25°C	kg/m³	1018
Vapor pressure at 20°C	hPa	<1.33
VOC	g/l	0
HEALTH & SAFETY		
HMIS classification	Flammability	1
	Health	0
	Physical hazard	0
Flash point	°C	171
Flash point method	-	PMCC
Animal testing, acute toxicity, Rat oral LD50	mg/kg	>5000
Animal testing, acute toxicity, Rat dermal LD50	mg/kg	>2000
Animal testing, acute toxicity, Rat inhalation LC50	mg/l	>1.03
First aid, eye		Get medical attention if symptoms occur. If in eyes, hold eyes open, flood with water for at least 15 minutes and see a doctor.
First aid, inhalation		Move into fresh air and keep at rest. Get medical attention if symptoms occur.
First aid, skin		Remove contaminated clothing and shoes. Wash skin thoroughly with soap and water. Get medical attention if symptoms occur.
USE & PERFORMANCE		
Manufacturer		Momentive

Silquest PA-1

PARAMETER	UNIT	VALUE
Outstanding features	acceptable at trace level for food contact; it is used to enhance the extrudability of high-viscosity polyolefins. Resin producers can add it to reactor fluff or can add it later during extrusion or compounding. Silquest PA-1 organosilicone is an excellent carrier and dispersant for other additives, and it can eliminate the need for metal stearates.	
Recommended for polymers	polyethylene	
Recommended for products	film	
Concentrations used	ppm	500-1000

Silquest RC-1

PARAMETER	UNIT	VALUE
GENERAL INFORMATION		
Name		Silquest RC-1
Composition		organosilane ester, <5% ethanol
Chemical class		silane
Functional organic group		vinyl/trialkoxy
Vinyl content	wt%	9.1
PHYSICAL PROPERTIES		
State	-	liquid
Odor	-	ester-like
Color	-	clear, pale yellow
Color, Gardner scale	-	4
Initial boiling point	°C	>160
Melting/freezing point	°C	<0
Density at 25°C	kg/m^3	954
Evaporation rate (butyl acetate=1)	-	<1
Solubility in water at 20°C		hydrolytic decomposition releasing methanol
Vapor pressure at 20°C	hPa	2.66
VOC	g/l	624
HEALTH & SAFETY		
HMIS classification	Flammability	2
	Health	1
	Reactivity	1
Carcinogenicity		not determined
Mutagenicity		not determined
DOT class		Flammable liquids, n.o.s. (Organosilane Esters) 3, III
ICAO/IATA class		Flammable liquids, n.o.s. (Organosilane Esters) 3,III
IMDG class		Flammable liquids, n.o.s. (Organosilane Esters) 3,III
Flash point	°C	47
Flash point method	-	PMCC
Hazardous combustion products		Carbon monoxide, carbon dioxide, nitrogen oxides
Agency rating, listed		TSCA USA, DSL Canada, AICS Australia, MITI Japan, EINECS Europe, ECL Korea, PICCS Philippines, Taiwan CSNN

Silquest RC-1

PARAMETER	UNIT	VALUE
Hazardous products of hydrolysis	methanol	
Effect of exposure, eye (human)	Not determined	
Effect of exposure, inhalation (human)	Respiratory tract irritation. Narcotic effects. Causes damage to central nervous system (CNS)	
Effect of repeated or overexposure (human)	May cause damage to organs through prolonged or repeated exposure to methanol (kidneys, liver)	
Exposure, personal protection	Safety glasses, protective clothing based on chemical resistance data, chemical-resistant gloves, general and local exhaust ventilation.	
First aid, eye	Rinse immediately with plenty of water, also under the eyelids, for at least 15 minutes. Get medical attention.	
First aid, inhalation	Move the exposed person to fresh air at once. If respiratory problems, artificial respiration/oxygen. Call a physician or poison control center immediately.	
First aid, skin	Wash off promptly and flush contaminated skin with water. Promptly remove clothing if soaked through and flush skin with water. Wash contaminated clothing before reuse. Get medical attention.	
NIOSH, REL	mg/m^3	1900/ethanol
OSHA, PEL	mg/m^3	1900/ethanol
ACGIH, TLV	ppm	1000/ethanol
NIOSH, REL	ppm	1000/ethanol
OSHA, PEL	ppm	1000/ethanol
UN/NA class	-	1993
ECOLOGICAL PROPERTIES		
Partition coefficient, log K_{oc}	-	-0.35 ethanol
USE & PERFORMANCE		
Manufacturer	Momentive	
Outstanding properties	improved modulus and tensile strength along with stable wet electrical properties in rubber compounds. It does not contain and will not liberate EGME (ethylene glycol monomethyl ether) or other glycol ethers during its use. It is recommended as an alternate to Silquest A-172 silane [vinyl-tris(2-methoxyethoxy)silane], which does liberate EGME upon contact with moisture.	

Silquest RC-1

PARAMETER	UNIT	VALUE
Recommended for polymers	EPM, EPDM	
Recommended applications		coupling agent, used in mineral reinforced peroxide-cured rubber wire and cable insulation formulations. It offers a balance of mechanical properties and stable wet electrical properties necessary to meet power cable specifications for EPM and/or EPDM rubber compounds containing clay, talc, ATH, silica, and other minerals.

Silquest VS-142

PARAMETER	UNIT	VALUE
GENERAL INFORMATION		
Name		Silquest VS-142
Composition		amino-alkyl-silane solution in water
Amine (NH$_2$)	%	2.9
PHYSICAL PROPERTIES		
State	-	liquid
Odor	-	amine-like
Color	-	clear, light straw
Initial boiling point	°C	>100
Melting point	°C	<-1
Density at 25°C	kg/m^3	1075
Solubility		toluene, xylene, acetone and many alcohols
Vapor pressure at 25°C	Pa	4.6
Viscosity at 25°C	cSt	3
HEALTH & SAFETY		
HMIS classification	Flammability	2
	Health	0
	Physical hazard	0
Flash point	°C	65
First aid, eye		In the event of contact with the eyes, rinse thoroughly with clean water. Get medical attention if any discomfort continues.
First aid, inhalation		Move to fresh air. Get medical attention if symptoms persist
First aid, skin		Wash area with soap and water. Get medical attention if symptoms occur.
UN #		1993
USE & PERFORMANCE		
Manufacturer		Momentive
Outstanding properties		prehydrolyzed form
Recommended for polymers		acrylics, phenolic resins
Recommended for products		acrylic emulsion based sealants and caulks, wool binders

Silquest VX-225

PARAMETER	UNIT	VALUE
GENERAL INFORMATION		
Name	Silquest VX-225	
CAS #	-	749886-39-3
Composition	amino-functional oligosiloxane	
Molecular mass	daltons	
Chemical class	silane	
Functional organic group	amino	
PHYSICAL PROPERTIES		
State	-	liquid
Color	-	clear
Color Index (Gardner)	-	1
Density	kg/m^3	990
Boiling point	°C	266
Refractive index at 25°C	-	1.42
Viscosity at 25°C	mPas	20
HEALTH & SAFETY		
Flash point	°C	80
Flash point method	-	PMCC
USE & PERFORMANCE		
Manufacturer	Momentive	
Outstanding properties	improved adhesion, especially under wet conditions. Reduced VOCs of the final formulation vs. typical amino silanes - reduced hazardous air pollutants (HAPs) emissions. Lower sealant viscosity to improve sealant application rate and lower sealant residual surface tack compared to typical amino silanes. Modified physical properties of hybrid and RTV sealants by increasing their elasticity lower modulus and decrease residual surface tack of the final cured product. It can also help reduce yellowing in sealant formulations.	

Silquest VX-225

PARAMETER	UNIT	VALUE
Recommended applications		adhesion promoter for use over a broad range of adhesives and sealants applications. Adhesion promoter on concrete and difficult, low polarity substrates. Improved the performance of moisture curable hybrid sealants, room temperature vulcanization (RTV) silicones, and polyurethane sealants, adhesives, and coatings.

Silquest Y-9627

PARAMETER	UNIT	VALUE
GENERAL INFORMATION		
Name	Silquest Y-9627	
General description	secondary aminofunctional bis-silanes	
Molecular mass	daltons	341.5
Chemical class	silane	
Active matter	wt%	>80
Functional organic group	secondary diamino	
PHYSICAL PROPERTIES		
State	-	liquid
Odor	-	ester-like
Color	-	clear, straw to dark
Boiling point at 4 mm Hg	°C	>100
Density at 25°C	kg/m³	1030
Refractive index at 25°C	-	1.423
HEALTH & SAFETY		
Flash point	°C	82
Flash point method	-	PMCC
Exposure, personal protection	Safety glasses, protective clothing based on chemical resistance data, chemical-resistant gloves, general and local exhaust ventilation.	
First aid, eye	Immediately flush with plenty of water for at least 15 minutes. If easy to do, remove contact lenses. Call a physician or poison control center immediately. In case of irritation from airborne exposure, move to fresh air. Get medical attention if symptoms persist	
First aid, inhalation	Remove to fresh air. If not breathing, give artificial respiration. If breathing is difficult, give oxygen. If irritation persists, obtain medical advice.	
First aid, skin	Immediately flush with plenty of water for at least 15 minutes while removing contaminated clothing and shoes. Wash contaminated clothing before reuse. Immediately call a POISON CENTER or physician	
USE & PERFORMANCE		
Manufacturer	Momentive	

Silquest Y-9627

PARAMETER	UNIT	VALUE
Outstanding properties		improved mechanical properties such as tensile and flexural strengths. Excellent stability in aggressive environments, such as high temperature and high humidity conditions. Greater bonding or adhesion promotion to inorganic substrates
Recommended for polymers		epoxy, urethane, melamine, polyimide, phenolic and furan thermosetting resins, many thermoplastics, such as PA and polyesters
Recommended for products		composites, paints, adhesives, and sealants
Recommended applications		adhesion promoter to improve adhesion in coatings, adhesives, and sealants between organic polymers and glass, metal, wood, or cast plastic substrates. Typically, it functions as coupling agents to particulate mineral fillers in composites, as in foundry molds and cores. Secondary amino functionality typically improves shelf stability in various resin systems.

Silquest Y-9669

PARAMETER	UNIT	VALUE
GENERAL INFORMATION		
Name		Silquest Y-9669
CAS #	-	3068-76-6, 62-53-3
EC number	-	221-328-2
Composition		50-100% n-phenyl-γ-aminopropyl trimethoxysilane, <5% aniline, <1.5% methanol, <1% toluene
Common synonym		phenylaminopropyltrimethoxysilane
Formula		
Molecular mass	daltons	255.00
Chemical class	silane	
Functional organic group	diamino/trimethoxy	
PHYSICAL PROPERTIES		
State	-	liquid
Odor	-	ester-like
Color	-	clear, pale yellow
Initial boiling point	°C	310
Melting point	°C	<0
Density at 25°C	kg/m³	1,070
Evaporation rate (butyl acetate=1)	-	<1
Solubility in water at 20°C		rapid hydrolytic decomposition releasing methanol
Vapor density	-	>1.0
Vapor pressure at 20°C	hPa	1.33
HEALTH & SAFETY		
HMIS classification	Flammability	2
	Health	2
	Reactivity	1
Carcinogenicity	not determined	
Mutagenicity	no data available	
Teratogenicity	no data available	
ICAO/IATA class	not regulated	
IMDG class	not regulated	
Flash point	°C	146
Flash point method	-	PMCC

Silquest Y-9669

PARAMETER	UNIT	VALUE
Hazardous combustion products	Carbon monoxide (CO), Carbon dioxide (CO2), Nitrogen Oxides	
Agency rating, listed	TSCA USA, DSL Canada, AICS Australia, MITI Japan, ECL Korea, IECSC China, CSNN Taiwan	
Hazardous products of hydrolysis	methanol	
Animal testing, acute toxicity, Rat oral LD50	mg/kg	1379/males, 2843/females
Animal testing, acute toxicity, Rat dermal LD50	mg/kg	>2000 (female)
Effect of exposure, eye (human)	Causes severe eye irritation.	
Effect of exposure, inhalation (human)	Harmful if inhaled. Causes damage to central nervous system (CNS), optic nerve, respiratory tract irritation. May cause allergy or asthma symptoms or breathing difficulties if inhaled.	
Effect of exposure, skin (human)	Harmful if absorbed through the skin.	
Effect of exposure, swallowing (human)	Harmful if swallowed. May cause blindness if swallowed. May cause dizziness and drowsiness	
Effect of repeated or overexposure (human)	May cause damage to organs through prolonged or repeated exposure (kidneys, liver, bladder)	
Exposure, personal protection	Safety glasses, protective clothing based on chemical resistance data, chemical-resistant gloves, general and local exhaust ventilation.	
First aid, eye	Immediately flush with plenty of water for at least 15 minutes. If easy to do, remove contact lenses. Get medical attention if symptoms persist.	
First aid, inhalation	If inhaled, remove to fresh air. If not breathing, give artificial respiration. If breathing is difficult, trained personnel should give oxygen. Get medical attention immediately.	
First aid, skin	Immediately flush with plenty of water for at least 15 minutes while removing contaminated clothing and shoes. Get medical attention if symptoms persist. Wash contaminated clothing before reuse. Destroy or thoroughly clean contaminated shoes.	
ACGIH, TLV	ppm	2/skin/aniline, 200/skin/methanol, 50/skin/toluene

Silquest Y-9669

PARAMETER	UNIT	VALUE
USE & PERFORMANCE		
Manufacturer		Momentive
Outstanding properties		excellent thermal, good wet properties. The outstanding high-temperature performance of Silquest Y-9669 silane was demonstrated by the improvement in durability of a glass-fiber-reinforced phenolic resin laminate. Composites prepared with Silquest Y-9669 silane maintained superior flexural strength even after prolonged exposure at 260°C compared to composites prepared with another silane
Recommended for polymers		acrylates, epoxy, phenolic resin, PU
Recommended for products		adhesives and sealants, glass fiber sizes and finishes, primers and foundry sand binders
Recommended applications		extremely effective adhesion promoter for many filled and reinforced resin systems. Particularly effective for resin systems that react with a secondary amino group. Applications include: adhesives and sealants, coatings, glass fiber sizes and finishes, primer and foundry-sand binder.

Silquest Y-9936

PARAMETER	UNIT	VALUE
GENERAL INFORMATION		
Name		Silquest Y-9936
CAS #	-	21142-29-0
Composition		50–100% γ-methacryloxypropyl-triethoxysilane
Formula		
Molecular mass	daltons	290.43
Purity	%	95
Acidity as chloride	ppm	1
PHYSICAL PROPERTIES		
State	-	liquid
Odor	-	ester-like
Color	-	clear, colorless
Initial boiling point	°C	130
Melting point	°C	<-45
Density at 25°C	kg/m³	986
Vapor pressure at 25°C	hPa	4.6
HEALTH & SAFETY		
HMIS classification	Flammability	1
	Health	0
	Physical hazard	1
Autoignition temperature	°C	>250
Flash point	°C	116
Flash point method	-	PMCC
Animal testing, acute toxicity, Rat oral LD50	mg/kg	5000
Animal testing, acute toxicity, Rat dermal LD50	mg/kg	>2000
First aid, eye		Get medical attention if symptoms occur. If in eyes, hold eyes open, flood with water for at least 15 minutes and see a doctor.
First aid, inhalation		Move to fresh air. Get medical attention if symptoms persist
First aid, skin		Remove contaminated clothing and shoes. Wash skin thoroughly with soap and water. Get medical attention if symptoms occur.

Silquest Y-9936

PARAMETER	UNIT	VALUE
ECOLOGICAL PROPERTIES		
Aquatic toxicity, *Algae*, 72-h EC50	mg/l	4.1
Aquatic toxicity, *Daphnia magna*, 48-h LC50	mg/l	5.2
Partition coefficient, log K_{ow}	-	3.6
USE & PERFORMANCE		
Manufacturer	Momentive	
Recommended for products	auto and protective coatings	

Silquest Y-11699

PARAMETER	UNIT	VALUE
GENERAL INFORMATION		
Name	Silquest Y-11699	
CAS #	-	13497-18-2
EC number	-	236-818-1
General description	bis-(γ-triethoxysilylpropyl)amine	
Empirical formula	NH[CH2CH2CH2Si(OCH2CH3)3]2	
Formula		
Molecular mass	daltons	425.5
Active matter	wt%	100
Functional organic group	diamino	
PHYSICAL PROPERTIES		
State	-	liquid
Odor	-	ester-like
Color	-	clear, pale
Boiling point	°C	>150
Melting point	°C	<0
Density at 25°C	kg/m^3	968
HEALTH & SAFETY		
Flash point	°C	>93
Flash point method	-	PMCC
Exposure, personal protection	Safety glasses, protective clothing based on chemical resistance data, chemical-resistant gloves, general and local exhaust ventilation.	
First aid, eye	Rinse immediately with plenty of water, also under the eyelids, for at least 15 minutes. Get medical attention.	
First aid, inhalation	Move the exposed person to fresh air at once. If respiratory problems, artificial respiration/oxygen. Call a physician or poison control center immediately.	
First aid, skin	Wash off promptly and flush contaminated skin with water. Promptly remove clothing if soaked through and flush skin with water. Wash contaminated clothing before reuse. Get medical attention.	
UN risk phrases, R	R34	

Silquest Y-11699

PARAMETER	UNIT	VALUE
US safety phrases, S		S26,S28,S36/37/39,S45
USE & PERFORMANCE		
Manufacturer		Momentive
Outstanding properties		improved mechanical properties such as tensile and flexural strengths. Excellent stability in aggressive environments, such as high temperature and high humidity conditions. Provides a controlled cure with a by-product of the coupling or crosslinking mechanisms that have a lower impact on the environment. Silquest Y-11699 silane generally provides a controlled cure with a by-product of the coupling or crosslinking mechanisms that have a lower impact on the environment
Recommended for polymers		epoxy, urethane, melamine, polyimide, phenolic and furan thermosetting resins, many thermoplastics, such as PA and polyesters
Recommended applications		adhesion promoter to improve adhesion in coatings, adhesives and sealants between organic polymers and glass, metal, wood or cast plastic substrates. Typically, it functions as coupling agents for particulate mineral fillers in composites, as in foundry molds and cores.

Silquest Y-15744

PARAMETER	UNIT	VALUE
GENERAL INFORMATION		
Name		Silquest Y-15744
Composition		amino-functional silane oligomer
Chemical class		silane
Active matter	wt%	100
PHYSICAL PROPERTIES		
State	-	liquid
Color	-	light yellow
Boiling point at 4 mm Hg	°C	>210
Density at 25°C	kg/m³	1010
HEALTH & SAFETY		
Flash point	°C	80
Flash point method	-	PMCC
USE & PERFORMANCE		
Manufacturer		Momentive
Outstanding properties		improved adhesion in wet environment. Improved wetting properties, even on substrates with low surface tension. Improved adhesion to damp masonry and concrete. Strong, weather-resistant bonds.
Recommended for polymers		ABS, PMMA, POM, PS
Recommended applications		adhesion promoter between organic/in-organic substrates and organic polymers and in moisture-curing formulations such as hybrid or silicone-based seal-ants, adhesives, and coatings. Provides weather and moisture-resistant bonds to substrates such as glass, metal, but also on more difficult surfaces like concrete, porous and plastics substrates.

Silquest Y-15866 Silane

PARAMETER	UNIT	VALUE
GENERAL INFORMATION		
Name		Silquest Y-15866 Silane
Composition		vinyl-functional silane that offers hydrolysis rates similar to vinyltrimethoxysilane, but is not based on methanol
Purity	%	95
Acidity as chloride	ppm	1
PHYSICAL PROPERTIES		
State	-	liquid
Odor	-	mild
Color	-	clear, colorless
Initial boiling point	°C	216.7
Melting point	°C	<-71
Density at 25°C	kg/m^3	1074
Solubility Hatch number	NTU	6
Vapor pressure at 25°C	Pa	4.6
HEALTH & SAFETY		
HMIS classification	Flammability	1
	Health	1
	Physical hazard	0
Autoignition temperature	°C	>250
Flash point	°C	104
Flash point method	-	FCC
Animal testing, acute toxicity, Rat oral LD50	mg/kg	>2000
Animal testing, acute toxicity, Rat dermal LD50	mg/kg	>2000
First aid, eye		In the event of contact with the eyes, rinse thoroughly with clean water. Get medical attention if any discomfort continues.
First aid, inhalation		Move to fresh air. Get medical attention if symptoms persist
First aid, skin		Wash area with soap and water. Get medical attention if symptoms occur.
ECOLOGICAL PROPERTIES		
Aquatic toxicity, *Daphnia magna*, 48-h LC50	mg/l	>139
Partition coefficient, log K_{ow}	-	<0.5-4.52; pH 7.3
USE & PERFORMANCE		
Manufacturer		Momentive

Silquest Y-15866 Silane

PARAMETER	UNIT	VALUE
Outstanding properties		substantially lower volatility than vinyltrimethoxysilane. Silquest Y-15866 silane is an excellent candidate for use as a moisture scavenger in hybrid sealant technology (e.g., SPUR+ prepolymer systems), latex modification, and wherever a more common vinyl silane may be used. Rapid hydrolysis without methanol generation. High boiling point
Recommended for polymers		silicone
Recommended for products		hybrid sealants

Silquest Wetlink 78

PARAMETER	UNIT	VALUE
GENERAL INFORMATION		
Name		Silquest Wetlink 78
CAS #	-	2897-60-1
Composition		3-glycidoxypropylmethyldiethoxysilane
Formula		
Molecular mass	daltons	248.39
PHYSICAL PROPERTIES		
State	-	liquid
Odor	-	ester-like
Color	-	pale yellow
Initial boiling point	°C	260
Melting point	°C	-80
Density at 25°C	kg/m³	940
Vapor pressure at 20°C	hPa	<1.33
HEALTH & SAFETY		
HMIS classification	Flammability	1
	Health	2
	Physical hazard	1
Flash point	°C	107
Animal testing, acute toxicity, Rat oral LD50	mg/kg	>2000
Animal testing, acute toxicity, Rabbit dermal LD50	mg/kg	>2000
First aid, eye		Immediately flush with plenty of water for up to 15 minutes. Remove any contact lenses and open eyes wide apart. Continue to rinse for at least 15 minutes. Get medical attention if symptoms occur.
First aid, inhalation		Move into fresh air and keep at rest. If breathing has stopped, trained personnel should begin artificial respiration immediately and if the heart has stopped, trained personnel should begin cardiopulmonary resuscitation immediately. Get medical attention.
First aid, skin		Wash area with soap and water. Get medical attention if symptoms persist.

Silquest Wetlink 78

PARAMETER	UNIT	VALUE
ECOLOGICAL PROPERTIES		
Partition coefficient log P_{ow}	-	2.7
USE & PERFORMANCE		
Manufacturer	Momentive	
Outstanding properties	shelf stable non-yellowing adhesion promoter performance while enhancing physical properties in latexes and waterborne adhesive and sealant systems. When Silquest Wetlink 78 silane is incorporated as a crosslinker or adhesion promoter, it provides improved water resistance and wet adhesion, with good shelf stability.	
Recommended for polymers	acrylics, SBR, polyurethanes	
Recommended for products	waterborne adhesives (automotive, furniture, and flocking) and sealant systems	

SiSiB PC1100

PARAMETER	UNIT	VALUE
GENERAL INFORMATION		
Name		SiSiB® PC1100
CAS #	-	919-30-2
EC number	-	213-048-4
Common synonym	γ-aminopropyltriethoxysilane	
Empirical formula	-	C9H23NO2Si
Formula	$H_3CCH_2O-\underset{\underset{OCH_2CH_3}{|}}{\overset{\overset{OCH_2CH_3}{|}}{Si}}-(CH_2)_3NH_2$	
Molecular mass	daltons	221.37
Functional organic group	-	primary amine
Purity	wt%	99; 95 (technical grade)
PHYSICAL PROPERTIES		
State	-	liquid
Color	-	colorless
Boiling point	°C	215
Kinematic viscosity at 25°C	cSt	2
Refractive index at 25°C	-	1.42
Solubility in water at 25°C	g/l	soluble
HEALTH & SAFETY		
Flash point	°C	96
USE & PERFORMANCE		
Manufacturer	SiSiB Silicones/PCC group	
Recommended for polymers	epoxy, phenolics, polyamides, poly(butylene terephthalate)	
Recommended for products	adhesives, coatings, inks, magnetic materials, pigment dispersion, rubber products	

SiSiB PC1120

PARAMETER	UNIT	VALUE
GENERAL INFORMATION		
Name	SiSiB® PC1120	
CAS #	-	3179-76-8
EC number	-	221-660-8
Common synonym	γ-aminopropylmethyldiethoxysilane	
Empirical formula	-	C8H21NO2Si
Formula		$H_3CCH_2O-\overset{\displaystyle CH_3}{\underset{\displaystyle OCH_2CH_3}{Si}}-(CH_2)_3NH_2$
Molecular mass	daltons	191.3
Functional organic group	-	primary amine
Purity	wt%	98
PHYSICAL PROPERTIES		
State	-	liquid
Color	-	colorless
Boiling point (15 mm Hg)	°C	88
Density at 25°C	kg/m³	916
Kinematic viscosity at 25°C	cSt	2
Refractive index at 25°C	-	1.4272
Solubility in water at 25°C	g/l	soluble
HEALTH & SAFETY		
Flash point	°C	68
USE & PERFORMANCE		
Manufacturer	SiSiB Silicones/PCC group	
Recommended for polymers	epoxy resins, polyurethanes, phenolic resins, furan resins, melamine resins, PA, PBT, PC, PEK, PE, EVA, PP, PVB, PVAc, PVC, acrylates and silicone	
Recommended for products	abrasives, adhesives and sealants, composites	
Guidelines for use	the hydrolysis of SiSiB® PC1120 takes place autocatalytically in a short time of about 5-10 minutes. Hydrolysates having a concentration of < 5% are stable for more than 72 hours. The pH is about 11.	

SiSiB PC1200

PARAMETER	UNIT	VALUE
GENERAL INFORMATION		
Name	SiSiB® PC1200	
CAS #	-	1760-24-3
EC number	-	217-164-6
Common synonym	N-β-(aminoethyl)-γ-aminopropyl-trimethoxysilane	
Empirical formula	-	C8H22N2O3Si
Formula	$H_3CO-\underset{\underset{OCH_3}{\vert}}{\overset{\overset{OCH_3}{\vert}}{Si}}-(CH_2)_3\overset{H}{N}(CH_2)_2NH_2$	
Molecular mass	daltons	222.4
Functional organic group	primary and secondary amines	
Purity	wt%	99
PHYSICAL PROPERTIES		
State	-	liquid
Color	-	colorless
Boiling point	°C	259
Density at 25°C	kg/m³	1025
Kinematic viscosity at 25°C	cSt	2
Refractive index at 25°C	-	1.446
Solubility in water at 25°C	g/l	soluble
HEALTH & SAFETY		
Flash point	°C	128
USE & PERFORMANCE		
Manufacturer	SiSiB Silicones/PCC group	
Outstanding features	modification of surfaces (corrosion prevention, component of primers) or silicone polymers or as crosslinker (moisture crosslinking of polymers). The application as a coupling agent leads to improvement of mechanical and electrical product properties above all under exposure to heat and/or moisture.	
Recommended for polymers	PA-6, PA-6/6 and polybutyleneterephthalate, natural and nitrile rubbers, phenolic, melamine and epoxy thermosets, polyurethanes, n polysulfide, polyvinylchloride	
Recommended for products	abrasives, adhesives and sealants, coatings, composites	

SiSiB PC1200

PARAMETER	UNIT	VALUE
Guidelines for use		Due to solution enthalpies mixing water is exothermic. Always stir SiSiB® PC1200 into water. With alcohols miscibility is, in general, possible with self-catalyzed exchange of the alkoxy-groups. In aliphatic and aromatic hydro-carbons and (moisture-free!) ethers or esters SiSiB® PC1200 is easily soluble at differing levels. With ketones and various halogenated compounds a slow reaction can occur. Towards acids, epoxides or isocyanates SiSiB® PC1200 shows typical amine function. Some nonferrous metals can discolor upon contact.

SiSiB PC1220

PARAMETER	UNIT	VALUE
GENERAL INFORMATION		
Name		SiSiB® PC1220
CAS #	-	3069-29-2
EC number	-	221-336-6
Common synonym		N-β-(aminoethyl)-γ-aminopropyl-methyl-dimethoxysilane
Empirical formula	-	C8H22N2O2Si
Formula		
Molecular mass	daltons	206.4
Functional organic group		primary and secondary amines
Purity	wt%	99
PHYSICAL PROPERTIES		
State	-	liquid
Color	-	colorless
Boiling point	°C	265
Density at 25°C	kg/m³	970-980
Refractive index at 25°C	-	1.448
Solubility in water at 25°C	g/l	soluble
HEALTH & SAFETY		
Autoignition temperature	°C	290
Flash point	°C	93
USE & PERFORMANCE		
Manufacturer		SiSiB Silicones/PCC group
Outstanding features		alkaline liquid with amine smell, very sensitive to hydrolysis, lower reactivity, and therefore higher stability in the aqueous environment
Recommended for polymers		PVC plastisols, polyurethanes, epoxy, phenolics, furan resins, polysulfides, siliconized urethanes
Recommended for products		adhesives and sealants, coatings, composites, foundry resins molding compounds

$$H_3CO-\underset{\underset{OCH_3}{|}}{\overset{\overset{CH_3}{|}}{Si}}-(CH_2)_3\underset{}{\overset{H}{N}}(CH_2)_2NH_2$$

SiSiB PC1220

PARAMETER	UNIT	VALUE
Guidelines for use		due to solution enthalpies mixing with water is exothermic. Always stir SiSiB® PC1220 into water. With alcohols miscibility, the self-catalyzed exchange of the alkoxy-groups is possible. Some nonferrous metals can discolor upon contact.

SiSiB PC1710

PARAMETER	UNIT	VALUE
GENERAL INFORMATION		
Name		SiSiB® PC1710
CAS #	-	77855-73-3
Common synonym		(N-phenylamino)methyltrimethoxysilane, anilinomethyltrimethoxysilane
Empirical formula	-	C10H17O3NSi
Formula		
Molecular mass	daltons	227.34
Functional organic group		secondary amine
Purity	wt%	97
PHYSICAL PROPERTIES		
State	-	liquid
Color		colorless to yellowish
Boiling point	°C	235
Density at 25°C	kg/m³	1080
Refractive index at 25°C	-	1.509
HEALTH & SAFETY		
Flash point	°C	110
USE & PERFORMANCE		
Manufacturer		SiSiB Silicones/PCC group
Outstanding features		close proximity of nitrogen atom to silicon atom accelerates hydrolysis reaction compared to (amino-propyl)silanes.
Recommended for products		production of silyl modified polymers, crosslinker, water scavenger, and adhesion promoter in silane-crosslinked formulations, adhesives, sealants and coatings, surface modifier for fillers, such as glass, metal oxides, aluminum hydroxide, kaolin, wollastonite, mica, and pigments.

SiSiB PC1711

PARAMETER	UNIT	VALUE
GENERAL INFORMATION		
Name		SiSiB® PC1711
CAS #	-	3473-76-5
Common synonym		(N-phenylamino)methyltriethoxysilane, anilinomethyltrimethoxysilane
Empirical formula	-	C13H23O3NSi
Formula		

$$H_5C_2O-\underset{\underset{OC_2H_5}{|}}{\overset{\overset{OC_2H_5}{|}}{Si}}-CH_2\underset{}{\overset{H}{N}}-\bigcirc$$

PARAMETER	UNIT	VALUE
Molecular mass	daltons	269.42
Functional organic group	secondary amine	
Purity	wt%	97
PHYSICAL PROPERTIES		
State	-	liquid
Color	colorless to yellowish	
Boiling point (4 mm Hg)	°C	136
Density at 25°C	kg/m³	1000
Refractive index at 25°C	-	1.485
HEALTH & SAFETY		
Flash point	°C	>110
USE & PERFORMANCE		
Manufacturer	SiSiB Silicones/PCC group	
Outstanding features	close proximity of nitrogen atom to silicon atom accelerates hydrolysis reaction compared to (amino-propyl)silanes.	
Recommended for products	production of silyl modified polymers, crosslinker, water scavenger, and adhesion promoter in silane-crosslinked formulations, adhesives, sealants and coatings, surface modifier for fillers, such as glass, metal oxides, aluminum hydroxide, kaolin, wollastonite, mica, and pigments.	

SiSiB PC1800

PARAMETER	UNIT	VALUE
GENERAL INFORMATION		
Name	SiSiB® PC1800	
CAS #	-	15180-47-9
Common synonym	diethylaminomethyltriethoxysilane	
Empirical formula	-	C11H27NO3Si
Formula		

$$H_5C_2O-\underset{\underset{OC_2H_5}{|}}{\overset{\overset{OC_2H_5}{|}}{Si}}-CH_2N\overset{C_2H_5}{\underset{C_2H_5}{\diagdown}}$$

PARAMETER	UNIT	VALUE
Molecular mass	daltons	249.42
Functional organic group	ethoxy only	
Purity	wt%	98; 75 (technical grade)
PHYSICAL PROPERTIES		
State	-	liquid
Color	colorless to yellowish	
Boiling point	°C	237
Density at 25°C	kg/m³	916-933
Refractive index at 25°C	-	1.432
HEALTH & SAFETY		
Flash point	°C	>110
USE & PERFORMANCE		
Manufacturer	SiSiB Silicones/PCC group	
Outstanding features	close proximity of nitrogen atom to silicon atom accelerates hydrolysis reaction compared to (amino-propyl)silanes.	
Recommended for polymers	silicone rubber	
Recommended for products	RTV sealants, fabrics	

SiSiB PC2000

PARAMETER	UNIT	VALUE
GENERAL INFORMATION		
Name	SiSiB® PC2000	
CAS #	-	40372-72-3
EC number	-	254-896-5
Common synonym	bis(3-triethoxysilylpropyl)tetrasulfide	
Empirical formula	-	C18H42O6S4Si2
Formula		
Molecular mass	daltons	539
Functional organic group	sulfidic	
Average chain length	-	3.75
Total sulfur	%	22.7
Secondary components	propyltriethoxysilane, chloropropyl-triethoxylsilane, ethanol	
PHYSICAL PROPERTIES		
State	-	liquid
Color	-	yellowish
Boiling point	°C	>250 (decomp)
Pour point	°C	-80
Density at 25°C	kg/m³	1100
Refractive index at 25°C	-	1.446
Solubility	primary alcohols, ketones, benzene, toluene, dimethylformaminde, chlori-nated hydrocarbons, cetonitrile, dime-thysulfoxide	
VOC	%	<=4
HEALTH & SAFETY		
Flash point	°C	100
USE & PERFORMANCE		
Manufacturer	SiSiB Silicones/PCC group	

SiSiB PC2000

PARAMETER	UNIT	VALUE
Outstanding features		carbon black has been used as a reinforcing filler for rubber because carbon black provides better reinforcement and abrasion resistance than other fillers. Because of the demand to save energy and resources, particularly to cut down the fuel consumption of automobiles, a decrease in heat buildup in rubber compositions is required. Silica provides decreased heat buildup, but it is difficult to disperse and causes problems with vulcanization because of its acidic character. SiSiB® PC2000 and SiSiB® PC2200 help to solve these problems.
Recommended for polymers		rubber (coupling agent for non-black pigments, cure equilibrium for reversion resistance, and curing agent for good heat aging)
Recommended for products		flat belts, footwear, hoses, molded goods, rollers, solid tires, tires, v belts
Concentrations used	phf	3-13 (silica), 0.5-1 (clay and talc)

SiSiB PC2200

PARAMETER	UNIT	VALUE
GENERAL INFORMATION		
Name	SiSiB® PC2200	
CAS #	-	56706-10-6
EC number	-	260-350-7
Common synonym	bis(3-triethoxysilylpropyl)disulfide	
Empirical formula	-	C18H42O6S2Si2
Formula	$H_5C_2O-\underset{\underset{OC_2H_5}{|}}{\overset{\overset{OC_2H_5}{|}}{Si}}-(CH_2)_3-S_2-(CH_2)_3-\underset{\underset{OC_2H_5}{|}}{\overset{\overset{OC_2H_5}{|}}{Si}}-OC_2H_5$	
Molecular mass	daltons	486
Functional organic group	sulfidic	
Average chain length	-	2.35
Total sulfur	%	15.2
Secondary components	propyltriethoxysilane, chloropropyl-triethoxylsilane, ethanol	
PHYSICAL PROPERTIES		
State	-	liquid
Color	-	light yellowish
Density at 25°C	kg/m³	1030
Solubility	primary alcohols, ketones, benzene, toluene, dimethylformaminde, chlorinated hydrocarbons, cetonitrile, dimethysulfoxide	
VOC	%	<=4
HEALTH & SAFETY		
Flash point	°C	>120
USE & PERFORMANCE		
Manufacturer	SiSiB Silicones/PCC group	

SiSiB PC2200

PARAMETER	UNIT	VALUE
Outstanding features		carbon black has been used as a reinforcing filler for rubber because carbon black provides better reinforcement and abrasion resistance than other fillers. Because of the demand to save energy and resources, particularly to cut down the fuel consumption of automobiles, a decrease in heat buildup in rubber compositions is required. Silica provides decreased heat buildup, but it is difficult to disperse and causes problems with vulcanization because of its acidic character. SiSiB® PC2000 and SiSiB® PC2200 help to solve these problems.
Recommended for polymers		rubber (coupling agent for non-black pigments, cure equilibrium for reversion resistance, and curing agent for good heat aging)
Recommended for products		flat belts, footwear, molded goods, rollers, solid tires, tires, v belts

SiSiB PC2300

PARAMETER	UNIT	VALUE
GENERAL INFORMATION		
Name		SiSiB® PC2300
CAS #	-	4420-74-0
EC number	-	224-588-5
Common synonym		γ-mercaptopropyltrimethoxysilane
Empirical formula	-	C6H16O3SSi
Formula		$H_3CO-\underset{\underset{OCH_3}{\vert}}{\overset{\overset{OCH_3}{\vert}}{Si}}-(CH_2)_3SH$
Molecular mass	daltons	196.4
Functional organic group	mercapto	
Purity	%	97
PHYSICAL PROPERTIES		
State	-	liquid
Odor	-	slight mercaptan
Color	colorless to light yellowish	
Boiling point	°C	212
Density at 25°C	kg/m³	1057
Refractive index	-	1.44
Solubility	alcohols, ketones and aliphatic or aromatic hydrocarbons	
HEALTH & SAFETY		
Flash point	°C	88
USE & PERFORMANCE		
Manufacturer	SiSiB Silicones/PCC group	
Outstanding features	bifunctional organosilane possessing a reactive organic mercapto and a hydrolyzable inorganic methoxysilyl group. It can be used with lower silane loadings. Improves low-rolling resistance in silica-reinforced tire tread compounds.	
Recommended for polymers	polysulfides, polyurethanes	
Recommended for products	sealants, shoe soles, rubber rollers and wheels, white sidewalls, and wire and cable insulation, pretreatment of minerals, tires	

SiSiB PC2310

PARAMETER	UNIT	VALUE
GENERAL INFORMATION		
Name		SiSiB® PC2310
CAS #	-	14814-09-6
EC number	-	238-883-1
Common synonym	γ-mercaptopropyltriethoxysilane	
Empirical formula	-	C9H22O3SSi
Formula	$H_5C_2O-\underset{\underset{OC_2H_5}{\mid}}{\overset{\overset{OC_2H_5}{\mid}}{Si}}-(CH_2)_3SH$	
Molecular mass	daltons	238.42
Functional organic group	mercapto	
Purity	%	98
PHYSICAL PROPERTIES		
State	-	liquid
Odor	-	slight mercaptan
Color	colorless to light yellowish	
Boiling point	°C	210
Density at 25°C	kg/m³	993
Refractive index	-	1.4331
Solubility	alcohols, ketones and aliphatic or aromatic hydrocarbons	
HEALTH & SAFETY		
Flash point	°C	88
USE & PERFORMANCE		
Manufacturer	SiSiB Silicones/PCC group	
Outstanding features	bifunctional organosilane possessing a reactive organic mercapto and a hydrolyzable inorganic ethoxysilyl group. It can be used with lower silane loadings. Improves low-rolling resistance in silica-reinforced tire tread compounds.	
Recommended for polymers	polysulfides, polyurethanes	
Recommended for products	sealants, shoe soles, rubber rollers and wheels, white sidewalls, and wire and cable insulation, pretreatment of minerals, tires	

SiSiB PC2320

PARAMETER	UNIT	VALUE
GENERAL INFORMATION		
Name		SiSiB® PC2320
CAS #	-	31001-77-1
EC number	-	250-426-8
Common synonym		γ-mercaptopropylmethyldimethoxysilane
Empirical formula	-	C6H16O2SSi
Formula		$H_3CO-\underset{\underset{OCH_3}{|}}{\overset{\overset{CH_3}{|}}{Si}}-(CH_2)_3SH$
Molecular mass	daltons	180.34
Functional organic group	mercapto	
Purity	%	98
PHYSICAL PROPERTIES		
State	-	liquid
Odor	-	slight mercaptan
Color		colorless to light yellowish
Boiling point (30 mm Hg)	°C	96
Density at 25°C	kg/m³	1000
Refractive index	-	1.4502
Solubility		alcohols, ketones and aliphatic or aromatic hydrocarbons
HEALTH & SAFETY		
Flash point	°C	93
USE & PERFORMANCE		
Manufacturer		SiSiB Silicones/PCC group
Outstanding features		bifunctional organosilane possessing a reactive organic mercapto and a hydrolyzable inorganic ethoxysilyl group. It can be used with lower silane loadings. Improves low-rolling resistance in silica-reinforced tire tread compounds.
Recommended for polymers		polysulfides, polyurethanes
Recommended for products		sealants, shoe soles, rubber rollers and wheels, white sidewalls, and wire and cable insulation, pretreatment of minerals, tires

SiSiB PC2521

PARAMETER	UNIT	VALUE
GENERAL INFORMATION		
Name	SiSiB® PC2521	
CAS #	116912-64-2 or 23779-32-0	
EC number	-	245-876-7
Common synonym	γ-ureidopropyltriethoxysilane	
Empirical formula	-	C10H24N2O4Si
Formula		
Molecular mass	daltons	264.4
Functional organic group	primary and secondary amino	
Purity	%	50 (in methanol)
PHYSICAL PROPERTIES		
State	-	liquid
Color		colorless
Density at 25°C	kg/m³	920
Refractive index	-	1.386
Solubility	alcohols	
VOC	%	<=4
HEALTH & SAFETY		
USE & PERFORMANCE		
Manufacturer	SiSiB Silicones/PCC group	
Outstanding features	bifunctional organosilane possessing a reactive organic ureido and a hydrolyzable inorganic ethoxysilyl group. The dual nature of its reactivity allows SiSiB® PC2521 to bind chemically to both inorganic materials (e.g., glass, metals, fillers) and organic polymers (e.g., thermosets, thermoplastics, elastomers), functioning as an adhesion promoter and as a surface modifier. SiSiB® PC2521 is a 50% solution of silane in methanol.	
Recommended for polymers	epoxy, phenolic, furan, and melamine resins, polyurethanes, PA, PBT, PE, EVA, PP, PVB, PVAc, acrylics, and silicones	
Recommended for products	glass fiber/glass fabric composites, foundry resins and abrasives, sealants and adhesives, paints and varnishes	

SiSiB PC2640

PARAMETER	UNIT	VALUE
GENERAL INFORMATION		
Name		SiSiB® PC2640
CAS #	-	34708-08-2
EC number	-	252-161-3
Common synonym		3-thiocyanatopropyltriethoxysilane
Empirical formula	-	C10H21NO3SSi
Formula		$H_5C_2O-\underset{\underset{OC_2H_5}{|}}{\overset{\overset{OC_2H_5}{|}}{Si}}-(CH_2)_3S-C\equiv N$
Molecular mass	daltons	263.43
Functional organic group	thiocyanate	
Purity	%	97
PHYSICAL PROPERTIES		
State	-	liquid
Odor	-	mild
Color		straw to yellowish
Boiling point (0.1 mm Hg)	°C	95
Density at 25°C	kg/m³	1030
Refractive index	-	1.446
Solubility	alcohols	
HEALTH & SAFETY		
Flash point	°C	138
USE & PERFORMANCE		
Manufacturer		SiSiB Silicones/PCC group
Outstanding features		improves physical and mechanical properties of vulcanizates. It can improve tensile strength, tearing strength, and abrasive resistance and reduce the compression set of vulcanizates. In addition, it can reduce the viscosity and improve the processability of rubber products.

SiSiB PC2720

PARAMETER	UNIT	VALUE
GENERAL INFORMATION		
Name		SiSiB® PC2720
CAS #	-	24801-88-5
EC number	-	246-467-6
Common synonym		3-isocyanatepropyltriethoxysilane
Empirical formula	-	C10H21NO4Si
Formula		$H_5C_2O-\overset{\displaystyle OC_2H_5}{\underset{\displaystyle OC_2H_5}{Si}}-(CH_2)_3NCO$
Molecular mass	daltons	247.36
Functional organic group	isocyanate	
Purity	%	98
PHYSICAL PROPERTIES		
State	-	liquid
Color		colorless
Boiling point	°C	238
Density at 25°C	kg/m³	1000
Refractive index	-	1.421
Solubility	alcohols	
HEALTH & SAFETY		
Flash point	°C	80
USE & PERFORMANCE		
Manufacturer		SiSiB Silicones/PCC group
Outstanding features		it is used for the functionalization of numerous compounds with active hydrogen atoms
Recommended for polymers		urethanes, silicones
Recommended for products		adhesives and sealants

SiSiB PC3100

PARAMETER	UNIT	VALUE
GENERAL INFORMATION		
Name		SiSiB® PC3100
CAS #	-	2530-83-8
EC number	-	219-784-2
Common synonym	γ-glycidoxypropyltrimethoxysilane	
Empirical formula	-	C9H20O5Si
Formula		
Molecular mass	daltons	236.34
Functional organic group	epoxy	
Purity	%	99
PHYSICAL PROPERTIES		
State	-	liquid
Color		light straw
Boiling point	°C	290
Density at 25°C	kg/m³	1070
Refractive index	-	1.427
Solubility	alcohols	
HEALTH & SAFETY		
Flash point	°C	110
USE & PERFORMANCE		
Manufacturer	SiSiB Silicones/PCC group	
Outstanding features	adhesion-promoting additive in water-borne systems	
Recommended for polymers	acrylics, polysulfides, polyurethanes	
Recommended for products	caulks and sealants, coatings, composites, electronic encapsulants, glass roving size-binders	

SiSiB PC3200

PARAMETER	UNIT	VALUE
GENERAL INFORMATION		
Name		SiSiB® PC3200
CAS #	-	2602-34-8
EC number	-	220-011-6
Common synonym		γ-glycidoxypropyltriethoxysilane
Empirical formula	-	C12H26O5Si
Formula		
Molecular mass	daltons	278.4
Functional organic group		epoxy
Purity	%	99
PHYSICAL PROPERTIES		
State	-	liquid
Odor	-	low
Color		colorless
Boiling point (3 mm Hg)	°C	124
Density at 25°C	kg/m³	1004
Refractive index	-	1.425
Solubility	alcohols	
HEALTH & SAFETY		
Flash point	°C	110
USE & PERFORMANCE		
Manufacturer		SiSiB Silicones/PCC group
Outstanding features		PC3200 may be used as an adhesion promoter (coupling agent) for organic/inorganic interfaces, as a surface modifier (e.g., regulating surface polarity), or as a crosslinking agent (moisture-curing of polymers). When used as coupling agent, it generally reduces the sensitivity of the products' mechanical and electrical properties to heat and/or moisture.
Recommended for polymers		acrylics, epoxy, polysulfides
Recommended for products		sealants and adhesives, electronic encapsulants, glass roving size-binders

SiSiB PC3300

PARAMETER	UNIT	VALUE
GENERAL INFORMATION		
Name	SiSiB® PC3300	
CAS #	-	2897-60-1
EC number	-	220-780-8
Common synonym	γ-glycidoxypropylmethyldiethoxysilane	
Empirical formula	-	C11H24O4Si
Formula		
Molecular mass	daltons	248.39
Functional organic group	epoxy	
Purity	%	98
PHYSICAL PROPERTIES		
State	-	liquid
Odor	-	low
Color		colorless
Boiling point (5 mm Hg)	°C	123
Density at 25°C	kg/m³	978
Refractive index	-	1.421
Solubility	alcohols	
HEALTH & SAFETY		
Flash point	°C	122
USE & PERFORMANCE		
Manufacturer	SiSiB Silicones/PCC group	
Outstanding features	PC3300 may be used as an adhesion promoter (coupling agent) for organic/inorganic interfaces, as a surface modifier (e.g., regulating surface polarity), or as a crosslinking agent (moisture-curing of polymers). When used as coupling agent, it generally reduces the sensitivity of the products' mechanical and electrical properties to heat and/or moisture.	
Recommended for polymers	acrylics, epoxy, polysulfides	
Recommended for products	sealants and adhesives, electronic encapsulants, composites reinforced with glass fiber rovings	

SiSiB PC4100

PARAMETER	UNIT	VALUE
GENERAL INFORMATION		
Name		SiSiB® PC4100
CAS #	-	2530-85-0
EC number	-	219-785-8
Common synonym		γ-methacryloxypropyltrimethoxysilane
Empirical formula	-	C10H20O5Si
Formula		
Molecular mass	daltons	248.35
Functional organic group	methoxyacrylo	
Purity	%	99
PHYSICAL PROPERTIES		
State	-	liquid
Color		colorless
Boiling point	°C	255
Density at 25°C	kg/m³	1045
Refractive index	-	1.430
Solubility		methanol, ethanol, isopropanol, acetone, benzene, toluene, and xylene
Viscosity at 25°C	cSt	2
HEALTH & SAFETY		
Flash point	°C	108
USE & PERFORMANCE		
Manufacturer		SiSiB Silicones/PCC group

SiSiB PC4100

PARAMETER	UNIT	VALUE
Outstanding features		SiSiB® PC4100 is used as an adhesion promoter for organic/inorganic interfaces, as a surface modifier (e.g., imparting water repellency, organophilic surface adjustment), or for crosslinking of polymers. It is used as a coupling agent to improve the physical and electrical properties of glass-reinforced and mineral-filled thermosetting resins under exposure to heat and/or moisture. It is typically employed as a blend additive in resin systems that cure *via* free radical mechanism (e.g., polyester, acrylic) and in filled or reinforced thermoplastic polymers, including polyolefins and polyurethanes. It is also used to functionalize resins *via* radical initiated processes – copolymerization or grafting – and to modify surfaces. SiSiB® PC4100 shows copolymerization or grafting reactions when catalyzed by (organic) initiator systems, e.g., peroxides or by radiation (e.g., UV).
Recommended for polymers		acrylics, epoxy, polyester, polyolefins, polyurethanes
Recommended for products		sealants and adhesives, electronic encapsulants, composites reinforced with glass fiber rovings

SiSiB PC6110

PARAMETER	UNIT	VALUE
GENERAL INFORMATION		
Name		SiSiB® PC6110
CAS #	-	2768-02-7
EC number	-	220-449-8
Common synonym	vinyltrimethoxysilane	
Empirical formula	-	C5H12O3Si
Formula	$H_3CO-\underset{\underset{OCH_3}{\vert}}{\overset{\overset{OCH_3}{\vert}}{Si}} - \overset{\overset{H}{\vert}}{C}=CH_2$	
Molecular mass	daltons	148.2
Functional organic group	methoxyacrylo	
Purity	%	99
PHYSICAL PROPERTIES		
State	-	liquid
Color		colorless
Boiling point	°C	122
Density at 25°C	kg/m³	960-970
Refractive index	-	1.3905
Solubility	methanol, ethanol, isopropanol, acetone, benzene, toluene, and xylene	
HEALTH & SAFETY		
Flash point	°C	28
USE & PERFORMANCE		
Manufacturer	SiSiB Silicones/PCC group	
Outstanding features	SiSiB® PC6110, vinyltrimethoxysilane, is used as a polymer modifier via grafting reactions. The resulting pendant trimethoxysilyl groups can function as moisture-activated crosslinking sites. The silane grafted polymer is processed as a thermoplastic, and its crosslinking occurs after fabrication of the finished article upon exposure to moisture.	
Recommended for polymers	acrylics, polyethylene	
Recommended for products	moisture-curing polymers, silane crosslinking, cable insulation, water/sanitary pipes, underfloor heating, sealants	

SiSiB PC6120

PARAMETER	UNIT	VALUE
GENERAL INFORMATION		
Name	SiSiB® PC6120	
CAS #	-	78-08-0
EC number	-	201-081-7
Common synonym	vinyltriethoxysilane	
Empirical formula	-	C8H18O3Si
Formula		H_5C_2O-Si with OC_2H_5 (top), OC_2H_5 (bottom), vinyl group
Molecular mass	daltons	190.4
Functional organic group	methoxyacrylo	
Purity	%	99
PHYSICAL PROPERTIES		
State	-	liquid
Color		colorless
Boiling point	°C	160
Density at 25°C	kg/m³	904-908
Refractive index	-	1.3965
Solubility	methanol, ethanol, isopropanol, acetone, benzene, toluene, and xylene	
HEALTH & SAFETY		
Flash point	°C	44
USE & PERFORMANCE		
Manufacturer	SiSiB Silicones/PCC group	
Outstanding features	SiSiB® PC6120, vinyltriethoxysilane, is a vinyl-functional silane that may be used to improve the bond between glass fiber or mineral fillers and resins that are reactive towards the vinyl group. It is also employed to functionalize resins *via* free radical mechanisms – copolymerization or grafting – and to modify surfaces.	
Recommended for polymers	acrylics, polyethylene	
Recommended for products	moisture-curing polymers, silane cross-linking, cable insulation, water/sanitary pipes, underfloor heating, sealants	

SiSiB PC6130

PARAMETER	UNIT	VALUE
GENERAL INFORMATION		
Name		SiSiB® PC6130
CAS #	-	1067-53-4
EC number	-	213-934-0
Common synonym		vinyltris(2-methoxyethoxy)silane
Empirical formula	-	C11H24O6Si
Formula		$H_3CO(CH_2)_2O-Si\begin{smallmatrix}O(CH_2)_2OCH_3\\\\O(CH_2)_2OCH_3\end{smallmatrix}$
Molecular mass	daltons	280.4
Functional organic group		methoxyacrylo
Purity	%	99
PHYSICAL PROPERTIES		
State	-	liquid
Color		colorless
Boiling point	°C	285
Density at 25°C	kg/m³	1035
Refractive index	-	1.427
HEALTH & SAFETY		
Flash point	°C	92
USE & PERFORMANCE		
Manufacturer		SiSiB Silicones/PCC group
Outstanding features		SiSiB® PC6130 is a vinyl-functional coupling agent that promotes adhesion among unsaturated, polyester-type resins or crosslinked polyethylene resins or elastomers and inorganic substrates, including fiber glass, silica, silicates, and many metal oxides. When used as a coupling agent, it reduces the sensitivity of the products' mechanical and electrical properties to heat and/or moisture.
Recommended for polymers		acrylics, polyethylene, unsaturated polyester
Recommended for products		moisture-curing polymers, silane crosslinking, composites

Wacker® Adhesion Promoter AMS 60

PARAMETER	UNIT	VALUE
GENERAL INFORMATION		
Name		Wacker® Adhesion Promoter AMS 60
CAS #	-	67923-07-3
General description		aminofunctional silane
Composition		>99% amino functional polydimethyl siloxan, >0.1% octamethyl cyclotetrasiloxane
Amine content	meq/g	4.2-4.7
PHYSICAL PROPERTIES		
State	-	liquid
Odor	-	faint
Color	-	colorless, dark
Boiling point	°C	>170
Density at 25°C	kg/m³	1000
Refractive index at 25°C	-	1.432
Vapor pressure at 20°C	hPa	0.0008
Viscosity at 25°C	mPas	20-40
HEALTH & SAFETY		
Autoignition temperature	°C	340
Flash point	°C	>100
Hazardous decomposition products		Methanol by hydrolysis. Measurements have shown the formation of small amounts of formaldehyde at temperatures above about 150°C through oxidation.
Animal testing, acute toxicity, Rat oral LD50	mg/kg	>2000
First aid, eye		If contact with eyes, immediately hold eyelids apart and flush with plenty of water for at least 15 min.
First aid, inhalation		If inhaled remove to fresh air. If not breathing, give artificial respiration. If breathing is difficult give oxygen
First aid, skin		For skin contact, immediately wipe away excess material. Use a waterless hand cleaner to remove as much of the remaining material as possible. Wash with soap and water.
ECOLOGICAL PROPERTIES		
Bioaccumulative potential		Bioaccumulation is not expected to occur

Wacker® Adhesion Promoter AMS 60

PARAMETER	UNIT	VALUE
USE & PERFORMANCE		
Manufacturer		Wacker Chemie AG
Recommended for products		used as a raw material for the production of RTV-1-silicone rubbers

Wacker® Adhesion Promoter AMS 70

PARAMETER	UNIT	VALUE
GENERAL INFORMATION		
Name		Wacker® Adhesion Promoter AMS 70
CAS #	-	67923-07-3
General description		aminofunctional silane
Composition		>75% 3-aminopropyl(methyl) silsesquioxanes, ethoxyterminated, <=20% 3-aminopropyltriethoxysilane
Amine content	meq/g	2.0-2.4
PHYSICAL PROPERTIES		
State	-	liquid
Odor	-	faint
Color	-	colorless
Melting point	°C	<-50
Boiling point	°C	>150
Density at 25°C	kg/m³	1040
Refractive index at 25°C	-	1.415
Vapor pressure at 20°C	hPa	<5
Viscosity at 25°C	mPas	7
HEALTH & SAFETY		
Autoignition temperature	°C	280
Flash point	°C	55
Hazardous decomposition products		Methanol by hydrolysis. Measurements have shown the formation of small amounts of formaldehyde at temperatures above about 150°C through oxidation.
Animal testing, acute toxicity, Rat oral LD50	mg/kg	>2000
First aid, eye		If contact with eyes, immediately hold eyelids apart and flush with plenty of water for at least 15 min.
First aid, inhalation		If inhaled remove to fresh air. If not breathing, give artificial respiration. If breathing is difficult give oxygen
First aid, skin		For skin contact, immediately wipe away excess material. Use a waterless hand cleaner to remove as much of the remaining material as possible. Wash with soap and water.

Wacker® Adhesion Promoter AMS 70

PARAMETER	UNIT	VALUE
ECOLOGICAL PROPERTIES		
Bioaccumulative potential		bioaccumulation is not expected to occur
USE & PERFORMANCE		
Manufacturer		Wacker Chemie AG
Recommended for products		used as a raw material for the production of RTV-1-silicone rubbers

Xiameter OFS-6011 Silane

PARAMETER	UNIT	VALUE
GENERAL INFORMATION		
Name	Xiameter™ OFS-6011 Silane	
CAS #	-	919-30-2
EC number	-	213-048-4
General description	amino functional alkoxysilane	
Composition	98-100% 3-aminopropyltriethoxysilane	
Acronym	-	APTES, APTS
Empirical formula	$H_2NC_3H_6-Si(OC_2H_5)_3$	
Formula		
Molecular mass	daltons	221.37
Chemical class	silane	
Mixture	-	yes
Active matter	wt%	>98.5
Functional organic group	primary amine	
PHYSICAL PROPERTIES		
State	-	liquid
Color	colorless to very pale yellow liquid	
Odor	-	fishy
APHA color	-	<25
Boiling point	°C	217
Kinematic viscosity at 25°C	cSt	1.6
Solubility (diluents)	alcohols, water	
Solubility in water at 25°C	g/l	hydrolytic decomposition releasing ethanol
Specific gravity at 25°C	-	0.946
Viscosity SUS at 25°C	s	1.6
HEALTH & SAFETY		
Flash point	°C	96
Flash point method	-	SCC
Animal testing, acute toxicity, Rat oral LD50	mg/kg	1479-2665
Animal testing, acute toxicity, Rabbit dermal LD50	mg/kg	4041
Carcinogenicity	Did not cause cancer in laboratory animals	

Xiameter OFS-6011 Silane

PARAMETER	UNIT	VALUE
Mutagenicity	In vitro genetic toxicity studies were negative. Animal genetic toxicity studies were negative.	
Teratogenicity	Did not cause birth defects in laboratory animals.	
Hazardous ingredients, labelling	Ethanol	
Exposure, personal protection	Safety glasses, protective clothing based on chemical resistance data, chemical-resistant gloves, general and local exhaust ventilation.	
Dow IHG, TWA	mg/m^3	0.5
NIOSH, REL	mg/m^3	1900/ethanol
OSHA, PEL	mg/m^3	1900/ethanol
ACGIH, TLV	ppm	1000/ethanol
NIOSH, REL	ppm	1000/ethanol
OSHA, PEL	ppm	1000/ethanol
ECOLOGICAL PROPERTIES		
Aquatic toxicity, *Green algae*, 72-h EC50	mg/l	>1000
Aquatic toxicity, *Daphnia magna*, 48-h LC50	mg/l	331
Aquatic toxicity, *Zebra fish*, 96-h LC50	mg/l	>934
Biodegradation	67%/28 d	
Partition coefficient, logP$_{ow}$	-	1.7
USE & PERFORMANCE		
Manufacturer	Dow	
Outstanding properties	improved adhesion of many plastics, resins, and elastomers to inorganic materials and surfaces, improving the properties of mineral-filled rubber, reinforced glass composite. Increased composite wet and dry tensile and flexural strength and modulus. Increased transparency of fiberglass composites.	
Recommended for polymers	WB & SB acrylic, SB PU	
Recommended applications	SB and WB coatings, pigment treatment in WB coating, coupling agent for thermoset resins with glass or mineral fillers	
Concentrations used	0.01 - 2.0% grind or let-down for each specific application, the optimum level of additive should be determined by testing.	
Food approval (FDA)	FDA 175.105	

Xiameter OFS-6030 Silane

PARAMETER	UNIT	VALUE
GENERAL INFORMATION		
Name		Xiameter™ OFS-6030 Silane
CAS #	-	2530-85-0
EC number	-	219-785-8
Composition		3-methacryloxypropyltrimethoxysilane
Empirical formula		H2C=CH(CH3)C(O)OC3H6-Si(OCH3)3
Formula		

PARAMETER	UNIT	VALUE
Molecular mass	daltons	248.35
RTECS number	-	UC0230000
Chemical class	silane	
Active matter	wt%	98
Functional organic group	methacryloxy	
Purity	%	98
PHYSICAL PROPERTIES		
State	-	liquid
Color		clear, white, light straw
Odor	-	aromatic
Boiling point	°C	190
Kinematic viscosity at 25°C	cSt	2.5
Refractive index at 20°C	-	1.43
Specific gravity at 25°C	-	1.045
HEALTH & SAFETY		
NFPA classification	Flammability	1
	Health	0
	Instability	0
HMIS classification	Flammability	1
	Health	0
	Physical hazard	0
Flash point	°C	138
Flash point method	-	OC
Auto-ignition temperature	°C	360
Exposure, personal protection		Safety glasses, protective clothing based on chemical resistance data, chemical-resistant gloves, general and local exhaust ventilation.

Xiameter OFS-6030 Silane

PARAMETER	UNIT	VALUE
Animal testing, acute toxicity, Rat oral LD50	mg/kg	>2000
Animal testing, acute toxicity, Rat dermal LD50	mg/kg	>2000
Animal testing, acute toxicity, Rat inhalation Lc50	mg/kg	>2.28/4H
ECOLOGICAL PROPERTIES		
Aquatic toxicity, *Green algae*, 72-h ErC50	mg/l	>100
Aquatic toxicity, *Zebra fish*, 96-h LC50	mg/l	>1042
Aquatic toxicity, *Daphnia magna*, 96-h LC50	mg/l	>876
USE & PERFORMANCE		
Manufacturer	Dow	
Outstanding properties	reinforce glass composite, reinforcement coupling. Increased composite tensile and flexural strength – both dry and wet. Improved chemical bonding. Increased transparency of polyester fiberglass composites	
Recommended for polymers	unsaturated resin systems, polyester	
Recommended applications	coupling agent for many thermoset and thermoplastics resins with glass or mineral fillers	
Food approval (FDA)	21 CFR Section 177.2465	
Alternative product(s)	XIAMETER OFS-6030, Geniosil GF 31, Dynasylant MEMO VP Si 123, KBM503, SiSiB PC4100	

Xiameter OFS-6032 Silane

PARAMETER	UNIT	VALUE
GENERAL INFORMATION		
Name		Xiameter™ OFS-6032 Silane
CAS #		171869-89-9, 67-56-1, 1760-24-3
Composition		27-37% aminosilane hydrochloride, 53-63% methanol, 8-12% N-(3-(trimethoxysilyl)propyl)ethylenediamine
Common synonym		N1-(vinylbenzyl)-N2-(3-(trimethoxysilyl) propyl)ethane-1,2-diamine hydrochloride
Empirical formula		(H2C=CHC6H4-CH2-NHC2H4NHC3H6-Si(OCH3)3)•HCl
Formula		
Molecular mass	daltons	374.5
RTECS number	-	DG0875000/ cas171869-89-9; PC1400/cas 67-56-1; KV7400000/ cas1760-24-3
Mixture	-	yes
Active matter	wt%	40
Functional organic group		vinylbenzyl-amino, cationic styrlamine
PHYSICAL PROPERTIES		
State	-	liquid
Odor	-	alcohol-like
Color		greenish yellow changing to reddish amber with time
Boiling point	°C	>65
Kinematic viscosity at 25°C	cSt	2
Refractive index at 20°C	-	1.395
Solubility in water at 25°C		hydrolytic decomposition releasing methanol
Specific gravity at 25°C	-	0.90
HEALTH & SAFETY		
NFPA classification	Flammability	3
	Health	3
	Instability	0

Xiameter OFS-6032 Silane

PARAMETER	UNIT	VALUE
HMIS classification	Flammability	3
	Health	4
	Physical hazard	0
Carcinogenicity	IARC, OSHA, NTP: no ingredient of this product present at levels greater than or equal to 0.1% is identified as probable, possible or confirmed human carcinogen	
TDG class	Flammable liquids. METHANOL SOLUTION, 3,II	
ICAO/IATA class	Flammable liquids, n.o.s. Methanol solution, 3, II	
IMDG class	Flammable liquids. METHANOL SOLUTION, 3,II	
Flash point	°C	12.7
Flash point method	-	PMCC
Animal testing, acute toxicity, Rat oral LD50	mg/kg	1897-2574
Animal testing, acute toxicity, Rabbit dermal LD50	mg/kg	>2000/N-(3-(Trimethoxysilyl) propyl)ethylenedi-amine
Animal testing, acute toxicity, Rat inhalation, LC50	mg/m^3	1.49-2.44/4H/ dust/mist/N-(3-(Trimethoxysilyl) propyl)ethylenedi-amine
Effect of exposure, eye (human)	Causes serious eye damage.	
Effect of exposure, inhalation (human)	May cause allergy or asthma symptoms or breathing difficulties if inhaled. May cause respiratory irritation, drowsiness or dizziness or organs damage. Estimate acute toxicity 4.55 mg/l/4H	
Effect of exposure, skin (human)	Causes skin irritation. May cause an allergic skin reaction. Estimate acute toxicity 454.55 mg/kg	
Effect of exposure, swallowing (human)	Harmful if swallowed. Estimated acute toxicity 454.55 mg/kg	
Exposure, personal protection	Safety glasses, protective clothing based on chemical resistance data, chemical-resistant gloves, general and local exhaust ventilation.	
First aid, eye	Flush eyes with plenty of water for at least 15 minutes. Remove contact lens, if worn.	

Xiameter OFS-6032 Silane

PARAMETER	UNIT	VALUE
First aid, inhalation	Remove to fresh air. If not breathing, give artificial respiration. If breathing is difficult, give oxygen. Get medical attention.	
First aid, skin	Flush with plenty of water for at least for 15 min. and removing contaminated clothing and shoes. Get medical attention.	
Specific target organ	eyes, central nervous system/methanol	
NIOSH, REL	mg/m^3	ST325/methanol
OSHA, PEL	mg/m^3	260/methanol
ACGIH, TLV	ppm	200/methanol
NIOSH, REL	ppm	ST250/methanol
OSHA, PEL	ppm	200/methanol
UN/NA class	-	1230
ECOLOGICAL PROPERTIES		
Aquatic toxicity, *Green algae*, 72-h EC50	mg/l	69
Aquatic toxicity, *Daphnia magna*, 48-h LC50	mg/l	74
Bioconcentration factor	BCF	<10/methanol/ Leuciscus idus (*Golden orfe*)
Biodegradation probability	95%/20d/methanol/readily biodegradable; 39%/N-(3-(Trimethoxysilyl)propyl) ethylenediamine(not readily biodegradable)	
Partition coefficient, log K$_{oc}$	-0.77/methanol; -0.3/N-(3-(Trimethoxysilyl)propyl)ethylenediamine;	
USE & PERFORMANCE		
Manufacturer	Dow	
Outstanding properties	organic and inorganic reactivity, improved chemical bonding, good resin wet-out, increased flexural and tensile strength	
Recommended for polymers	epoxies for PCBs, polyolefins, all polymer types	
Recommended applications	coupling agent for many resin systems; especially useful for fiberglass-reinforced printed circuit boards	

Xiameter OFS-6032 Silane

PARAMETER	UNIT	VALUE
Concentrations used		The total concentration of silane should be varied according to the surface area of the inorganic substrate. For example, fillers with high surface areas require more silane than those with relatively low surface areas. Typical concentrations range from 0.1 to 1.0% active ingredients based on the weight of the inorganic material.

Xiameter OFS-6040 Silane

PARAMETER	UNIT	VALUE
GENERAL INFORMATION		
Name		Xiameter™ OFS-6040 Silane
CAS #	-	2530-83-8
EC number	-	219-784-2
Composition	98-100% glycidoxypropyltrimethoxysilane	
Common synonym	3-glycidyloxypropyltrimethoxysilane	
Acronym	-	TMSPGE/GLYMO
Empirical formula	CH2(O)CHCH2OC3H6-Si(OCH3)3	
Formula		
Molecular mass	daltons	236.34
RTECS number	-	VV4025000
Chemical class	silane	
Purity	%	>98.5
PHYSICAL PROPERTIES		
State	-	liquid
Odor	-	aromatic
Color	colorless to pale yellow	
Color, Platinum-cobalt scale	-	50
Boiling point	°C	>250
Melting point	°C	-70
Kinematic viscosity at 25°C	cSt	3.09
Refractive index at 20°C	-	1.428
Solubility in water at 25°C	rapid hydrolytic decomposition releasing methanol	
Specific gravity at 25°C	-	1.07
Vapor pressure at 20°C	kPa	0.0011
Viscosity at 20°C	mPas	2.9
HEALTH & SAFETY		
NFPA classification	Flammability	1
	Health	3
	Instability	0
HMIS classification	Flammability	1
	Health	3
	Physical hazard	0

Xiameter OFS-6040 Silane

PARAMETER	UNIT	VALUE
Carcinogenicity		IARC, OSHA, NTP: no ingredient of this product present at levels greater than or equal to 0.1% is identified as probable, possible or confirmed human carcinogen
Mutagenicity		not classified based on available information.
DOT class		not regulated as a dangerous goods
TDG class		not regulated as a dangerous goods
ICAO/IATA class		not regulated as a dangerous goods
IMDG class		not regulated as a dangerous goods
Flash point	°C	>94
Flash point method	-	SCC
Animal testing, acute toxicity, Rat oral, LC50	mg/kg	8025
Animal testing, acute toxicity, Rabbit dermal LD50	mg/kg	3970
Animal testing, acute toxicity, Rat inhalation, LC50	mg/m^3	>5300/4H
Effect of exposure, eye (human)		Serious eye damage/eye irritation.
Effect of exposure, skin (human)		Does not cause skin sensitization.
Effect of exposure, swallowing (human)		Harmful if swallowed.
Exposure, personal protection		Safety glasses, protective clothing based on chemical resistance data, chemical-resistant gloves, general and local exhaust ventilation.
First aid, eye		Flush eyes with plenty of water for at least 15 minutes. Remove contact lens, if worn.
First aid, inhalation		Remove to fresh air. If not breathing, give artificial respiration. If breathing is difficult, give oxygen. Get medical attention.
First aid, skin		Flush with plenty of water for at least for 15 min. and removing contaminated clothing and shoes. Get medical attention.
Specific target organ		eyes, central nervous system/methanol
TWA Dow IHG	ppm	0.5
NIOSH, REL	mg/m^3	ST325/methanol
OSHA, PEL	mg/m^3	260/methanol
ACGIH, TLV	ppm	200/methanol

Xiameter OFS-6040 Silane

PARAMETER	UNIT	VALUE
NIOSH, REL	ppm	ST250/methanol
OSHA, PEL	ppm	200/methanol
ECOLOGICAL PROPERTIES		
Aquatic toxicity, *Green algae*, 96-h EC50	mg/l	350
Aquatic toxicity, *Daphnia magna*, 48-h LC50	mg/l	324
Aquatic toxicity *Fathead minnow*, 96-h LC50	mg/l	<100
Aquatic toxicity, *Zebra fish*, 96-h LC50	mg/l	<100
Bioconcentration factor	BCF	bioconcentration potential is low (BCF < 100 or Log Pow < 3).
Biodegradation probability	37%/28 d (not readily biodegradable)	
Partition coefficient, log K_{oc}	-	-2.6
Partition coefficient, log K_{ow}	-	-0.9
USE & PERFORMANCE		
Manufacturer	Dow	
Outstanding properties	improved adhesion. Increases composite strength properties. Increased composite wet and dry tensile strength and modulus. Increased composite wet and dry flexural strength and modulus.	
Recommended for polymers	acrylics, EMC, epoxy, PA, PU, polysulfide	
Recommended applications	enhanced the bonding of a polymer coating, paint, or adhesive to glass, metals, and polymer surfaces, used for silica treatment for EMC application	
Alternative product(s)	Geniosil GF 80, Dynasylan GLMO, KBM403, SiSiB PC3100	

Xiameter OFS-6062 Silane

PARAMETER	UNIT	VALUE
GENERAL INFORMATION		
Name	Xiameter OFS-6062 Silane	
CAS #	-	4420-74-0
Composition	3-mercaptopropyltrimethoxysilane	
Empirical formula	HS(CH2)3Si(OCH3)3	
Formula		

$$\text{H}_3\text{CO}-\underset{\underset{\text{OCH}_3}{|}}{\overset{\overset{\text{OCH}_3}{|}}{\text{Si}}}\diagup\diagdown\diagup\text{SH}$$

RTECS number	-	TZ7800000
Chemical class	silane	
Functional organic group	mercaptan	
Purity	wt%	98
PHYSICAL PROPERTIES		
State	-	liquid
Odor	-	mercaptan
Color	-	colorless
Kinematic viscosity at 25°C	cSt	2.0
Solubility in water at 25°C	hydrolytic decomposition releasing methanol	
Specific gravity at 25°C	-	1.05
Thermal decomposition products	formaldehyde	
Viscosity SUS at 25°C	s	2.0
HEALTH & SAFETY		
NFPA classification	Flammability	3
	Health	3
	Reactivity	0
HMIS classification	Flammability	2
	Health	2
	Reactivity	0
Carcinogenicity	IARC, OSHA, NTP: no ingredient of this product present at levels greater than or equal to 0.1% is identified as probable, possible or confirmed human carcinogen	
Mutagenicity	not classified based on available information.	
TDG class	ENVIRONMENTALLY HAZARDOUS SUBSTANCE, LIQUID, N.O.S.(3-Mercaptopropyl trimethoxysilane) 9, III	

Xiameter OFS-6062 Silane

PARAMETER	UNIT	VALUE
ICAO/IATA class	Environmentally hazardous substance, liquid, n.o.s. (3-Mercaptopropyl trimethoxysilane), 9, III	
IMDG class	ENVIRONMENTALLY HAZARDOUS SUBSTANCE, LIQUID, N.O.S.(3-Mercaptopropyl trimethoxysilane), 9, III, Marine pollutant	
Flash point	°C	75
Flash point method	-	TCC
Effect of exposure, inhalation (human)	May cause skin irritation. Low to moderate skin sensitization rate in humans. Estimate acute toxicity 2,243 mg/kg	
Effect of exposure, skin (human)	May cause skin irritation. Low to moderate skin sensitization rate in humans. Estimate acute toxicity 2,243 mg/kg	
Effect of exposure, swallowing (human)	Harmful if swallowed. Estimated acute toxicity 454.55 mg/kg	
Exposure, personal protection	Safety glasses, protective clothing based on chemical resistance data, chemical-resistant gloves, general and local exhaust ventilation.	
First aid, eye	Flush eyes with plenty of water for at least 15 minutes. Remove contact lens, if worn.	
First aid, inhalation	Remove to fresh air. If not breathing, give artificial respiration. If breathing is difficult, give oxygen. Get medical attention.	
First aid, skin	Flush with plenty of water for at least for 15 min. and removing contaminated clothing and shoes. Get medical attention.	
Specific target organ	eyes, central nervous system/methanol	
NIOSH, REL	mg/m^3	ST325/methanol
OSHA, PEL	mg/m^3	260/methanol
ACGIH, TLV	ppm	200/methanol
NIOSH, REL	ppm	ST250/methanol
OSHA, PEL	ppm	200/methanol
UN/NA class	-	3082

Xiameter OFS-6062 Silane

PARAMETER	UNIT	VALUE
ECOLOGICAL PROPERTIES		
Aquatic toxicity, *Green algae*, 96-h EC50	mg/l	931.0/72H/3-mercaptopropyl trimethoxysilane; 22000/96H/methanol
Aquatic toxicity, *Bluegill sunfish*, 96-h LC50	mg/l	15400/methanol
Aquatic toxicity, *Daphnia magna*, 48-h LC50	mg/l	6.7/ 3-mercaptopropyl trimethoxysilane; 10000/methanol
Aquatic toxicity, *Zebra fish*, 96-h LC50	mg/l	439.00/ 3-mercaptopropyl trimethoxysilane
Bioconcentration factor	BCF	<10/methanol/ *Leuciscus idus* (*Golden orfe*)
Biodegradation probability		95%/20d/methanol/readily biodegradable; 51%/28d/3-Mercaptopropyl trimethoxysilane (not readily biodegradable)
Partition coefficient, log K_{oc}		-0.77/methanol
USE & PERFORMANCE		
Manufacturer		Dow
Outstanding properties		provides coupling between inorganic surfaces (such as clay, glass, etc.) and sulfur cured elastomers
Recommended for polymers		epoxy, polysulfide, EPDM, natural rubber, SBR, nitrile rubber
Recommended applications		coupling agent to improve the adhesion of sulfur-cured elastomers to inorganic fillers, fiberglass, and surfaces; treated fillers compatible with epoxy, polysulfide, EPDM, natural

Xiameter OFS-6075 Silane

PARAMETER	UNIT	VALUE
GENERAL INFORMATION		
Name		Xiameter OFS-6075 Silane
CAS #	-	4130-08-9
Composition		90-<100% vinyltriacetoxysilane + 4-6% Dimer of vinylacetoxysilane + 2% Acetic anhydride
Formula		
Chemical class		silane
Mixture	-	yes
Functional organic group		acetoxy
Purity	wt%	98
PHYSICAL PROPERTIES		
State	-	liquid
Odor	-	acetic acid
Color	-	colorless
Boiling point	°C	220
Kinematic viscosity at 25°C	cSt	1.0
Specific gravity at 25°C	-	1.65
HEALTH & SAFETY		
NFPA classification	Flammability	1
	Health	3
	Instability	0
HMIS classification	Flammability	1
	Health	3
	Physical hazard	4
Carcinogenicity		IARC, OSHA, NTP: no ingredient of this product present at levels greater than or equal to 0.1% is identified as probable, possible or confirmed human carcinogen
TDG class		CORROSIVE LIQUID, ACIDIC, ORGANIC, N.O.S. (Vinyltriacetoxysilane, Dimer of Vinylacetoxysilane), 8, II
ICAO/IATA class		Corrosive liquid, acidic, organic, n.o.s. (Vinyltriacetoxysilane, Dimer of Vinylacetoxysilane) 8, II
IMDG class		CORROSIVE LIQUID, ACIDIC, ORGANIC, N.O.S. (Vinyltriacetoxysilane, Dimer of Vinylacetoxysilane), 8, II

Xiameter OFS-6075 Silane

PARAMETER	UNIT	VALUE
Autoignition temperature	°C	415
Flash point	°C	112
Flash point method	-	PMCC
Hazardous combustion products		carbon oxides, SiO_x, formaldehyde
Hazardous ingredients, labelling		vinyltriacetoxysilane, dimer of vinylacetoxysilane
Animal testing, acute toxicity, Rat oral, LD50	mg/kg	>3000
Animal testing, acute toxicity, Rat inhalation, LC50	mg/m^3	<1670/4H/ acetic anhydride, 2300ppm/7H/vinyl-triacetoxysilane,
Effect of exposure, eye (human)		Causes serious eye damage.
Effect of exposure, skin (human)		Causes severe skin burns.
Effect of expcsure, swallowing (human)		Harmful if swallowed. Causes digestive tract burns.
Exposure, personal protection		Safety glasses, protective clothing based on chemical resistance data, chemical-resistant gloves, general and local exhaust ventilation.
First aid, eye		Immediately flush the contaminated eye(s) with lukewarm, gently flowing water for 15 - 20 minutes while holding the eyelid(s) open. If contact lens is present, DO NOT delay irrigation or attempt to remove the lens. Take care not to rinse contaminated water into the unaffected eye or onto the face. Immediately obtain medical attention.
First aid, inhalation		Remove to fresh air. If not breathing, give artificial respiration. If breathing is difficult, give oxygen. Get medical attention.
First aid, skin		Flush skin with soap and plenty of water for at least for 15 min. and removing contaminated clothing and shoes. Get medical attention.
NIOSH, REL	mg/m^3	25/acetic anhydride
OSHA, PEL	mg/m^3	25/acetic anhydride
ACGIH, TLV	ppm	10/acetic anhydride
NIOSH, REL	ppm	10/acetic anhydride
OSHA, PEL	ppm	10/acetic anhydride
UN/NA class	-	3265

Xiameter OFS-6075 Silane

PARAMETER	UNIT	VALUE
ECOLOGICAL PROPERTIES		
Aquatic toxicity, *Green algae*, 72-h EC50	mg/l	22.83
Aquatic toxicity, *Daphnia magna*, 48-h LC50	mg/l	83.81
Bioconcentration factor	BCF	
Biodegradation probability	96%/20d/acetic anhydride	
Partition coefficient, log K_{oc}	-0.27 acetic anhydride	
USE & PERFORMANCE		
Manufacturer	Dow	
Recommended for products	wire and cable	
Alternative product(s)	chemical equivalent of Dowsil Z-6075 Silane	

Xiameter OFS-6076 Silane

PARAMETER	UNIT	VALUE
GENERAL INFORMATION		
Name		Xiameter™ OFS-6076 Silane
CAS #	-	2530-87-2
EC number	-	219-787-9
Composition	89-100% chloropropyltrimethoxysilane	
Empirical formula	ClC3H6-Si(OCH3)3	
Formula		

PARAMETER	UNIT	VALUE
Molecular mass	daltons	198.5
Chemical class	silane	
Functionality	-	chloroalkyl
Mixture	-	yes
Purity	wt%	97
PHYSICAL PROPERTIES		
State	-	liquid
Odor	-	alcohol-like
Color	-	clear, light yellow
Boiling point	°C	196
Kinematic viscosity at 25°C	cSt	1.4
Refractive index at 20°C	-	1.418
Solubility in water at 25°C	hydrolytic decomposition releasing methanol	
Specific gravity at 25°C	-	1.09
Thermal decomposition products	formaldehyde	
HEALTH & SAFETY		
NFPA classification	Flammability	2
	Health	0
	Instability	0
HMIS classification	Flammability	2
	Health	0
	Physical hazard	0
Carcinogenicity	IARC, OSHA, NTP: no ingredient of this product present at levels greater than or equal to 0.1% is identified as probable, possible or confirmed human carcinogen	
TDG class	FLAMMABLE LIQUID, N.O.S. (Chloropropyltrimethoxysilane) 3,III	

Xiameter OFS-6076 Silane

PARAMETER	UNIT	VALUE
ICAO/IATA class		Flammable liquid. n.o.s. (Chloropropyltrimethoxysilane), 3,III
Flash point	°C	51
Flash point method	-	CC
Hazardous combustion products		Carbon oxides, SiOx, chlorine compounds, formaldehyde
Hazardous ingredients, labelling		Chloropropyltrimethoxysilane
Hazardous products of hydrolysis		methanol
Animal testing, acute toxicity, Rat oral LD50	mg/kg	>2000
Animal testing, acute toxicity, Rat dermal LD50	mg/kg	>2000
Effect of exposure, eye (human)		Avoid contact with eyes. Based on available information not classified as a eye irritant. Rabbit irritation to eyes, reversing within 21 days.
Effect of exposure, inhalation (human)		Based on available information not classified as a irritant. No significant health effects observed in animals at concentrations of 1 mg/l/6h/d or less.
Effect of exposure, skin (human)		Based on available information not classified as a skin irritant. Does not cause skin sensitization
Effect of exposure, swallowing (human)		The substance or mixture has no acute oral toxicity.
Exposure, personal protection		Safety glasses, protective clothing based on chemical resistance data, chemical-resistant gloves, general and local exhaust ventilation.
First aid, eye		Flush eyes with plenty of water for at least 15 minutes. Remove contact lens, if worn. Get medical attention if irritation develops and persists
First aid, inhalation		Remove to fresh air. If not breathing, give artificial respiration. If breathing is difficult, give oxygen. Get medical attention.
First aid, skin		Flush skin with soap and plenty of water for at least for 15 min. and removing contaminated clothing and shoes. Get medical attention.
Specific target organ		eyes, central nervous system/methanol
TWA Dow IHG	ppm	0.25
NIOSH, REL	mg/m^3	260/chloropropyl-trimethoxysilane, ST325/methanol

Xiameter OFS-6076 Silane

PARAMETER	UNIT	VALUE
OSHA, PEL	mg/m³	260/chloropropyl-trimethoxysilane, 260/methanol
ACGIH, TLV	ppm	260/chloropropyl-trimethoxysilane, 260/methanol
NIOSH, REL	ppm	260/chloropropyl-trimethoxysilane, 260/methanol
OSHA, PEL	ppm	260/chloropropyl-trimethoxysilane, 260/methanol
UN risk phrases, R	R10,R36/37/38,R20	
US safety phrases, S	S26,S36/37/39	
UN/NA class	-	1993
ECOLOGICAL PROPERTIES		
Aquatic toxicity, *Green algae*, 72-h EC50	mg/l	883/chloropropyltri-methoxysilane
Aquatic toxicity, *Daphnia magna*, 48-h LC50	mg/l	869/chloropropyltri-methoxysilane
Aquatic toxicity, *Zebra fish*, 96-h LC50	mg/l	>100/chloropropyl-trimethoxysilane
Biodegradation probability	chloropropyltrimethoxysilane/ not readily biodegradable	
Partition coefficient, log K_{oc}	-	-1.12/chloropropyl-trimethoxysilane
Stability in water (half-life)	-	0.89 h/7pH 7
USE & PERFORMANCE		
Manufacturer	Dow	
Outstanding properties	improved adhesion, increased dry and wet flexural strength, versatile chloroalkyl functionality	
Recommended for polymers	epoxy, PS, PU	
Recommended for products	adhesives, glass fabric	
Recommended applications	effective coupling agent for treating glass fabric used in polystyrene laminates.	
Alternative product(s)	SiSiB PC 5011, KBM703	

Xiameter OFS-6094 Silane

PARAMETER	UNIT	VALUE
GENERAL INFORMATION		
Name	Xiameter OFS-6094 Silane	
CAS #	-	1760-24-3
EC number		217-164-6
General description	high purity version of Xiameter OFS-6020 Silane	
Composition	88%-<100% N-(3-(trimethoxysilyl)pro-pyl)ethylenediamine + 0.77%-<1.04% methanol	
Common synonym	N-(2-aminoethyl)-3-aminopropyltrime-thoxysilane	
Acronym	-	DAMO
Empirical formula	H2NC2H4NHC3H6-Si(OCH3)3	
Formula		
Molecular mass	daltons	222.36
RTECS number	-	KV7400000
Chemical class	silane	
Mixture	-	yes
Functional organic group	amine	
Purity	wt%	98
PHYSICAL PROPERTIES		
State	-	liquid
Color	-	colorless
Odor	-	amine-like
Boiling point	°C	264
Refractive index	-	1.445
Kinematic viscosity at 25°C	cSt	4.2
Specific gravity at 25°C	-	1.02
HEALTH & SAFETY		
NFPA classification	Flammability	1
	Health	3
	Instability	0
HMIS classification	Flammability	1
	Health	3
	Physical hazard	0

Xiameter OFS-6094 Silane

PARAMETER	UNIT	VALUE
Carcinogenicity		IARC, OSHA, NTP: no ingredient of this product present at levels greater than or equal to 0.1% is identified as probable, possible or confirmed human carcinogen
TDG class		not regulated as a dangerous goods
ICAO/IATA class		not regulated as a dangerous goods
IMDG class		not regulated as a dangerous goods
Autoignition temperature	°C	320
Flash point	°C	94
Flash point method	-	SCC
Hazardous combustion products		Carbon oxides, SiO, NOx, formaldehyde
Animal testing, acute toxicity, Rat oral LD50	mg/kg	>2300/ N-(3-(trimethoxysilyl)propyl)ethylenedi-amine, 300/metha-nol
Animal testing, acute toxicity, Rabbit dermal LD50	mg/kg	>2000/N-(3-(trimethoxysilyl)propyl)ethylenedi-amine
Animal testing, acute toxicity, Rat inhalation, LC50	mg/m^3	>1490/4H/N-(3-(trimethoxysilyl)propyl)ethylenedi-amine, 3000/4h/methanol
Effect of exposure, eye (human)		Causes serious eye damage.
Effect of exposure, inhalation (human)		Acute toxicity estimated: 1.49 mg/l/4H
Effect of exposure, skin (human)		May cause an allergic skin reaction. Acute toxicity estimated: > 5,000 mg/kg
Effect of exposure, swallowing (human)		Harmful if swallowed. Acute toxicity estimated: 2,188 mg/kg
Exposure, personal protection		Safety glasses, protective clothing based on chemical resistance data, chemical-resistant gloves, general and local exhaust ventilation.
First aid, eye		Immediately flush the contaminated eye(s) with lukewarm, gently flowing water for 15 - 20 minutes while hold-ing the eyelid(s) open. If contact lens is present, DO NOT delay irrigation or attempt to remove the lens. Take care not to rinse contaminated water into the unaffected eye or onto the face. Immedi-ately obtain medical attention.

Xiameter OFS-6094 Silane

PARAMETER	UNIT	VALUE
First aid, inhalation	Remove to fresh air. If not breathing, give artificial respiration. If breathing is difficult, give oxygen. Get medical attention.	
First aid, skin	Flush skin with soap and plenty of water for at least for 15 min. and removing contaminated clothing and shoes. Get medical attention.	
NIOSH, REL	mg/m³	ST325/methanol
OSHA, PEL	mg/m³	260/methanol
ACGIH, TLV	ppm	200/methanol
NIOSH, REL	ppm	ST250/methanol
OSHA, PEL	ppm	200/methanol
ECOLOGICAL PROPERTIES		
Aquatic toxicity, *Green algae*, 96-h EC50	mg/l	8.8/72/N-(3-(trimethoxysilyl)propyl)ethylenediamine
Aquatic toxicity, *Daphnia magna*, 48-h LC50	mg/l	81/N-(3-(trimethoxysilyl)propyl)ethylenediamine, >10000/methanol
Aquatic toxicity, *Zebra fish*, 96-h LC50	mg/l	597/N-(3-(trimethoxysilyl)propyl)ethylenediamine
Biodegradation probability	39% N-(3-(Trimethoxysilyl)propyl)ethylenediamine/not readily biodegradable	
Partition coefficient, log K_{oc}	-	-0.30 N-(3-(trimethoxysilyl)propyl)ethylenediamine
USE & PERFORMANCE		
Manufacturer	Dow	
Outstanding properties	high purity version of OFS-6020, coupling agent. Improved adhesion, increased wet and dry tensile strength and modulus to the composite, increased wet and dry flexural strength and modulus to the composite, improved compatibility between inorganic filler and organic polymer, reinforced glass composite.	

Xiameter OFS-6094 Silane

PARAMETER	UNIT	VALUE
Recommended for polymers		acrylic, PA, epoxy, phenolics, PVC, urethanes, melamines, nitrile rubber
Recommended applications		coupling agent for clay reinforced elastomers and for phenolic, melamine, and other organic resins used as binders for glass and mineral wood insulation, abrasives, and molding components
Alternative product(s)		chemical equivalent of Dow Corning Z-6094 Silane

Xiameter OFS-6106 Silane

PARAMETER	UNIT	VALUE
GENERAL INFORMATION		
Name	Xiameter OFS-6106 Silane	
CAS #	-	2530-83-8
General description	mixture of glycidoxypropyltrimethoxysilane (Z-6040) with a melamine resin	
Composition	40%-60% melamine-formaldehyde resin, 30-40% hexamethyoxymethyl-melamine, 8%-12% glycidoxypropyl trimethoxysilane	
Formula		
Chemical class	silane	
Mixture	-	yes
Active matter	wt%	100
Functional organic group	epoxy	
PHYSICAL PROPERTIES		
State	-	liquid
Color	white to slightly yellow	
Odor	-	amine-like
Boiling point	°C	>250
Kinematic viscosity at 25°C	cSt	725
Refractive index at 20°C	-	1.5104
Solubility (diluents)	solvents: methanol, isopropanol, methoxypropanol, xylene, and aqueous alcohol mixture	
Solubility in water at 25°C	hydrolytic decomposition releasing ethanol	
Specific gravity at 25°C	-	1.19
Thermal decomposition products	formaldehyde	
HEALTH & SAFETY		
NFPA classification	Flammability	2
	Health	3
	Instability	0
Carcinogenicity	IARC, OSHA, NTP: no ingredient of this product present at levels greater than or equal to 0.1% is identified as probable, possible or confirmed human carcinogen	
Mutagenicity	not classified based on available information.	

Xiameter OFS-6106 Silane

PARAMETER	UNIT	VALUE
DOT class	not regulated as a dangerous goods	
TDG class	FLAMMABLE LIQUID, N.O.S. (Methanol) 3,III	
ICAO/IATA class	Flammable liquid, n.o.s. (methanol), 3, III	
IMDG class	FLAMMABLE LIQUID, N.O.S. (Methanol), 3,III	
Flash point	°C	48.8
Flash point method	-	PMCC
Hazardous combustion products	Carbon oxides, SiO, NOx, formaldehyde	
Hazardous ingredients, labelling	Methanol	
Hazardous products of hydrolysis	Methanol	
Animal testing, acute toxicity, Rat oral LD50	mg/kg	1606 (hexa-methyoxymethyl-melamine), 8025 (glycidoxypropyltri-methoxysilane)
Animal testing, acute toxicity, Rabbit dermal LD50	mg/kg	>3500
Animal testing, acute toxicity, Rat inhalation, LC50	mg/m^3	>5300/4H
Effect of exposure, eye (human)	Avoid contact with eyes. Direct contact may cause severe irritation.	
Effect of exposure, inhalation (human)	Avoid inhalation of vapor or mist. Acute inhalation toxicity estimate: >20 mg/l/4H	
Effect of exposure, skin (human)	Avoid prolonged or repeated contact with skin. Acute dermal toxicity estimate: >2000 mg/kg	
Effect of exposure, swallowing (human)	Harmful if swallowed.	
Effect of repeated or overexposure (human)	Damage to health by prolonged exposure through inhalation and in contact with skin and if swallowed.	
Exposure, personal protection	Safety glasses, protective clothing based on chemical resistance data, chemical-resistant gloves, general and local exhaust ventilation.	
First aid, eye	Flush eyes with plenty of water for at least 15 minutes. Remove contact lens, if worn.	
First aid, inhalation	Remove to fresh air. If not breathing, give artificial respiration. If breathing is difficult, give oxygen. Get medical attention.	

Xiameter OFS-6106 Silane

PARAMETER	UNIT	VALUE
First aid, skin	Flush with plenty of water for at least for 15 min. and removing contaminated clothing and shoes. Get medical attention.	
Specific target organ	eyes, central nervous system/methanol	
TWA Dow IHG (glycidoxypropyltrimethoxysil)	ppm	0.5
NIOSH, REL	mg/m^3	ST325/methanol
OSHA, PEL	mg/m^3	260/methanol
ACGIH, TLV	ppm	200/methanol
NIOSH, REL	ppm	ST250/methanol
OSHA, PEL	ppm	200/methanol
UN/NA class	-	1993
ECOLOGICAL PROPERTIES		
Aquatic toxicity, *Green algae*, 96-h EC50	mg/l	>10-100/72H
Aquatic toxicity, *Daphnia magna*, 96-h LC50	mg/l	>10-100/melamine-formaldehyde resin; >10-100/methylated melamine
Aquatic toxicity, *Daphnia magna*, 48-h LC50	mg/l	710/glycidoxypropyl trimethoxysilane; >10-100/ melamine-formaldehyde resin; >10-100/methylated melamine
Bioconcentration factor	BCF	<10/methanol/Leuciscus idus (Golden orfe)
Biodegradation probability	95%/20d/methanol/readily biodegradable; 37%/28d/Glycidoxypropyl trimethoxysilane (not readily biodegradable)	
Partition coefficient, log K_{oc}	-	-0.77/ methanol; -2.6/P70Glycidoxypropyl trimethoxysilane
USE & PERFORMANCE		
Manufacturer	Dow	
Outstanding properties	improved adhesion of the organic polymer to inorganic substrate or filler, improved wet and dry physical properties of the composite, improved mixing and compatibility of filled systems	

Xiameter OFS-6106 Silane

PARAMETER	UNIT	VALUE
Recommended for polymers		acrylics, epoxy, EVA, fluoropolymers, urethane, phenolic, PEEK, rubber, thermoplastics polyester, polysulfones, poly(phenylene sulfide)s, melamines, polyimides
Recommended for products		adhesives, coatings, paints, fiberglass, filler treatment
Recommended applications		provides coupling for inorganic materials, such as glass and metals, to many organic polymers and engineering plastics
Concentrations used		mix 1-2% in the resin during compounding or before final fabrication to impart unprimed adhesion.
Guidelines for use		make a 5% solution of XIAMETER OFS-6106 in a solvent and wipe it on the surface of an inorganic substrate or on a polymer surface as a primer. Allow the solvent to evaporate, or heat (to not greater than 110°C) to remove solvent.

Xiameter OFS-6224 Silane

PARAMETER	UNIT	VALUE
GENERAL INFORMATION		
Name		Xiameter™ OFS-6224 Silane
CAS #		171869-90-2, 67-56-1
Composition		30-50% (N((vinylphenyl)methyl)(ethyl-enediaminepropyl))trimethoxysilane hydrolyzed derivatives, 50-70% methanol, (impurity: 1-5% chloride salts of amino silanes)
Common synonym		N1-(vinylbenzyl)-N2-(3-(trimethoxysilyl) propyl)ethane-1,2-diamine hydrochloride
Empirical formula		(H2C=CHC6H4-CH2-NHC2H4NHC3H6-Si(OCH3)3)
Formula		
Molecular mass	daltons	374.5
RTECS number	-	DG0875000/ cas171869-89-9; PC1400/cas 67-56-1; KV7400000/ cas1760-24-3
Chemical class	silane	
Mixture	-	yes
Functional organic group	vinylbenzyl-amine	
Organoreactive group	-(CH2)3NR(CH2)N-R2; R is either H or vinylbenzene	
PHYSICAL PROPERTIES		
State	-	liquid
Odor	-	alcohol-like
Color	greenish yellow changing to reddish amber with time	
Boiling point	°C	>65
Kinematic viscosity at 25°C	cSt	1.00 - 3.00
Refractive index at 20°C	-	1.388
Solubility in water at 25°C	hydrolytic decomposition releasing methanol	
Specific gravity at 25°C	-	0.88
Viscosity SUS at 25°C	s	1.0 - 3.0

Xiameter OFS-6224 Silane

PARAMETER	UNIT	VALUE
HEALTH & SAFETY		
NFPA classification	Flammability	3
	Health	3
	Reactivity	0
HMIS classification	Flammability	3
	Health	4
	Reactivity	0
Carcinogenicity	IARC, OSHA, NTP: no ingredient of this product present at levels greater than or equal to 0.1% is identified as probable, possible or confirmed human carcinogen	
Mutagenicity	not classified based on available information.	
TDG class	Flammable liquids. METHANOL SOLUTION, 3,II	
ICAO/IATA class	Flammable liquids, n.o.s. Methanol solution, 3, II	
IMDG class	Flammable liquids. METHANOL SOLUTION, 3,II	
Flash point	°C	13.0
Flash point method	-	CC
Animal testing, acute toxicity, Rabbit dermal LD50	mg/kg	not irritating/mercaptopropyltrimethoxysilane
Animal testing, acute toxicity, Rat inhalation, LC50	mg/m^3	1.49/4H/dust/mist/N-(3-(trimethoxysilyl)propyl)ethylenediamine
Effect of exposure, eye (human)	Causes serious eye damage.	
Effect of exposure, inhalation (human)	May cause allergy or asthma symptoms or breathing difficulties if inhaled. May cause respiratory irritation, drowsiness or dizziness or organs damage. Estimate acute toxicity 4.55 mg/l/4H	
Effect of exposure, skin (human)	Causes skin irritation. May cause an allergic skin reaction. Estimate acute toxicity 454.55 mg/kg	
Effect of exposure, swallowing (human)	Harmful if swallowed. Estimated acute toxicity 454.55 mg/kg	

Xiameter OFS-6224 Silane

PARAMETER	UNIT	VALUE
Exposure, personal protection	Safety glasses, protective clothing based on chemical resistance data, chemical-resistant gloves, general and local exhaust ventilation.	
First aid, eye	Flush eyes with plenty of water for at least 15 minutes. Remove contact lens, if worn.	
First aid, inhalation	Remove to fresh air. If not breathing, give artificial respiration. If breathing is difficult, give oxygen. Get medical attention.	
First aid, skin	Flush with plenty of water for at least for 15 min. and removing contaminated clothing and shoes. Get medical attention.	
Specific target organ	eyes, central nervous system/methanol	
NIOSH, REL	mg/m^3	ST325/methanol
OSHA, PEL	mg/m^3	260/methanol
ACGIH, TLV	ppm	200/methanol
NIOSH, REL	ppm	ST250/methanol
OSHA, PEL	ppm	200/methanol
UN/NA class	-	1223
ECOLOGICAL PROPERTIES		
Aquatic toxicity, *Daphnia magna*, 48-h LC50	mg/l	10000/methanol
Bioconcentration factor	BCF	<10/methanol/ *Leuciscus idus* (*Golden orfe*)
Biodegradation probability	95%/20d/methanol/readily biodegradable	
Partition coefficient, log K$_{oc}$	-0.77/methanol	
USE & PERFORMANCE		
Manufacturer	Dow	
Outstanding properties	improved adhesion of the organic polymer to inorganic substrate or filler, improved wet and dry physical properties of the composite, improved mixing and compatibility of filled system	
Recommended for polymers	epoxies for PCBs, polyolefins, all polymer types	

Xiameter OFS-6224 Silane

PARAMETER	UNIT	VALUE
Recommended applications		coupling agent for many resin systems; especially useful for fiberglass-reinforced printed circuit boards, finish on woven glass fabric for the reinforcement of resins to improve physical strength properties of the composites

Xiameter OFS-6610 Silane

PARAMETER	UNIT	VALUE
GENERAL INFORMATION		
Name		Xiameter™ OFS-6610 Silane
CAS #	-	919-30-2
EC number	-	213-048-4
General description		amino functional alkoxysilane
Composition		98-100% 3-aminopropyltriethoxysilane
Acronym	-	APTES, APTS
Empirical formula		H2NC3H6-Si(OC2H5)3
Formula		
Molecular mass	daltons	221.37
Chemical class	silane	
Mixture	-	yes
Active matter	wt%	>98.5
Functional organic group		primary amine
PHYSICAL PROPERTIES		
State	-	liquid
Color		colorless to very pale yellow liquid
Odor	-	fishy
APHA color	-	<25
Boiling point	°C	217
Kinematic viscosity at 25°C	cSt	1.6
Solubility (diluents)	alcohols, water	
Solubility in water at 25°C	g/l	hydrolytic decomposition releasing ethanol
Specific gravity at 25°C	-	0.946
HEALTH & SAFETY		
Flash point	°C	96
Flash point method	-	SCC
Animal testing, acute toxicity, Rat oral LD50	mg/kg	1479-2665
Animal testing, acute toxicity, Rabbit dermal LD50	mg/kg	4041
Carcinogenicity		Did not cause cancer in laboratory animals

Xiameter OFS-6610 Silane

PARAMETER	UNIT	VALUE
Mutagenicity		In vitro genetic toxicity studies were negative. Animal genetic toxicity studies were negative.
Teratogenicity		Did not cause birth defects in laboratory animals.
Hazardous ingredients, labelling		Ethanol
Exposure, personal protection		Safety glasses, protective clothing based on chemical resistance data, chemical-resistant gloves, general and local exhaust ventilation.
Dow IHG, TWA	mg/m^3	0.5
NIOSH, REL	mg/m^3	1900/ethanol
OSHA, PEL	mg/m^3	1900/ethanol
ACGIH, TLV	ppm	1000/ethanol
NIOSH, REL	ppm	1000/ethanol
OSHA, PEL	ppm	1000/ethanol
ECOLOGICAL PROPERTIES		
Aquatic toxicity, *Green algae*, 72-h EC50	mg/l	>1000
Aquatic toxicity, *Daphnia magna*, 48-h LC50	mg/l	331
Aquatic toxicity, *Zebra fish*, 96-h LC50	mg/l	>934
Biodegradation	67%/28 d	
Partition coefficient, logP$_{ow}$	-	1.7
USE & PERFORMANCE		
Manufacturer		Dow
Outstanding properties		improved adhesion of many plastics, resins, and elastomers to inorganic materials and surfaces, improving the properties of mineral-filled rubber, reinforced glass composite. Increased composite wet and dry tensile and flexural strength and modulus. Increased transparency of fiberglass composites.
Recommended for polymers		WB & SB acrylic, SB PU
Recommended applications		SB and WB coatings, pigment treatment in WB coating, coupling agent for thermoset resins with glass or mineral fillers
Concentrations used		0.01 - 2.0% grind or let-down for each specific application, the optimum level of additive should be determined by testing.
Food approval (FDA)		FDA 175.105

Xiameter OFS-6697 Silane

PARAMETER	UNIT	VALUE
GENERAL INFORMATION		
Name		Xiameter™ OFS-6697 Silane
CAS #	-	78-10-4
Composition		90-100% tetraethoxysilane
Acronym	-	TEOS
Formula		
Chemical class	silane	
Mixture	-	yes
Active matter	wt%	99
Purity	wt%	99
SiO$_2$ content	%	28.7
PHYSICAL PROPERTIES		
State	-	liquid
Odor	-	aromatic
Color	-	clear
Boiling point	°C	168.3
Density at 25°C	kg/m³	930
Kinematic viscosity at 25°C	cSt	0.72
Sulfur oxide (SO$_2$) content	wt%	28.7/SO$_2$
HEALTH & SAFETY		
NFPA classification	Flammability	2
	Health	3
	Reactivity	0
DOT class	Combustible liquid, n.o.s. (Tetraethyl orthosilicate, ethanol) 3, III	
ICAO/IATA class	Flammable liquid. n.o.s. (Tetraethyl orthosilicate, Ethanol) 3, III	
IMDG class	Flammable liquid. n.o.s. (Tetraethyl orthosilicate, Ethanol) 3, III	
Autoignition temperature	°C	235
Flash point	°C	54
Flash point method	-	CC
Hazardous ingredients, labelling	Tetraethyl orthosilicate, ethanol	
Animal testing, acute toxicity, Rat oral LD50	mg/kg	>2500
Animal testing, acute toxicity, Rabbit dermal LD50	mg/kg	5878

Xiameter OFS-6697 Silane

PARAMETER	UNIT	VALUE
Effect of exposure, eye (human)		Direct contact may cause severe irritation. Vapor may cause eye irritation
Effect of exposure, inhalation (human)		Direct contact may cause severe irritation. Vapor may cause eye irritation
Effect of exposure, skin (human)		May cause mild irritation. Repeated exposure may cause defatting and drying of skin which can result in skin irritation and dermatitis
Effect of exposure, swallowing (human)		May cause vomiting
Effect of repeated or overexposure (human)		Overexposure may cause pulmonary edema. Overexposure by inhalation cause drowsiness, dizziness, confusion or loss of coordination. Repeated exposure may cause defatting and drying of skin which can result in skin irritation and dermatitis. Overexposure by inhalation and ingestion may injure kidneys, liver, lung, blood.
Exposure, personal protection		Safety glasses, protective clothing based on chemical resistance data, chemical-resistant gloves, general and local exhaust ventilation.
First aid, eye		Flush eyes with plenty of water for at least 15 minutes. Remove contact lens, if worn. Get medical attention if irritation develops and persists
First aid, inhalation		Remove to fresh air. If not breathing, give artificial respiration. If breathing is difficult, give oxygen. Get medical attention.
First aid, skin		Flush skin with soap and plenty of water for at least for 15 min. and removing contaminated clothing and shoes. Get medical attention.
OSHA, PEL	mg/m^3	850/ tetraethoxysilane
ACGIH, TLV	ppm	10/ tetraethoxysilane
NIOSH, REL	ppm	10/ tetraethoxysilane
OSHA, PEL	ppm	100/ tetraethoxysilane
UN/NA class	-	1993
ECOLOGICAL PROPERTIES		
Aquatic toxicity, *Green algae*, 96-h EC50	mg/l	>100

Xiameter OFS-6697 Silane

PARAMETER	UNIT	VALUE
Aquatic toxicity, *Zebra fish*, 96-h LC50	mg/l	>245
Aquatic toxicity, *Daphnia magna*, 48-h LC50	mg/l	>75
USE & PERFORMANCE		
Manufacturer	Dow	
Outstanding properties	increases substrate strength, good penetration due to small molecular size, UV stable	
Recommended for products	diluent for zinc-rich primers, additive for other coupling agents, consolidation of construction materials such as stones in buildings or monuments.	
Recommended applications	applied to natural stones or other construction materials, forming silica-gel-like binder (SiO_2) that increases the substrate strength	
Guidelines for use	the product should be applied between 5 and 20°C and relative humidity not below 40%. During the first 2-3 days after the treatment, the substrate should be protected against rain and direct sunlight. The reaction will be complete after 2-4 weeks. Methods of application include spraying, brushing, or dipping.	

Xiameter OFS-6920 Silane

PARAMETER	UNIT	VALUE
GENERAL INFORMATION		
Name		Xiameter™ OFS-6920 Silane
CAS #	-	56706-10-6
EC number	-	260-350-7
General description		mixture of polysulfidosilane species with an average sulfur chain length of 2.20
Composition		<100% bis-(triethoxysilylpropyl)disulfide, 1.5-3%triethoxy(3-mercaptopropyl)silane
Acronym	-	TSPD
Empirical formula		C18H42O6S2Si2/(OC2H5)3Si-(C2H3)-S2-(C2H3)-Si(OC2H5)3
Formula		
Molecular mass	daltons	474.00
Chemical class	silane	
Mixture	-	yes
Active matter	wt%	99
Functional organic group	sulfido	
PHYSICAL PROPERTIES		
State	-	liquid
Color	-	pale yellow
Odor	-	slight
Boiling point	°C	>150
Density	kg/m^3	1027
Kinematic viscosity at 25°C	cSt	1.5
Solubility in water at 25°C		hydrolytic decomposition releasing ethanol
Thermal decomposition products	formaldehyde	
Sulfur content	%	14.4
Average sulfur chain length	-	2.2
HEALTH & SAFETY		
NFPA classification	Flammability	1
	Health	0
	Reactivity	0
HMIS classification	Flammability	1
	Health	0
	Reactivity	0

Xiameter OFS-6920 Silane

PARAMETER	UNIT	VALUE
Carcinogenicity	IARC, OSHA, NTP: no ingredient of this product present at levels greater than or equal to 0.1% is identified as probable, possible or confirmed human carcinogen	
Mutagenicity	not classified based on available information.	
Teratogenicity	not classified based on available information.	
DOT class	not regulated as a dangerous goods	
TDG class	not regulated as a dangerous goods	
ICAO/IATA class	not regulated as a dangerous goods	
IMDG class	not regulated as a dangerous goods	
Flash point	°C	105.56
Flash point method	-	SCC
Hazardous combustion products	Carbon oxides, SiO, NOx, formaldehyde	
Hazardous products of hydrolysis	Methanol	
Effect of exposure, eye (human)	Avoid contact with eyes. Direct contact may cause severe irritation.	
Effect of exposure, swallowing (human)	Harmful if swallowed.	
Exposure, personal protection	Safety glasses, protective clothing based on chemical resistance data, chemical-resistant gloves, general and local exhaust ventilation.	
First aid, eye	Flush eyes with plenty of water for at least 15 minutes. Remove contact lens, if worn.	
First aid, inhalation	Remove to fresh air. If not breathing, give artificial respiration. If breathing is difficult, give oxygen. Get medical attention.	
First aid, skin	Flush with plenty of water for at least for 15 min. and removing contaminated clothing and shoes. Get medical attention.	
OSHA, PEL	ppm	200/methanol
ECOLOGICAL PROPERTIES		
Aquatic toxicity, *Daphnia magna*, 48-h LC50	mg/l	6.7/triethoxy(3-mercaptopropyl) silane
USE & PERFORMANCE		
Manufacturer	Dow	

Xiameter OFS-6920 Silane

PARAMETER	UNIT	VALUE
Outstanding properties		improved properties of silica-filled tire rubber, improved processability versus Xiameter OFS-6940 Silane. Low sulfur content allows higher rubber processing temperatures without excessive scorch.
Recommended for polymers		silica-filled rubber, reactive in sulfur-vulcanized rubber.
Recommended for products		tires, filler treatment
Recommended applications		provides compatibilization and coupling between inorganic fillers and organic rubber (e.g., silica and SBR)

Xiameter OFS-6925 Silane

PARAMETER	UNIT	VALUE
GENERAL INFORMATION		
Name	Xiameter™ OFS-6925 Silane	
CAS #	56706-10-6, 1333-86-4, 14814-09-6	
EC number	-	260-350-7
General description	mixture of polysulfidosilane species with an average sulfur chain length of 2.20 on carbon black, N300 type carrier. It delivers polysulfidosilanes active ingredient in an easy to handle solid form	
Composition	50% Disulfidosilane, 43-58% carbon black, 0.9-1.2% triethoxy(3-mercaptopropyl)silane	
Common synonym	bis(triethoxysilylpropyl) disulfide	
Acronym	-	TESPD
Empirical formula	(EtO)3Si(CH2)3-Sx-(CH2)3Si(OEt)3 combined with N300 type carbon black	
Formula		
Molecular mass	daltons	474.82
Chemical class	silane	
Active matter	wt%	50
Functional organic group	sulfido	
PHYSICAL PROPERTIES		
State	small, irregular pellets	
Color	-	black
Ash content	wt%	12.5
Density	kg/m³	1350
Sulfur oxide (SO₂) content	wt%	7.2
Average sulfur chain length	-	2.2
HEALTH & SAFETY		
NFPA classification	Flammability	1
	Health	0
	Reactivity	0
HMIS classification	Flammability	1
	Health	0
	Reactivity	0

Xiameter OFS-6925 Silane

PARAMETER	UNIT	VALUE
Carcinogenicity	IARC Group 2B: Possibly carcinogenic to humans, OSHA, NTP: no ingredient of this product present at levels greater than or equal to 0.1% is identified as probable, possible or confirmed human carcinogen	
DOT class	not regulated as a dangerous goods	
TDG class	not regulated as a dangerous goods	
ICAO/IATA class	not regulated as a dangerous goods	
IMDG class	not regulated as a dangerous goods	
Hazardous combustion products	When heated to temperatures above 150°C in the presence of air, product can form formaldehyde vapors.	
Animal testing, acute toxicity, Rat oral LD50	mg/kg	>5000
Animal testing, acute toxicity, Rabbit dermal LD50	mg/kg	>2000
Effect of exposure, eye (human)	Dust contact with the eyes can lead to mechanical irritation.	
Effect of exposure, inhalation (human)	Carbon black is inextricably bound in the product and therefore do not contribute to a dust inhalation hazard	
Effect of exposure, skin (human)	Contact with dust may cause mechanical irritation or drying of the skin. Avoid prolonged or repeated contact with skin	
Exposure, personal protection	Safety glasses, protective clothing based on chemical resistance data, chemical-resistant gloves, general and local exhaust ventilation.	
First aid, eye	Flush eyes with plenty of water for at least 15 minutes. Remove contact lens, if worn.	
First aid, inhalation	Remove to fresh air. If not breathing, give artificial respiration. If breathing is difficult, give oxygen. Get medical attention.	
First aid, skin	Flush with plenty of water for at least for 15 min. and removing contaminated clothing and shoes. Get medical attention.	
Specific target organ	eyes, skin	
NIOSH, REL	mg/m^3	max 3.5/ carbon black
OSHA, PEL	mg/m^3	max 3.5/ carbon black

Xiameter OFS-6925 Silane

PARAMETER	UNIT	VALUE
ECOLOGICAL PROPERTIES		
Aquatic toxicity, *Daphnia magna*, 48-h LC50	mg/l	>6.7
USE & PERFORMANCE		
Manufacturer	Dow	
Outstanding properties	provides compatibilization and coupling between inorganic fillers and organic rubber (SBR). Low sulfur content allows higher rubber processing temperatures without excessive scorch	
Recommended for polymers	inorganic fillers in organic rubber (such as silica in black rubber used in tire treads)	
Recommended for products	tires, filler treatment	
Recommended applications	coupling agent for inorganic fillers in organic rubber (such as silica in black rubber used in tire treads)	
Processing methods	can be incorporated in-situ in a black rubber formulation. The sulfur portion of the silane will then participate in the sulfur vulcanization of the rubber to couple the filler to the organic matrix.	

Xiameter OFS-6940 Silane

PARAMETER	UNIT	VALUE
GENERAL INFORMATION		
Name		Xiameter™ OFS-6940 Silane
CAS #		40372-72-3, 5089-70-3
EC number	-	260-350-7
General description		mixture of polysulfidosilane species with an average sulfur chain length of 2.20 on carbon black, N300 type carrier. It delivers polysulfidosilanes active ingredient in an easy to handle solid form
Composition		bis-(triethoxysilylpropyl)tetrasulfidosilane, 1-5% chloropropyl triethoxysilane
Acronym	-	TESPT
Empirical formula		(EtO)3Si(CH2)3-Sx(CH2)3Si(OEt)3
Formula		
Chemical class		silane
Functional organic group		sulfido
Purity	wt%	100
PHYSICAL PROPERTIES		
State	-	liquid
Color	-	clear, amber
Boiling point	°C	>65
Kinematic viscosity at 25°C	cSt	10
Specific gravity at 25°C	-	1.085
Sulfur oxide (SO_2) content	wt%	22.7
Average sulfur chain length	-	3.75
HEALTH & SAFETY		
NFPA classification	Flammability	1
	Health	0
	Instability	0
HMIS classification	Flammability	1
	Health	0
	Physical hazard	0
Carcinogenicity		IARC, OSHA, NTP: no ingredient of this product present at levels greater than or equal to 0.1% is identified as probable, possible or confirmed human carcinogen

Xiameter OFS-6940 Silane

PARAMETER	UNIT	VALUE
TDG class	not regulated as a dangerous goods	
ICAO/IATA class	not regulated as a dangerous goods	
IMDG class	not regulated as a dangerous goods	
Flash point	°C	101.67
Flash point method	-	CC
Hazardous combustion products	Carbon oxides, Chlorine compounds, SiO, NOx, formaldehyde, Sulfur oxides	
Agency rating, listed		
Animal testing, acute toxicity, Rat oral LD50	mg/kg	>5000
Animal testing, acute toxicity, Rabbit dermal LD50	mg/kg	>2000
Effect of exposure, eye (human)	Avoid contact with eyes. Direct contact may cause severe irritation.	
Effect of exposure, inhalation (human)	Carbon black is inextricably bound in the product and therefore do not contribute to a dust inhalation hazard	
Effect of exposure, skin (human)	Contact with dust may cause mechanical irritation or drying of the skin. Avoid prolonged or repeated contact with skin	
Exposure, personal protection	Safety glasses, protective clothing based on chemical resistance data, chemical-resistant gloves, general and local exhaust ventilation.	
ACGIH, TLV		3.0 inhalable fraction/carbon black
UN risk phrases, R	R20/21/22	
US safety phrases, S	S26,S36/37/39	
ECOLOGICAL PROPERTIES		
Aquatic toxicity, *Green algae*, 96-h EC50	mg/l	819/72H/chloropropyl triethoxysilane
Aquatic toxicity, *Daphnia magna*, 48-h LC50	mg/l	21.2/chloropropyl triethoxysilane
Aquatic toxicity, *Zebra fish*, 96-h LC50	mg/l	80/chloropropyl triethoxysilane
Biodegradation probability	46% chloropropyl triethoxysilane/not readily biodegradable	
Partition coefficient, log K_{oc}	-	3.13/chloropropyl triethoxysilane

Xiameter OFS-6940 Silane

PARAMETER	UNIT	VALUE
USE & PERFORMANCE		
Manufacturer		Dow
Outstanding properties		provides compatibilization and coupling between inorganic fillers and organic rubber. Low sulfur content allows higher rubber processing temperatures without excessive scorch
Recommended for polymers		silica-filled rubber
Recommended for products		tires, filler treatment
Recommended applications		coupling agent for inorganic fillers in organic rubber (such as silica in black rubber used in tire treads)
Processing methods		can be incorporated in-situ in a black rubber formulation, reactive in sulfur-vulcanized rubber

Xiameter OFS-6945 Silane

PARAMETER	UNIT	VALUE
GENERAL INFORMATION		
Name	Xiameter OFS-6945 Silane	
CAS #	40372-72-3, 1333-86-4	
EC number	-	254-896-5
General description	mixture of polysulfidosilane species with an average sulfur chain length of 3.75 on an HAF carbon black, N330 type carrier.	
Composition	50% bis-(triethoxysilylpropyl)tetrasulfide, 50% carbon black	
Acronym	-	TESPT
Empirical formula	C18H42O6S4Si2	
Formula		
Molecular mass	daltons	538.95
Chemical class	silane	
Active matter	wt%	50
Functional organic group	sulfido	
PHYSICAL PROPERTIES		
State	small, irregular pellets	
Color	-	black
Ash content	wt%	11.5
Density	kg/m³	1350
Sulfur oxide (SO$_2$) content	wt%	11.5
HEALTH & SAFETY		
Carcinogenicity	IARC, OSHA, NTP: no ingredient of this product present at levels greater than or equal to 0.1% is identified as probable, possible or confirmed human carcinogen	
DOT class	not regulated as a dangerous goods	
TDG class	not regulated as a dangerous goods	
ICAO/IATA class	not regulated as a dangerous goods	
IMDG class	not regulated as a dangerous goods	
Hazardous combustion products	carbon oxides, SiO, NO$_x$, formaldehyde, sulfur oxides	
Animal testing, acute toxicity, Rat oral LD50	mg/kg	>5000 est.

Xiameter OFS-6945 Silane

PARAMETER	UNIT	VALUE
Animal testing, acute toxicity, Rabbit dermal LD50	mg/kg	>5000 est.
Effect of exposure, eye (human)		Avoid contact with eyes. Direct contact may cause severe irritation.
Effect of exposure, inhalation (human)		Carbon black is inextricably bound in the product and therefore do not contribute to a dust inhalation hazard
Effect of exposure, skin (human)		Contact with dust may cause mechanical irritation or drying of the skin. Avoid prolonged or repeated contact with skin
Effect of exposure, swallowing (human)		Harmful if swallowed.
Exposure, personal protection		Safety glasses, protective clothing based on chemical resistance data, chemical-resistant gloves, general and local exhaust ventilation.
ACGIH, TLV		3.0 inhalable fraction/carbon black
UN risk phrases, R	R20/21/22	
US safety phrases, S	S26,S36/37/39	
ECOLOGICAL PROPERTIES		
Aquatic toxicity, *Green algae*, 96-h EC50	mg/l	10000/72H/ carbon black
Aquatic toxicity. *Daphnia magna*, 48-h LC50	mg/l	>5600/24H/ carbon black
USE & PERFORMANCE		
Manufacturer	Dow	
Outstanding properties		provides compatibilization and coupling between inorganic fillers and organic rubber (e.g., silica and SBR), sulfur functional silane. It delivers polysulfidosilanes active ingredient in an easy to handle solid form
Recommended for polymers		organic rubber-like SBR, silica-filled rubber
Recommended for products		tire treads, filler treatment
Recommended applications		coupling agent for inorganic fillers in organic rubber (such as silica in black rubber used in tire treads)
Processing methods		can be incorporated in-situ in a black rubber formulation, solid form facilitates easy handling and addition to the mixing system

3.28 Silane+silica
Coupsil 6109

PARAMETER	UNIT	VALUE
GENERAL INFORMATION		
Name		Coupsil 6109
CAS #	-	367510-46-1
General description		precipitated silica, surface-modified with organosilane Si 69® for application in the rubber industry.
Sulfur content	wt%	1.85
PHYSICAL PROPERTIES		
State	-	powder
Bulk density at 20°C	kg/m³	220
pH	-	7.2
Volatile content	%	3.5
USE & PERFORMANCE		
Manufacturer		Evonik
Outstanding properties		Coupsil® 6109 is used as a reinforcing filler in rubber compounds. It reacts with unsaturated polymers during vulcanization under the formation of covalent chemical bonds. This imparts greater tensile strength, higher moduli, reduced compression set, increased abrasion resistance, and optimized dynamic properties. It is used in many fields of the rubber industry in combination with white fillers and where optimum technical properties are required.
Recommended for polymers		rubber

Coupsil 8113/8113 GR

PARAMETER	UNIT	VALUE
GENERAL INFORMATION		
Name		Coupsil 8113/8113 GR
CAS #	-	367510-46-1
General description		Precipitated silica, surface-modified with organosilane Si 69® for application in the rubber industry. Granulated form has extension GR.
Sulfur content	wt%	2.6
PHYSICAL PROPERTIES		
State	-	powder
Bulk density at 20°C	kg/m³	310
pH	-	6.6
Volatile content	%	3.5
USE & PERFORMANCE		
Manufacturer		Evonik
Outstanding properties		Coupsil® 8113/8113 GR is used as a reinforcing filler in rubber compounds. They react with unsaturated polymers during vulcanization under the formation of covalent chemical bonds. This imparts greater tensile strength, higher moduli, reduced compression set, increased abrasion resistance, and optimized dynamic properties.
Recommended for polymers		rubber
Recommended applications		low rolling resistant tires with high wet grip and low heat generation, mechanical rubber goods, shoe soles
Guidelines for use		products are used in many fields of the rubber industry in combination with white fillers and where optimum technical properties are required.

Dynasylan SIVO 160

PARAMETER	UNIT	VALUE
GENERAL INFORMATION		
Name		Dynasylan SIVO 160
General description		bifunctional
Composition		stable silane-based sol-gel, impurity <5% ethanol
Chemical class	-	silane+silica
Amine number	-	9.0
Solids content	wt%	8.5-9.5
PHYSICAL PROPERTIES		
State	-	liquid
Odor	-	odorless
Color		clear to slightly opaque, colorless to slightly yellow
Boiling point	°C	100
Density at 20°C	kg/m^3	1,025
pH at 20°C (1:1 with H$_2$O)	-	4-4.6
Solubility in water at 25°C	hydrolysis and condensation	
Viscosity at 20°C	mPas	9.1
Volatility	-	nearly VOC-free
HEALTH & SAFETY		
NFPA classification	Flammability	1
	Health	1
	Reactivity	0
HMIS classification	Flammability	1
	Health	1
	Reactivity	0
Carcinogenicity	IARC, OSHA, NTP: no ingredient of this product present at levels greater than or equal to 0.1% is identified as probable, possible or confirmed human carcinogen	
DOT class	not regulated	
TDG class	not regulated	
ICAO/IATA class	not regulated	
IMDG class	not regulated	
Flash point	°C	>95
Flash point method	-	PMCC
Agency rating, listed	EINECS Europe	
Animal testing, acute toxicity, Rat oral LD50	mg/kg	>2000

Dynasylan SIVO 160

PARAMETER	UNIT	VALUE
Exposure personal protection	Safety glasses, protective clothing based on chemical resistance data, chemical-resistant gloves, general and local exhaust ventilation.	
First aid, eye	Rinse cautiously with water for several minutes. Remove contact lenses, if present and easy to do. Continue rinsing. If eye irritation persists: Get medical advice/attention.	
First aid, inhalation	Remove to fresh air. If not breathing, give artificial respiration. If breathing is difficult, give oxygen. If irritation persists, obtain medical advice.	
First aid, skin	Immediately wipe away excess material. Use a waterless hand cleaner to remove as much of the remaining material as possible. Wash with soap and water.	
NIOSH, REL	mg/m^3	1900/ethanol
OSHA, PEL	mg/m^3	1900/ethanol
ACGIH, TLV	ppm	1000/ethanol
NIOSH, REL	ppm	1000/ethanol
OSHA, PEL	ppm	1000/ethanol
USE & PERFORMANCE		
Manufacturer	Evonik Industries	
Outstanding properties	Dynasylan SIVO 160 is highly reactive, binds strongly to the metal through the reactive silanol groups of the molecule while being able to interact with a great variety of organic resins with its organofunctional groups. Due to its high degree of polymerization and high reactivity, it has low VOC; it is heavy metal and fluoride-free, forms very thin/highly cross-linking layers, resistant to boiling water, thermally stable up to 220°C, excellent wetting properties, and extraordinary low curing temperature of 20 to 80°C.	

Dynasylan SIVO 160

PARAMETER	UNIT	VALUE
Recommended applications		used as an additive to formulate primers and wash-coats as conversion layers, sealers, or adhesion promotion primers on metal, glass, or ceramic surfaces. It can be used to formulate new conversion coatings for metal surfaces (stainless steel, HDG steel, zinc, aluminum, and magnesium). The performance of Dynasylan 160 can be optimized by special formulations with other waterborne sol-gel systems or additives like Dynasylan SIVO 112 or/and Dynasylan SIVO 110.
Guidelines for use		formulations can be sprayed, dipped or applied with doctor blade

3.29 Silane+silicate
Geniosil CS 2

PARAMETER	UNIT	VALUE
GENERAL INFORMATION		
Name		Geniosil CS 2
CAS #	-	78-10-4, 780-69-8
EC number		201-083-8, 212-305-8
Composition		<50% tetraethyl silicate + <3% triethoxyphenylsilane
Chemical class		silane+silicate
Mixture	-	yes
Functional organic group	-	alkoxy
PHYSICAL PROPERTIES		
State	-	liquid
Color	-	clear, colorless
Boiling point	°C	163
Density at 20°C	kg/m³	985
Kinematic viscosity at 25°C	cSt	1.29
pH	-	4-5 aq solution
Solubility (diluents)	-	organic solvents
Solubility in water at 25°C	insoluble in neutral water	
Vapor pressure at 20°C	kPa	0.26
Vapor pressure at 50°C	kPa	1.41
HEALTH & SAFETY		
Mutagenicity	negative	
TDG class	Dangerous Goods, (Tetraethyl silicate solution) 3,III	
ICAO/IATA class	Dangerous Goods, (Tetraethyl silicate solution) 3,III	
IMDG class	Dangerous Goods, (Tetraethyl silicate solution) 3,III	
Autoignition temperature	°C	255
Flash point	°C	50
Flash point method	-	PMCC
Hazardous combustion products	Carbon oxides, SiOx, formaldehyde.	
Agency rating, listed	EINECS Europe, ECL Korea, ENCS Japan, AICS Australia, IECSC China, DSL Canada, PICCS Philippines, TSCA USA, NZIoC-New Zealand, TCSI Taiwan, REACH-EU	
Hazardous ingredients, labelling	Tetraethyl silicate solution	
Hazardous products of hydrolysis	methanol	

3.29 Silane+silicate
Geniosil CS 2

PARAMETER	UNIT	VALUE
Animal testing, acute toxicity, Rat oral LD50	mg/kg	>2500/ tetraethyl silicate
Animal testing, acute toxicity, Rabbit dermal LD50	mg/kg	no skin irritation
Animal testing, acute toxicity, Rat inhalation, LC50	mg/m³	>168000/4H/female/tetraethyl silicate, >10000/4H/male/ tetraethyl silicate
Effect of exposure, eye (human)		Causes serious eye irritation.
Effect of exposure, inhalation (human)		May cause respiratory irritation. Inhalation of aerosol spray may damage health.
Effect of repeated or overexposure (human)		Target organs: respiratory system.
Exposure, personal protection		Safety glasses, protective clothing based on chemical resistance data, chemical-resistant gloves, general and local exhaust ventilation.
First aid, eye		Rinse cautiously with water for several minutes. Remove contact lenses, if present and easy to do. Continue rinsing. If eye irritation persists: Get medical advice/attention.
First aid, inhalation		Keep the patient calm. Protect against loss of body heat. Seek medical advice and clearly identify substance.
First aid, skin		Remove contaminated or soaked clothing. Wash off with plenty of water or water and soap immediately for 10-15 minutes. In serious cases, use emergency shower immediately. Seek medical advice and clearly identify substance.
NIOSH, REL	mg/m³	ST325/methanol
OSHA, PEL	mg/m³	260/methanol
ACGIH, TLV	ppm	200/methanol
NIOSH, REL	ppm	ST250/methanol
OSHA, PEL	ppm	200/methanol
UN/NA class	-	1292
ECOLOGICAL PROPERTIES		
Aquatic toxicity, *Daphnia magna*, 48-h LC50	mg/l	>75/ tetraethyl silicate
Aquatic toxicity, *Zebra fish*, 96-h LC50	mg/l	>245/ tetraethyl silicate

3.29 Silane+silicate
Geniosil CS 2

PARAMETER	UNIT	VALUE
Biodegradation probability		98%/28d/tetraethyl silicate/readily biodegradable
USE & PERFORMANCE		
Manufacturer		Wacker Chemie AG
Outstanding properties		improved the adhesion between the organic polymer and the inorganic fillers. This imparts significantly improved flexural strength and impact strength to the composites. It also reduces water absorption and dirt pick-up significantly. Geniosil CS 2 also liberates about 50% less methanol than conventional adhesion promoters.
Recommended for polymers		unsaturated polyester resin or polyacrylate composite
Recommended applications		acts as a coupling agent by forming a chemical bonding between the organic resin and the inorganic fillers or pigments used, e.g. quartz, granite or even glass. Construction materials
Concentrations used		1-3% added to the organic resin along with conventional additives
Conditions to avoid		contact with moisture must be avoided during processing to prevent undesired hydrolysis

3.30 Silane+titanate
Dowsil P5200

PARAMETER	UNIT	VALUE
GENERAL INFORMATION		
Name		Dowsil P5200
CAS #		107-51-7, 18407-95-9, 5593-70-4
Composition		70-90% octamethyltrisiloxane, >5 -< 10% 1-methoxyisopropyl orthosilicate, 1-5% Titanium tetrabutanolate
Chemical class		silane+titanate
Mixture	-	yes
Functional organic group		octamethyl
PHYSICAL PROPERTIES		
State	-	liquid
Odor	-	slight
Color		colorless to pale yellow or red
Boiling point	°C	>150
Kinematic viscosity at 25°C	cSt	1.0
Solubility in water at 25°C		incompatible with water
Specific gravity at 25°C	-	0.82
Thermal decomposition products		formaldehyde
Volatility		77 g/l/OS fluid exempt, 517 g/l/OS fluid non-exempt
HEALTH & SAFETY		
NFPA classification	Flammability	3
	Health	3
	Reactivity	0
HMIS classification	Flammability	3
	Health	3
	Reactivity	0
Carcinogenicity		IARC, OSHA, NTP: no ingredient of this product present at levels greater than or equal to 0.1% is identified as probable, possible or confirmed human carcinogen
TDG class		FLAMMABLE LIQUID, N.O.S. (Octamethyltrisiloxane, Organo Titanate), 3,III
ICAO/IATA class		Flammable liquids, n.o.s. (Octamethyltrisiloxane, Organo Titanate) 3,III
IMDG class		FLAMMABLE LIQUID, N.O.S. (Octamethyltrisiloxane, Organo Titanate) 3,III
Flash point	°C	30
Flash point method	-	CC

3.30 Silane+titanate Dowsil P5200

PARAMETER	UNIT	VALUE
Hazardous combustion products	Carbon oxides, SiOx, formaldehyde, metal oxides	
Hazardous ingredients, labelling	Octamethyltrisiloxane, Organo Titanate	
Hazardous products of hydrolysis	propan-1-ol	
Animal testing, acute toxicity, Rat oral LD50	mg/kg	>2000/octamethyltrisiloxane, 1530/1-methoxy-isopropyl ortho-silicate, >2000/titanium tetrabutanolate
Animal testing, acute toxicity, Rabbit dermal LD50	mg/kg	not irritating
Animal testing, acute toxicity, Rat dermal LD50	mg/kg	>2000/octamethyltrisiloxane
Animal testing, acute toxicity, Rat inhalation, LC50	mg/m^3	>2350ppm/4H/octamethyltrisiloxane
Effect of exposure, eye (human)	Direct contact may cause irritation.	
Effect of exposure, inhalation (human)	Vapor and/or mist may irritate respiratory tract.	
Effect of exposure, skin (human)	Does not cause skin sensitization.	
Effect of exposure, swallowing (human)	Low ingestion hazard in normal use. Acute toxicity estimated: >5000	
Effect of repeated or overexposure (human)	Overexposure by inhalation may cause drowsiness, dizziness, confusion or loss of coordination. Prolonged or repeated exposure by inhalation or ingestion may injure internally.	
Exposure, personal protection	Safety glasses, protective clothing based on chemical resistance data, chemical-resistant gloves, general and local exhaust ventilation.	
First aid, eye	Immediately flush the contaminated eye(s) with lukewarm, gently flowing water for 15 - 20 minutes while holding the eyelid(s) open. If contact lens is present, DO NOT delay irrigation or attempt to remove the lens. Take care not to rinse contaminated water into the unaffected eye or onto the face. Immediately obtain medical attention.	
First aid, inhalation	Remove to fresh air. If not breathing, give artificial respiration. If breathing is difficult, give oxygen. If irritation persists, obtain medical advice	

3.30 Silane+titanate
Dowsil P5200

PARAMETER	UNIT	VALUE
First aid, skin		Remove to fresh air. If not breathing, give artificial respiration. If breathing is difficult, give oxygen. If irritation persists, obtain medical advice
NIOSH, REL	mg/m^3	500/propan-1-ol, 150/butan-1-ol
OSHA, PEL	mg/m^3	500/propan-1-ol, 300/butan-1-ol
ACGIH, TLV	ppm	200/based on DCC OEL/octamethyl-trisiloxane, 100/propan-1-ol, 20/butan-1-ol
NIOSH, REL	ppm	200/propan-1-ol, 500/propan-1-ol
OSHA, PEL	ppm	200/propan-1-ol, 100/butan-1-ol
UN/NA class	-	1993
ECOLOGICAL PROPERTIES		
Aquatic toxicity, *Green algae*, 96-h EC50	mg/l	>0.0094/72H/octa-methyltrisiloxane
Aquatic toxicity, *Daphnia magna*, 48-h LC50	mg/l	>0.020/octamethyl-trisiloxane
Aquatic toxicity, *Rainbow trout*, 96-h LC50	mg/l	>0.019/octamethyl-trisiloxane
Bioconcentration factor	BCF	>500/octamethyl-trisiloxane
Biodegradation probability	octamethyltrisiloxane/not readily biode-gradable	
Partition coefficient, log K$_{oc}$	-	>4.00/octamethyl-trisiloxane, 0.88/titanium tetrabuta-nolate
USE & PERFORMANCE		
Manufacturer	Dow	
Outstanding properties	versatile adhesion enhancing, clear primer dispersed in low molecular weight silicone fluid.	
Recommended applications	enhanced adhesion/bonding RTV and heat cure silicones to: metals, ceramics, glass, wood, masonry, structural plastics	

3.30 Silane+titanate
Dowsil P5200

PARAMETER	UNIT	VALUE
Processing methods		applied in every light, even coat by: wiping, dipping, spraying. Diluting by factor 2 to 4 with additional solvent may prevent excessive build-up

Dowsil PR-2260

PARAMETER	UNIT	VALUE
GENERAL INFORMATION		
Name		Dowsil PR-2260
CAS #		142-82-5, 2768-02-7, 5593-70-4
General description		dilute solution of silane coupling agents and other active ingredients.
Composition		67-81% heptane, 14-22% vinyltrime-thoxysilane, 2.3-3.1% titanium tetrabu-tanolate
Chemical class		silane+titanate
Mixture	-	yes
Functional organic group		methoxy
PHYSICAL PROPERTIES		
State	-	liquid
Odor	-	solvent
Color	-	light straw
Boiling point	°C	98
Kinematic viscosity at 25°C	cSt	1
Solubility in water at 25°C	incompatible with water	
Specific gravity at 25°C	-	0.75
Thermal decomposition products	formaldehyde	
HEALTH & SAFETY		
NFPA classification	Flammability	3
	Health	2
	Instability	0
HMIS classification	Flammability	3
	Health	3
	Physical hazard	0
Carcinogenicity	IARC, OSHA, NTP: no ingredient of this product present at levels greater than or equal to 0.1% is identified as probable, possible or confirmed human carcinogen	
TDG class	FLAMMABLE LIQUID, N.O.S. (Heptane, Alkoxysilane),3,III	
ICAO/IATA class	Flammable liquids, n.o.s. (Heptane, Alkoxysilane) 3,III Flammable Liquids	
IMDG class	FLAMMABLE LIQUID, N.O.S. (Heptane, Alkoxysilane) 3,III	
Flash point	°C	9
Flash point method	-	PMCC

Dowsil PR-2260

PARAMETER	UNIT	VALUE
Hazardous combustion products		Carbon oxides, SiO_x, formaldehyde, metal oxides
Hazardous ingredients, labelling		heptane, alkoxysilane
Hazardous products of hydrolysis		propan-1-ol
Animal testing, acute toxicity, Rat oral LD50	mg/kg	> 5000/heptane, 7236/vinyltrimethoxysilane, 4220/titanium tetrabutanolate
Animal testing, acute toxicity, Rabbit dermal LD50	mg/kg	>2000
Effect of exposure, inhalation (human)		May cause respiratory irritation. May cause drowsiness or dizziness. Acute toxicity estimated: 98.82 mg/l/4H
Effect of exposure, skin (human)		Causes skin irritation.
Effect of exposure, swallowing (human)		Acute toxicity estimated > 5,000 mg/kg. May be fatal if swallowed and enters airways.
Effect of repeated or overexposure (human)		May cause damage to organs through prolonged or repeated exposure if swallowed.
Exposure, personal protection		Safety glasses, protective clothing based on chemical resistance data, chemical-resistant gloves, general and local exhaust ventilation.
First aid, eye		Immediately flush the contaminated eye(s) with lukewarm, gently flowing water for 15 - 20 minutes while holding the eyelid(s) open. If contact lens is present, DO NOT delay irrigation or attempt to remove the lens. Take care not to rinse contaminated water into the unaffected eye or onto the face. Immediately obtain medical attention.
First aid, inhalation		Remove to fresh air. If not breathing, give artificial respiration. If breathing is difficult, give oxygen. If irritation persists, obtain medical advice
First aid, skin		Remove to fresh air. If not breathing, give artificial respiration. If breathing is difficult, give oxygen. If irritation persists, obtain medical advice
NIOSH, REL	mg/m^3	350/heptane
OSHA, PEL	mg/m^3	2000/heptane

Dowsil PR-2260

PARAMETER	UNIT	VALUE
ACGIH, TLV	ppm	400/heptane, 500/STEL/heptane
NIOSH, REL	ppm	85/heptane
OSHA, PEL	ppm	500/heptane
UN/NA class	-	1993
ECOLOGICAL PROPERTIES		
Aquatic toxicity, *Daphnia magna*, 48-h LC50	mg/l	0.2/heptane, 168.7/vinyltrimethoxysilane
Aquatic toxicity, *Rainbow trout*, 96-h LC50	mg/l	191/vinyltrimethoxysilane
Aquatic toxicity, *Green algae*, 72-h LC50	mg/l	>89/vinyltrimethoxysilane
Biodegradation probability		51% vinyltrimethoxysilane/not readily biodegradable
Partition coefficient, log K_{oc}	-	4.5/heptane, 0.88/titanium tetrabutanolate, -2.0/Vinyltrimethoxysilane
USE & PERFORMANCE		
Manufacturer		Dow
Outstanding properties		enhanced bonding/adhesion of RTV and heat cure silicones to many metals, silicones and some plastics
Recommended applications		used for wide variety of surfaces including FR-4, ceramics, and metals. Not recommended for plastics
Processing methods		applied as light, even coat by: wiping, dipping, spraying. Diluting by factor 2 to 4 with additional solvent may prevent excessive build-up.
Guideline for use		Apply as a very light, even coat by wiping, dipping or spraying. Excess material should be wiped off to avoid over-application. Diluting by a factor of 2 to 4 with additional solvent may avoid excessive build-up

3.31 Silicate+silica
Dynasylan® 40

PARAMETER	UNIT	VALUE
GENERAL INFORMATION		
Name		Dynasylan® 40
Common synonym		ethyl silicate containing silica
Composition		silicon dioxide content of approximately 40-42% upon complete hydrolysis (the Si content of Dynasylan® 40 was calculated as SiO_2)
PHYSICAL PROPERTIES		
State	-	liquid
Color	-	colorless
Boiling point	°C	254-271
Density at 20°C	kg/m³	1050-1070
Solubility (diluents)		alcohols, aromatic and aliphatic hydrocarbons
Viscosity at 20°C	mPas	5
HEALTH & SAFETY		
Flash point	°C	62
USE & PERFORMANCE		
Manufacturer		Dow
Outstanding properties		the resulting silicic acid bonds well to many inorganic substrates, such as glass, ceramic, metal, fillers, pigments and synthetic fibers. The deposition of a thin SiO_2 layer improves the chemical and thermal stability and mechanical properties.
Recommended for products		binder component for 1- and 2-pack zinc dust paints, crucial for corrosion protection on steel, crosslinker, dentistry, foundry molds, highly scratch-, abrasion- and chemical-resistant coatings
Processing methods		in sol-gel processes, it is usually used in conjunction with alkylsilanes (e.g., Dynasylan® PTEO) organofunctional silanes and/or organic precursors(e.g. organic resins) to form siloxane networks.
Guidelines for use		the hydrolysis and condensation reaction of Dynasylan® 40 is catalyzed through via basic compounds such as diethanolamine.

3.32 Silicic acid ester
Dynasylan® A

PARAMETER	UNIT	VALUE
GENERAL INFORMATION		
Name		Dynasylan® A
Common synonym		ethyl ester of orthosilicic acid
Composition		SiO_2 content of 28.3-29.1%
Functionality	-	tetrafunctional
Functional groups	-	ethoxy
PHYSICAL PROPERTIES		
State	-	liquid
Odor	-	odorless
Color	-	colorless
Boiling point	°C	167 (initial)
Density at 20°C	kg/m³	940
Viscosity at 20°C	mPas	0.75
HEALTH & SAFETY		
Flash point	°C	45
USE & PERFORMANCE		
Manufacturer		Dow
Outstanding properties		Dynasylan® A is a ready source of silicic acid for many applications. Silicic acid is usually obtained by hydrolysis or thermally by condensation at elevated temperature. The resulting silicic acid bonds well to many inorganic substrates and can be deposited in situ in a controlled manner. The surfaces of glass, metals, pigments, fillers, and synthetic fibers can be coated with a very thin SiO_2 layer in order to improve chemical and thermal stability and mechanical properties.
Recommended for polymers		silicone rubber
Recommended for products		coatings, crosslinker, drying agent
Processing		condensation starts before hydrolysis is complete. During storage of these hydrolyzates condensation, continues until a gel is formed. The rate of gelation depends on the concentration of water.

Dynasylan® A SQ

PARAMETER	UNIT	VALUE
GENERAL INFORMATION		
Name		Dynasylan® A SQ
Common synonym		ethyl ester of orthosilicic acid
Composition		SiO_2 content of 28.3-29.1%
Functionality	-	tetrafunctional
Functional groups	-	ethoxy
Purity	%	>99.9
PHYSICAL PROPERTIES		
State	-	liquid
Odor	-	odorless
Color	-	colorless
Boiling point	°C	167 (initial)
Density at 20°C	kg/m³	940
Viscosity at 20°C	mPas	0.75
HEALTH & SAFETY		
Flash point	°C	45
USE & PERFORMANCE		
Manufacturer		Dow
Outstanding properties		Dynasylan® A SQ can replace Dynasylan® A in every application where higher purity is required.
Recommended for polymers		silicone rubber
Recommended for products		coatings (scratch- and abrasion-resistant coatings), crosslinker (electronic industry), drying agent, potting systems
Processing		The surfaces of glass, metals, pigments, fillers, and synthetic fibers can be coated with a very thin SiO_2 layer in order to improve chemical and thermal stability and mechanical properties. Dynasylan® A SQ is an important starting material for sol-gel processes. Hydrolysis leads to silanol groups which, in a subsequent condensation reaction, form very stable siloxane bonds (-Si-O-Si-).
Guideline for use		Suitable catalysts are acids or bases like mineral acids or ammonia, or even acetic acid and amines.

Dynasylan® AR

PARAMETER	UNIT	VALUE
GENERAL INFORMATION		
Name		Dynasylan® AR
Common synonym		pre-hydrolyzed, ready-for-use silicic acid ester – hybrid binder with additional colloidal SiO_2 particles. It contains ethanol as solvent and sulfuric acid as catalyst
Composition		SiO_2 content of 19-21%
Functionality	-	tetrafunctional
Functional groups	-	ethoxy
Acidity (H_2SO_4)	wt%	0.07-0.14
PHYSICAL PROPERTIES		
State	-	liquid
Odor	-	odorless
Color	-	milky
Density at 20°C	kg/m^3	930
Gel time	min	3-10
Viscosity at 20°C	mPas	4
HEALTH & SAFETY		
Flash point	°C	<21
USE & PERFORMANCE		
Manufacturer		Dow
Outstanding properties		the degree of hydrolysis and acidity have been optimized for the required reactivity and sufficient storage stability. Through the addition of alcohol and/or water, it is possible to vary the SiO_2-content and the curing properties.
Recommended for products		ceramic casting molds, refractory materials, 2-pack zinc dust anti-corrosion paints
Processing		the hydrolysis and condensation have been started during the production of the binder. Through a shift in the pH-value, this process is accelerated. This shift is achieved by the addition of fillers, pigments, additives, or through the evaporation of solvent or exposure to atmospheric moisture. The resulting silicic acid gel cures rapidly at ambient temperatures in the air. The process of curing can be accelerated through the addition of alkali catalysts.

Dynasylan® AR

PARAMETER	UNIT	VALUE
Guideline for use		the curing process is much faster than when Dynasylan® XAR is used. This can lead to the formation of cracks (mud cracking) in the thicker coating. Thus Dynasylan® AR is recommended for thinner coatings such as shop-primers.

Dynasylan® M

PARAMETER	UNIT	VALUE
GENERAL INFORMATION		
Name	Dynasylan® M	
Common synonym	methyl ester of orthosilicic acid	
Common name	tetramethoxysilane	
Composition	SiO_2 content of 39.5%	
Acronym	-	TMOS
Functionality	-	tetrafunctional
Functional groups	-	methoxy
PHYSICAL PROPERTIES		
State	-	liquid
Odor	-	odorless
Color	-	slightly yellow
Boiling point	°C	122 (initial)
Density at 20°C	kg/m³	1030
Viscosity at 20°C	mPas	0.7
HEALTH & SAFETY		
Flash point	°C	113
USE & PERFORMANCE		
Manufacturer	Dow	
Outstanding properties	silicic acid is usually obtained by hydrolysis or thermally by condensation at elevated temperatures. The resulting silicic acid bonds well to many inorganic substrates and can be deposited in situ in a controlled manner. The surfaces of glass, metals, pigments, fillers, and synthetic fibers can be coated with a very thin SiO_2 layer in order to improve chemical and thermal stability and mechanical properties.	
Recommended for products	scratch- and abrasion-resistant coatings, sol-gel systems	
Guideline for use	activity and shelf life are inversely proportional. The correct choice of the amount of water can give hydrolysates which have a shelf life of a few months.	

Dynasylan® MKS

PARAMETER	UNIT	VALUE
GENERAL INFORMATION		
Name		Dynasylan® MKS
Common synonym		silicic acid ester
Common name		tetramethoxysilane
Composition		SiO_2 content of 39.5%
Acronym	-	TMOS
Functionality	-	tetrafunctional
Functional groups	-	methoxy
PHYSICAL PROPERTIES		
State	-	liquid
Odor	-	odorless
Boiling point	°C	122 (initial)
Density at 20°C	kg/m³	900
Dilution		miscible with aliphatic and aromatic solvents
Viscosity at 20°C	mPas	135
HEALTH & SAFETY		
Flash point	°C	2
USE & PERFORMANCE		
Manufacturer		Dow
Outstanding properties		paint will cure to a hard coating. Temperature and humidity are important parameters of the curing process.
Recommended for products		zinc dust paints
Guideline for use		use of basic media accelerates hydrolysis and condensation of silicates. Dynasylan® MKS contains an aminic curing agent.
Paint formulation (wt%)		MKS – 17.1, Aerosil R 972 – 0.5, zinc dust – 77.5, Solvesso 100 – 1, xylol – 1.9, special benzin 100/140 – 2

Dynasylan® P

PARAMETER	UNIT	VALUE
GENERAL INFORMATION		
Name		Dynasylan® P
Common synonym		tetra-n-propylsilicate
PHYSICAL PROPERTIES		
State	-	liquid
Boiling point	°C	226
Density at 20°C	kg/m³	920
Viscosity at 20°C	mPas	1
HEALTH & SAFETY		
Flash point	°C	88
USE & PERFORMANCE		
Manufacturer		Dow
Recommended for polymers		silicone
Recommended for products		crosslinker, sol-gel formulations

3.33 Silicic acid ester+SiO$_2$
Dynasylan® XAR

PARAMETER	UNIT	VALUE
GENERAL INFORMATION		
Name		Dynasylan® XAR
Composition		pre-hydrolyzed, ready-for-use silicic acid ester – hybrid binder with additional colloidal SiO$_2$ particles. it contains a mixture of ethanol and isopropanol as solvents and sulfuric acid as catalyst.
SiO$_2$ content	wt%	19-21
Acidity (H$_2$SO$_4$)	%	0.07-0.14
PHYSICAL PROPERTIES		
State	-	liquid
Color		milky
Density at 20ºC	kg/m^3	910-935
Gel time	min	2-10
Viscosity at 20ºC	mPas	3-4
HEALTH & SAFETY		
Flash point	ºC	<21
USE & PERFORMANCE		
Manufacturer		Dow
Recommended for products		ceramic casting molds, 2-pack zinc dust anti-corrosion paints
Guidelines for use		the reactivity depends on the age of the binder which needs to be considered during use. Through the addition of alcohol and/or water it is possible to vary the SiO$_2$-content and the curing properties.

3.34 Sucrose acetate isobutyrate
Sucrose Acetate Isobutyrate 90

PARAMETER	UNIT	VALUE
GENERAL INFORMATION		
Name		Sucrose Acetate Isobutyrate 90
CAS #	-	27216-37-1
Composition		SAIB-90 is a sucrose-based adhesion promoter and plasticizer supplied as a solution of 90% sucrose acetate isobutyrate (SAIB) and 10% denatured ethyl alcohol by weight.
Acronym	-	SAIB
Empirical formula	-	C40H62O19
Molecular mass	daltons	846.91
PHYSICAL PROPERTIES		
State	-	viscous liquid
Odor	-	ester-like
Color	-	yellow
Color, Platinum-cobalt scale	-	1
Boiling point	°C	105
Acid number	mg KOH/g	0.2
Density at 20°C	kg/m³	1100
Vapor density	-	1.5
Vapor pressure at 20°C	mbar	53.2
Viscosity at 25°C	cP	770
HEALTH & SAFETY		
NFPA classification	Flammability	3
	Health	1
	Instability	0
HMIS classification	Flammability	1
	Health	3
	Physical hazard	0
Autoignition temperature	°C	402
Flash point	°C	17.1
Flash point method	-	TCC
First aid, eye		In case of contact, immediately flush eyes with plenty of water for at least 15 minutes. Get medical advice/ attention. Remove person to fresh air. If signs/ symptoms continue, get medical attention.
First aid, inhalation		Remove to fresh air. Treat symptomatically. Get medical advice/ attention

3.34 Sucrose acetate isobutyrate
Sucrose Acetate Isobutyrate 90

PARAMETER	UNIT	VALUE
First aid, skin	Remove contaminated clothing and shoes. Get medical attention if symptoms occur.	
UN/NA #	-	1993
USE & PERFORMANCE		
Manufacturer	Eastman	
Recommended for products	coatings, nail polish lacquers, personal care ingredients, printing inks	

Sustane™ SAIB

PARAMETER	UNIT	VALUE
GENERAL INFORMATION		
Name		Sustane™ SAIB
CAS #	-	27216-37-1
Composition		100% sucrose acetate isobutyrate
Acronym	-	SAIB
Empirical formula	-	C40H62O19
Molecular mass	daltons	846.91
PHYSICAL PROPERTIES		
State	-	viscous liquid
Odor	-	ester-like
Color	-	yellow
Color, Platinum-cobalt scale	-	200
Boiling point	°C	288 (decomp.)
Density at 20°C	kg/m^3	1146
Viscosity at 30°C	mPas	100,000
HEALTH & SAFETY		
NFPA classification	Flammability	1
	Health	1
	Instability	0
HMIS classification	Flammability	1
	Health	1
	Physical hazard	0
Autoignition temperature	°C	399
Flash point	°C	227
Flash point method	-	TCC
Carcinogenecity	No ingredient of this product present at levels greater than or equal to 0.1% is	
First aid, eye	identified as probable, possible or confirmed human carcinogen by IARC.	
First aid, inhalation	Move to fresh air. Treat symptomatically. If symptoms persist, call a physician.	
First aid, skin	Wash off with soap and water. If symptoms persist, call a physician.	
ECOLOGICAL IMPACT		
Partition coefficient: n-octanol/water, log P$_{ow}$	-	6
USE & PERFORMANCE		
Manufacturer	Eastman	

Sustane™ SAIB

PARAMETER	UNIT	VALUE
Recommended for products		viscous liquid used primarily in citrus beverages as a weighting agent or flavor emulsion stabilizer to prevent separation of essential citrus oils, personal care ingredients

Sustane™ SAIB MCT

PARAMETER	UNIT	VALUE
GENERAL INFORMATION		
Name		Sustane™ SAIB MCT
CAS #	-	27216-37-1
Composition		blend of 20% medium chain triglycerides (73398-61-5) and 80% SAIB
Acronym	-	SAIB
Empirical formula	-	C40H62O19
Molecular mass	daltons	846.91
Medium chain triglycerides	%	18-221
PHYSICAL PROPERTIES		
State	-	viscous liquid
Odor	-	slight
Color	-	colorless
Color, Platinum-cobalt scale	-	200
Acid number	mg KOH/g	0.3
Density at 20°C	kg/m^3	1100
Viscosity at 25°C	cP	5000
HEALTH & SAFETY		
NFPA classification	Flammability	1
	Health	1
	Instability	0
HMIS classification	Flammability	1
	Health	1
	Physical hazard	0
Flash point	°C	>99
Flash point method	-	TCC
Carcinogenicity		No ingredient of this product present at levels greater than or equal to 0.1% is identified as probable, possible or confirmed human carcinogen by IARC.
Animal testing, acute toxicity, Rat oral LD50	mg/kg	>25600
Animal testing, acute toxicity, Mouse oral LD50	mg/kg	>25600
Animal testing, acute toxicity, RAt dermal LD50	mg/kg	>22000
Animal testing, acute toxicity, Guinea pig dermal LD50	mg/kg	>22000
First aid, eye		In the case of contact with eyes, rinse immediately with plenty of water and seek medical advice.

Sustane™ SAIB MCT

PARAMETER	UNIT	VALUE
First aid, inhalation		Move to fresh air. Treat symptomatically. If symptoms persist, call a physician.
First aid, skin		Wash off with soap and water. If symptoms persist, call a physician.
ECOLOGICAL PROPERTIES		
Aquatic toxicity, *Daphnia magna*, 48-h LC50	mg/l	>1.11
Aquatic toxicity *Fathead minnow*, 96-h LC50	mg/l	>1.82
Biological oxygen demand, 5 days	mg/g	560-1400
Partition coefficient: n-octanol/water, log P_{ow}	-	6
USE & PERFORMANCE		
Manufacturer		Eastman
Outstanding properties		pours readily at room temperature. Combining SAIB with MCT provides enhanced cloud properties with maintaining a neutral taste, resistance to air oxidation.
Recommended for products		beverages, food ingredients, nail polish lacquers, personal care ingredients

3.35 Sulfur compounds
Duralink HTS

PARAMETER	UNIT	VALUE
GENERAL INFORMATION		
Name	Duralink HTS	
CAS #	-	5719-73-3
General description	sulfur based vulcanization systems to generate hybrid crosslinks	
Chemical name	hexamethylene-1,6-bis(thiosulphate), disodium salt, dihydrate	
Composition	98-99% disodium S,S'-hexane-1,6-diyldi(thiosulphate) dihydrate, 1-2% mineral spirit	
Common synonym	disodium S,S'-hexane-1,6-diyldi(thiosulphate) dihydrate	
Formula		
Molecular mass	daltons	390
Chemical class	sulfur compound	
Active matter	wt%	98-99
PHYSICAL PROPERTIES		
State	-	fine powder
Odor	-	mild, musty
Color	-	off-white
Melting point	°C	127-137 (decomp)
Density at 25°C	kg/m³	1,390
Particle size, residue on 150 μM sieve	%	<0.05
Solubility in water at 25°C	g/l	307
HEALTH & SAFETY		
HMIS classification	Flammability	1
	Health	2
	Reactivity	0
DOT class	not regulated	
ICAO/IATA class	not regulated	
Animal testing, acute toxicity, Rat oral LD50	mg/kg	>5000
Animal testing, acute toxicity, Rabbit dermal LD50	mg/kg	>5000
ACGIH, TLV	mg/m³	3-10
OSHA, PEL	mg/m³	5-15

3.35 Sulfur compounds
Duralink HTS

PARAMETER	UNIT	VALUE
ECOLOGICAL PROPERTIES		
Aquatic toxicity, *Bluegill sunfish*, 96-h LC50	mg/l	>1000
Aquatic toxicity, *Daphnia magna*, 48-h LC50	mg/l	250
Partition coefficient, log K_{oc}	-	-0.19
USE & PERFORMANCE		
Manufacturer	Eastman Chemical Company	
Outstanding properties	Duralink™ HTS is used in sulfur-based vulcanization systems to generate hybrid crosslinks, which provide increased retention of physical and dynamic properties when exposed to anaerobic conditions at elevated temperatures such as those experienced during overcure when using high curing temperatures, or during product service life.	
Recommended for polymers	NR, IR, SBR, BR	
Concentrations used	phr	1-3

Tytan AP20

PARAMETER	UNIT	VALUE
GENERAL INFORMATION		
Name		Tytan AP20
CAS #	-	68586-02-7
EC number	-	271-603-6
Composition		65-70% butyl(dialkyloxy(dibutoxyphosphoryloxy))titanium(trialkyloxy)titanium phosphate, 5-10% ethanol, 20-25% 2-propanol, 15% acetylacetone
Common synonym		ethoxybis(pentane-2,4-dionato-O,O') (propan-2-olato)titanium
Common name		titanium phosphate complex
Titanium content	wt%	8.6
PHYSICAL PROPERTIES		
State	-	liquid
Odor	-	alcohol-like
Color	-	orange to red
Initial boiling point	°C	78
Melting point	°C	<-20
Density at 20°C	kg/m^3	970
Vapor pressure at 25°C	Pa	5800
Viscosity at 25°C	mPas	5.5
HEALTH & SAFETY		
Flash point	°C	17
Flash point method	-	CC
Animal testing, acute toxicity, Rat oral LD50	mg/kg	>2000
First aid, eye		Immediately irrigate with eyewash solution or clean water, holding the eyelids apart, for at least 15 minutes. Obtain medical attention.
First aid, inhalation		Remove the victim from exposure to fresh air and keep at rest in a position comfortable for breathing. If breathing is difficult give oxygen, if breathing stops or shows signs of failing give artificial respiration. Do not use mouth-to-mouth method. Seek medical attention immediately.
First aid, skin		Immediately flush the affected area with soap and water using a brush to remove any precipitated solids. Seek medical attention if irritation or other symptoms occur.

Tytan AP20

PARAMETER	UNIT	VALUE
UN/NA #	-	1993
ECOLOGICAL PROPERTIES		
Aquatic toxicity, *Daphnia magna*, 48-h LC50	mg/l	45
Bioaccumulative potential		the organic hydrolysis products do not have bioaccumulation and bioconcentration potential.
Biodegradation probability		hydrolysis half-life ($t_{1/2}$) < 30 minutes, at 30°C, pH 7. Method: OECD 111. Main hydrolysis products include ethanol, 2-propanol, acetylacetone and TiO_2.
USE & PERFORMANCE		
Manufacturer		Borica Co., Ltd.
Outstanding properties		Tytan AP20 was designed to strongly increase adhesion of solvent based inks to a variety of plastics used for food packaging or other purposes, but also pigment distribution and heat sealing resistance are improved allowing perfect hiding power and appearance of printing on otherwise problematic substrates. Tytan AP20 is user friendly as the addition of the ethanol strongly reduces the pour point and as such the possibility for crystallization at low to normal temperatures.
Recommended for products		solvent based inks and coatings

Tytan AP100

PARAMETER	UNIT	VALUE
GENERAL INFORMATION		
Name		Tytan AP100
CAS #	-	109037-78-7
EC number	-	401-100-0
Composition		65% butyl(dialkyloxy(dibutoxyphosphoryloxy))titanium(trialkyloxy)titanium phosphate, 10% ethanol, 25% 2-propanol
Common synonym		butyl(dialkyloxy(dibutoxyphosphoryloxy))titanium(trialkyloxy)titanium phosphate
Common name		titanium phosphate complex
Titanium content	wt%	8.6
PHYSICAL PROPERTIES		
State	-	liquid
Odor	-	alcohol-like
Color		colorless to pale yellow
Initial boiling point	°C	80
Melting point	°C	<-20
Density at 20°C	kg/m^3	1000
Vapor pressure at 25°C	Pa	5500
Viscosity at 25°C	mPas	20
HEALTH & SAFETY		
Autoignition temperature	°C	386
Flash point	°C	12
Flash point method	-	CC
Animal testing, acute toxicity, Rat oral LD50	mg/kg	>5000
Animal testing, acute toxicity, Rat dermal LD50	mg/kg	>5000
First aid, eye		Immediately rinse eyes with eyewash solution or clean water, holding eyelids apart, for at least 10 minutes. Removal of any precipitated solids from the eye must only be attempted by a qualified medical person. In case of contact with eyes, rinse immediately with plenty of water and seek medical advice.
First aid, inhalation		Remove patient from exposure. If breathing is difficult give oxygen, if breathing stops or shows signs of failing give artificial respiration. Do not use mouth to mouth method. Seek medical attention.

Tytan AP100

PARAMETER	UNIT	VALUE
First aid, skin	Immediately wash affected area with soap and water using a brush to remove any precipitated solids. Seek medical attention if irritation occurs.	
UN/NA #	-	1993
ECOLOGICAL PROPERTIES		
Aquatic toxicity, *Daphnia magna*, 48-h LC50	mg/l	>10
Bioaccumulative potential	the substance is rapidly hydrolyzed and readily biodegradable. The substance is not expected to have bioaccumulation and bioconcentration potential.	
Biodegradation probability	readily biodegradable. 96% degradation after 28 days	
Partition coefficient, log K_{oc}	-	9.39
USE & PERFORMANCE		
Manufacturer	Borica Co., Ltd.	
Outstanding properties	Tytan AP100 is a Titanium Phosphate Complex of which the alkoxy groups can react with resins to promote adhesion and/or cross linking while the chelating agents stabilize the molecule.	
Recommended for products	solvent based inks use in food packaging	

Tytan AP110

PARAMETER	UNIT	VALUE
GENERAL INFORMATION		
Name	Tytan AP110	
CAS #	-	109037-78-7
EC number	-	401-100-0
Composition	50-60% butyl(dialkyloxy(dibutoxyphos phoryloxy))titanium(trialkyloxy)titanium phosphate, 25-35% ethanol, 5-10% 2-propanol, 5-10% alkylbenzensulfonic acid	
Common synonym	butyl(dialkyloxy(dibutoxyphosphoryloxy)) titanium(trialkyloxy)titanium phosphate	
Common name	titanium phosphate complex	
Titanium content	wt%	7.1
PHYSICAL PROPERTIES		
State	-	liquid
Odor	-	alcohol-like
Color	colorless to pale yellow	
Initial boiling point	°C	80
Melting point	°C	<-20
Density at 20°C	kg/m³	999
Vapor pressure at 25°C	Pa	5500
Viscosity at 25°C	mPas	20
HEALTH & SAFETY		
Autoignition temperature	°C	386
Flash point	°C	12
Flash point method	-	CC
Animal testing, acute toxicity, Rat oral LD50	mg/kg	>5000
Animal testing, acute toxicity, Rat dermal LD50	mg/kg	>2000
First aid, eye	Immediately rinse eyes with eyewash solution or clean water, holding eyelids apart, for at least 10 minutes. Removal of any precipitated solids from the eye must only be attempted by a qualified medical person. In case of contact with eyes, rinse immediately with plenty of water and seek medical advice.	

Tytan AP110

PARAMETER	UNIT	VALUE
First aid, inhalation	Remove patient from exposure. If breathing is difficult give oxygen, if breathing stops or shows signs of failing give artificial respiration. Do not use mouth to mouth method. Seek medical attention.	
First aid, skin	Immediately wash affected area with soap and water using a brush to remove any precipitated solids. Seek medical attention if irritation occurs.	
UN/NA #	-	1993
ECOLOGICAL PROPERTIES		
Aquatic toxicity, *Daphnia magna*, 48-h LC50	mg/l	>10
Bioaccumulative potential	the substance is rapidly hydrolyzed and readily biodegradable. The substance is not expected to have bioaccumulation and bioconcentration potential.	
Biodegradation probability	readily biodegradable. 96% degradation after 28 days	
Partition coefficient, log K_{oc}	-	9.39
USE & PERFORMANCE		
Manufacturer	Borica Co., Ltd.	
Outstanding properties	Tytan AP110 is modified Tytan AP100, which is aiming at better compatibility with various ink formulations. It has the same adhesion performance as Tytan AP100, but prevents viscosity change during printing process.	
Recommended for products	inks	

Tytan AP120

PARAMETER	UNIT	VALUE
GENERAL INFORMATION		
Name	Tytan AP120	
CAS #	-	109037-78-7
EC number	-	401-100-0
Composition	65% butyl(dialkyloxy(dibutoxyphosphoryloxy))titanium(trialkyloxy)titanium phosphate, 10% ethanol, 25% 2-propanol	
Common synonym	butyl(dialkyloxy(dibutoxyphosphoryloxy))titanium(trialkyloxy)titanium phosphate	
Common name	titanium phosphate complex	
Titanium content	wt%	8.6
PHYSICAL PROPERTIES		
State	-	liquid
Odor	-	alcohol-like
Color	colorless to pale yellow	
Initial boiling point	°C	80
Melting point	°C	<-20
Density at 20°C	kg/m^3	1000
Vapor pressure at 25°C	Pa	5500
Viscosity at 25°C	mPas	20
HEALTH & SAFETY		
Autoignition temperature	°C	386
Flash point	°C	12
Flash point method	-	CC
Animal testing, acute toxicity, Rat oral LD50	mg/kg	>5000
Animal testing, acute toxicity, Rat dermal LD50	mg/kg	>2000
First aid, eye	Immediately rinse eyes with eyewash solution or clean water, holding eyelids apart, for at least 10 minutes. Removal of any precipitated solids from the eye must only be attempted by a qualified medical person. In case of contact with eyes, rinse immediately with plenty of water and seek medical advice.	
First aid, inhalation	Remove patient from exposure. If breathing is difficult give oxygen, if breathing stops or shows signs of failing give artificial respiration. Do not use mouth to mouth method. Seek medical attention.	

Tytan AP120

PARAMETER	UNIT	VALUE
First aid, skin		Immediately wash affected area with soap and water using a brush to remove any precipitated solids. Seek medical attention if irritation occurs.
UN/NA #	-	1993
ECOLOGICAL PROPERTIES		
Aquatic toxicity, *Daphnia magna*, 48-h LC50	mg/l	>10
Bioaccumulative potential		the substance is rapidly hydrolyzed and readily biodegradable. The substance is not expected to have bioaccumulation and bioconcentration potential.
Biodegradation probability		readily biodegradable. 96% degradation after 28 days
Partition coefficient, log K_{oc}	-	9.39
USE & PERFORMANCE		
Manufacturer		Borica Co., Ltd.
Outstanding properties		Tytan AP120 is a Titanium Phosphate Complex of which the alkoxy groups can react with resins to promote adhesion and/or cross linking while the chelating agents stabilize the molecule. It is transparent and therefore most suitable for modification of ink for very clear color and high gloss printing.
Recommended for products		solvent based inks used for food packaging or other purposes.

Tytan AP130D

PARAMETER	UNIT	VALUE
GENERAL INFORMATION		
Name		Tytan AP130D
Composition		35-40% titanium tetraisopropanolate, 10-15% ethanol, 45-50% proprietary hydroxy acid ester
Common synonym		butyl(dialkyloxy(dibutoxyphosphoryloxy)) titanium(trialkyloxy)titanium phosphate
Common name		titanium phosphate complex
Titanium content	wt%	6
PHYSICAL PROPERTIES		
State	-	liquid
Odor	-	alcohol-like
Color		almost colorless to pale yellow. Yellow with age. This color change does not affect the products reactivity.
Initial boiling point	°C	78
Density at 20°C	kg/m³	1030
Viscosity at 25°C	mPas	40
HEALTH & SAFETY		
Autoignition temperature	°C	386
Flash point	°C	14
Flash point method	-	CC
First aid, eye		Immediately rinse eyes with eyewash solution or clean water, holding eyelids apart, for at least 10 minutes. Removal of any precipitated solids from the eye must only be attempted by a qualified medical person. In case of contact with eyes, rinse immediately with plenty of water and seek medical advice.
First aid, inhalation		Remove patient from exposure. If breathing is difficult give oxygen, if breathing stops or shows signs of failing give artificial respiration. Do not use mouth to mouth method. Seek medical attention. In case of accident by inhalation: remove casualty to fresh air and keep at rest.
First aid, skin		Immediately wash affected area with soap and water using a brush to remove any precipitated solids. Seek medical attention if irritation occurs.
UN/NA #	-	1993

Tytan AP130D

PARAMETER	UNIT	VALUE
USE & PERFORMANCE		
Manufacturer		Borica Co., Ltd.
Outstanding properties		Tytan AP310 is a mixture of titanium alkoxide and hydroxy acid ester that can be used as an adhesion promoter for printing, crosslinker for paints and as a catalyst for esterification. Tytan AP310 shows delayed reaction at ambient temperature and thus facilitates storage stability of inks.
Recommended for products		inks

Tytan AP310

PARAMETER	UNIT	VALUE
GENERAL INFORMATION		
Name		Tytan AP310
Composition		35-40% titanium tetraisopropanolate, 10-15% ethanol, 45-50% proprietary hydroxy acid ester
Common synonym		butyl(dialkyloxy(dibutoxyphosphoryloxy)) titanium(trialkyloxy)titanium phosphate
Common name		titanium phosphate complex
Titanium content	wt%	6
PHYSICAL PROPERTIES		
State	-	liquid
Odor	-	alcohol-like
Color		almost colorless to pale yellow. Yellow with age. This color change does not affect the products reactivity.
Initial boiling point	°C	78
Density at 20°C	kg/m³	1030
Viscosity at 25°C	mPas	40
HEALTH & SAFETY		
Flash point	°C	14
Flash point method	-	CC
First aid, eye		Immediately rinse eyes with eyewash solution or clean water, holding eyelids apart, for at least 10 minutes. Removal of any precipitated solids from the eye must only be attempted by a qualified medical person. In case of contact with eyes, rinse immediately with plenty of water and seek medical advice.
First aid, inhalation		Remove patient from exposure. If breathing is difficult give oxygen, if breathing stops or shows signs of failing give artificial respiration. Do not use mouth to mouth method. Seek medical attention. In case of accident by inhalation: remove casualty to fresh air and keep at rest.
First aid, skin		Immediately wash affected area with soap and water using a brush to remove any precipitated solids. Seek medical attention if irritation occurs.
UN/NA #	-	1993

Tytan AP310

PARAMETER	UNIT	VALUE
USE & PERFORMANCE		
Manufacturer		Borica Co., Ltd.
Outstanding properties		Tytan AP310 is a mixture of titanium alkoxide and hydroxy acid ester that can be used as an adhesion promoter for printing, crosslinker for paints and as a catalyst for esterification. Tytan AP310 shows delayed reaction at ambient temperature and thus facilitates storage stability of inks.
Recommended for products		inks

Tytan TAA

PARAMETER	UNIT	VALUE
GENERAL INFORMATION		
Name		Tytan TAA
CAS #	-	17927-72-9
EC number	-	241-866-1
Composition		75-80% bis(pentane-2,4-dionato-O,O') bis(propan-2-olato)titanium, 10-15% 2-propanol, 1-5% acetylacetone
Common synonym		bis(pentane-2,4-dionato-O,O') bis(propan-2-olato)titanium
Common name		titanium acetylacetonate
Formula		
Molecular mass	daltons	364.27
Titanium content	wt%	9.9
PHYSICAL PROPERTIES		
State	-	liquid
Odor	-	alcohol-like
Color	-	orange to red
Initial boiling point	°C	83
Melting point	°C	<0
Density at 20°C	kg/m³	1000
Vapor pressure at 25°C	Pa	47500
Viscosity at 25°C	mPas	8
HEALTH & SAFETY		
Flash point	°C	17
Flash point method	-	CC
Animal testing, acute toxicity, Rat oral LD50	mg/kg	2870
First aid, eye		Immediately irrigate with eyewash solution or clean water, holding the eyelids apart, for at least 15 minutes. Obtain medical attention.

Tytan TAA

PARAMETER	UNIT	VALUE
First aid, inhalation		Remove the victim from exposure to fresh air and keep at rest in a position comfortable for breathing. If breathing is difficult give oxygen, if breathing stops or shows signs of failing give artificial respiration. Do not use mouth-to-mouth method. Seek medical attention immediately.
First aid, skin		Immediately flush the affected area with soap and water using a brush to remove any precipitated solids. Seek medical attention if irritation or other symptoms occur.
UN/NA #	-	1993
ECOLOGICAL PROPERTIES		
Aquatic toxicity, *Daphnia magna*, 48-h LC50	mg/l	5790
Bioaccumulative factor, BCF		3.162 based on modeling. No bioaccumulation potential
Biodegradation probability		readily biodegradable. Half-life: 37.5 days based on modeling. Biodegradation in soil: half-lie: 75 days based on modeling.
Partition coefficient, K_{oc}	-	1093000
USE & PERFORMANCE		
Manufacturer		Borica Co., Ltd.
Outstanding properties		Tytan TAA is a titanium chelate of which the alkoxy groups can react with resins to promote adhesion and/or cross linking while the chelating agents stabilize the molecule. Tytan TAA has an excellent balance between reactivity and cross linking/adhesion improvement properties which makes it also very suitable to strongly improve performance properties of industrial coatings on difficult modern polymeric substrates, glass and some metals.
Recommended for products		coatings, inks

Tyzor AA-65

PARAMETER	UNIT	VALUE
GENERAL INFORMATION		
Name		Tyzor AA-65
CAS #	-	68568-02-7
Composition		15% (TiO_2), titanium acetylacetonate, 25% isopropyl alcohol, 10% ethanol
Active matter	wt%	65
PHYSICAL PROPERTIES		
State	-	liquid
Color	-	yellow to red
Boiling point	°C	78
Pour point	°C	-70
Density at 25°C	kg/m³	980
Solubility (diluents)		most organic solvents
Viscosity at 25°C	mPas	8
HEALTH & SAFETY		
Flash point	°C	14
Effect of exposure, swallowing (human)		Harmful if swallowed.
Exposure, personal protection		Safety glasses, protective clothing based on chemical resistance data, chemical-resistant gloves, general and local exhaust ventilation.
First aid, eye		Immediately flush eyes with plenty of water, occasionally lifting the upper and lower eyelids. Check for and remove any contact lenses. Rinse opened eye for several minutes under running water. Obtain medical attention if irritation develops.
First aid, inhalation		Supply fresh air. If required, provide artificial respiration. Keep patient warm. Consult doctor if symptoms persist. In case of unconsciousness place patient stably in side position for transportation.
First aid, skin		Immediately rinse with water. If skin irritation continues, consult a doctor.
USE & PERFORMANCE		
Manufacturer		Dorf Ketal Chemicals, LLC

Tyzor AA-65

PARAMETER	UNIT	VALUE
Outstanding properties		printing inks: improved the drying rate, solvent resistance, heat resistance, and adhesion to substrates. Coatings: increased surface hardness, adhesion promotion, scratch resistance, coloring effects, heat and light reflection, iridescence, and corrosion resistance.
Recommended for products		printing inks, coatings, paint additives.
Recommended applications		adhesion promotion, crosslinking of various functional polymers can occur, or to form polymeric titanium dioxide layers as a binder or coating. Used in solvent-based printing inks such as those based on nitrocellulose, glass, metals, fillers, pigments, and paints. Enhances performance of printing inks in a variety of applications, specifically flexible packaging.

Tyzor AA-105

PARAMETER	UNIT	VALUE
GENERAL INFORMATION		
Name		Tyzor AA-105
CAS #	-	445398-76-5
Composition		23% (TiO_2), bis(acetylactonate) ethoxide isopropoxide titanium
Empirical formula		C15H26O8Ti
Formula		
Molecular mass	daltons	382.3
Chemical class		titanate
PHYSICAL PROPERTIES		
State	-	liquid
Color	-	yellow to red
Pour point	°C	-25
Density at 25°C	kg/m³	1,120
Viscosity at 25°C	mPas	90
HEALTH & SAFETY		
Flash point	°C	90
Effect of exposure, swallowing (human)		Harmful if swallowed.
Exposure, personal protection		Safety glasses, protective clothing based on chemical resistance data, chemical-resistant gloves, general and local exhaust ventilation.
USE & PERFORMANCE		
Manufacturer		Dorf Ketal Chemicals, LLC
Outstanding properties		improved adhesion of the ink to the substrate, enhanced curing and heat resistance, improved chemical, solvent, and water resistance, decreased drying times or lower cure temperatures, increased lamination bond strength.
Recommended for polymers		PE, PP, PA, cellophane, PVC, metalized plastic
Recommended for products		printing inks, coatings, paint additives.
Recommended applications		adhesion promoters between inorganic (e.g., aluminum) and organic substrates, e.g., flexographic printing inks for flexible packaging.

Tyzor GBA

PARAMETER	UNIT	VALUE
GENERAL INFORMATION		
Name		Tyzor GBA
CAS #		17927-72-9, 67-63-0, 67-56-1, 70-36-3
Composition		16.40% (TiO_2), bis(acetylacetonate) diisopropoxide titanium, 13-30% propan-2-ol, 3-7% methanol, 3-7% butan-1-ol
Empirical formula		C16H28O6Ti
Formula		
Molecular mass	daltons	364.26
Chemical class	titanate	
Active matter	wt%	75
PHYSICAL PROPERTIES		
State	-	liquid
Color	-	yellow to red
Boiling point	°C	232
Pour point	°C	<-42
Density at 25°C	kg/m³	1,020
Solubility (diluents)	most organic solvents	
Viscosity at 25°C	mPas	60
HEALTH & SAFETY		
NFPA classification	Flammability	3
	Health	2
	Reactivity	0
HMIS classification	Flammability	3
	Health	2
	Reactivity	0
DOT class	Flammable liquids, n.o.s. (Propan-2-ol, Methanol)	
ICAO/IATA class	FLAMMABLE LIQUID, N.O.S. (Propan-2-ol, Methanol)	
IMDG class	FLAMMABLE LIQUID, N.O.S. (Propan-2-ol, Methanol)	
Flash point	°C	12
Flash point method	-	PMCC

Tyzor GBA

PARAMETER	UNIT	VALUE
Animal testing, acute toxicity, Rat oral LD50	mg/kg	5045/propan-2-ol, 5628/methanol, 790/butan-1-ol
Animal testing, acute toxicity, Rabbit dermal LD50	mg/kg	12800/propan-2-ol, 15800/methanol, 3400/butan-1-ol
Effect of exposure, eye (human)	Causes serious eye damage.	
Effect of exposure, swallowing (human)	Harmful if swallowed.	
Effect of repeated or overexposure (human)	May cause damage to organs.	
Exposure, personal protection	Safety glasses, protective clothing based on chemical resistance data, chemical-resistant gloves, general and local exhaust ventilation.	
First aid, eye	Immediately flush eyes with plenty of water, occasionally lifting the upper and lower eyelids. Check for and remove any contact lenses. Rinse opened eye for several minutes under running water. Obtain medical attention if irritation develops.	
First aid, inhalation	Supply fresh air. If required, provide artificial respiration. Keep patient warm. Consult doctor if symptoms persist. In case of unconsciousness place patient stably in side position for	
First aid, skin	Immediately rinse with water. If skin irritation continues, consult a doctor.	
ACGIH, TLV	mg/m³	984/ST/propan-2-ol, 328/ST/methanol
OSHA, PEL	mg/m³	1225/ST/propan-2-ol, 325/ST/methanol, 300/LT/butan-1-ol
ACGIH, TLV	ppm	400/ST/propan-2-ol, 250/ST/methanol
OSHA, PEL	ppm	400/ST/propan-2-ol, 250/ST/methanol, 100/LT/butan-1-ol
UN/NA class	-	1993

Tyzor GBA

PARAMETER	UNIT	VALUE
USE & PERFORMANCE		
Manufacturer		Dorf Ketal Chemicals, LLC
Outstanding properties		printing inks: improved the drying rate, solvent resistance, heat resistance, and adhesion to substrates. Coatings: increased surface hardness, adhesion promotion, scratch resistance, coloring effects, heat and light reflection, iridescence, and corrosion resistance.
Recommended for products		printing inks, coatings, paint additives.
Recommended applications		excellent adhesion promoters and crosslinkers in solvent- based printing inks (e.g. based on nitrocellulose), used as an additive in paints to crosslink −OH and −COOH functional polymers or binders, promote adhesion. Enhanced the performance of printing inks in a variety of applications, specifically flexible packaging.
Concentrations used	wt%	1-4

Tyzor GBO

PARAMETER	UNIT	VALUE
GENERAL INFORMATION		
Name	Tyzor GBO	
CAS #	-	17927-72-9
Composition	12.7% (TiO_2), titanium chelates with acetylacetone as a chelating agent, alcohol	
Empirical formula	C18H28O6Ti	
Formula		

Chemical class	titanate	
Active matter	wt%	60
PHYSICAL PROPERTIES		
State	-	liquid
Color	-	yellow to red
Boiling point	°C	232
Density at 25°C	kg/m³	950
Viscosity at 25°C	mPas	10
HEALTH & SAFETY		
NFPA classification	Flammability	3
	Health	2
	Reactivity	0
HMIS classification	Flammability	3
	Health	2
	Reactivity	0
Flash point	°C	12
Flash point method	-	PMCC
Animal testing, acute toxicity, Rat oral LD50	mg/kg	5045/propan-2-ol, 5628/methanol, 790/butan-1-ol
Animal testing, acute toxicity, Rabbit dermal LD50	mg/kg	12800/propan-2-ol, 15800/methanol, 3400/butan-1-ol
Effect of exposure, eye (human)	Causes serious eye damage.	
Effect of exposure, swallowing (human)	Harmful if swallowed.	
Effect of repeated or overexposure (human)	May cause damage to organs.	

Tyzor GBO

PARAMETER	UNIT	VALUE
Exposure, personal protection		Safety glasses, protective clothing based on chemical resistance data, chemical-resistant gloves, general and local exhaust ventilation.
First aid, eye		Immediately flush eyes with plenty of water, occasionally lifting the upper and lower eyelids. Check for and remove any contact lenses. Rinse opened eye for several minutes under running water. Obtain medical attention if irritation develops.
First aid, inhalation		Supply fresh air. If required, provide artificial respiration. Keep patient warm. Consult doctor if symptoms persist. In case of unconsciousness place patient stably in side position for
First aid, skin		Immediately rinse with water. If skin irritation continues, consult a doctor.
USE & PERFORMANCE		
Manufacturer		Dorf Ketal Chemicals, LLC
Outstanding properties		printing inks: improved the drying rate, solvent resistance, heat resistance, and adhesion to substrates. Coatings: increased surface hardness, adhesion promotion, scratch resistance, coloring effects, heat and light reflection, iridescence, and corrosion resistance.
Recommended for products		printing inks, coatings, paint additives.
Recommended applications		excellent adhesion promoters and cross-linkers in solvent-based printing inks (e.g., based on nitrocellulose), used as an additive in paints to crosslink −OH and −COOH functional polymers or binders, promote adhesion.

Tyzor IAM

PARAMETER	UNIT	VALUE
GENERAL INFORMATION		
Name		Tyzor IAM
CAS #	-	109037-78-7
Composition		12% (TiO_2), Bu phosphate Et alc. iso-Pr alc. complexes titanium-based phosphate complex
Chemical class		titanate
PHYSICAL PROPERTIES		
State	-	liquid
Odor	-	sweet
Color		colorless to light yellow
Boiling point	°C	80
Pour point	°C	< -50
Density at 25°C	kg/m³	960
Refractive index at 20°C	-	1.477
Solubility in water at 25°C	g/l	decomposes
Viscosity at 25°C	mPas	20
HEALTH & SAFETY		
NFPA classification	Flammability	3
	Health	2
	Reactivity	0
HMIS classification	Flammability	3
	Health	2
	Reactivity	0
DOT class	Flammable liquids, n.o.s.(Titanium, Bu phosphate Et alc. isoPr alc. Complexes) 3, II	
ICAO/IATA class	FLAMMABLE LIQUID, N.O.S. (Titanium, Bu phosphate Et alc. iso-Pr alc. Complexes) 3, II	
IMDG class	FLAMMABLE LIQUID, N.O.S. (Titanium, Bu phosphate Et alc. iso-Pr alc. Complexes) 3, II	
Autoignition temperature	°C	386
Flash point	°C	12
Hazardous combustion products	Carbon monoxide, carbon dioxide, metal oxides	

Tyzor IAM

PARAMETER	UNIT	VALUE
Hazardous ingredients, labelling		Flammable liquids, n.o.s.(Titanium, Bu phosphate Et alc. iso-Pr alc. Complexes), special provision 640D, ENVIRONMENTALLY HAZARDOUS
Animal testing, acute toxicity, Rat dermal LD50	mg/kg	>5000
Effect of exposure, eye (human)		Causes serious eye irritation.
Effect of exposure, inhalation (human)		May cause drowsiness or dizziness.
Exposure, personal protection		Safety glasses, protective clothing based on chemical resistance data, chemical-resistant gloves, general and local exhaust ventilation.
First aid, eye		Immediately flush eyes with plenty of water, occasionally lifting the upper and lower eyelids. Check for and remove any contact lenses. Rinse opened eye for several minutes under running water. Obtain medical attention if irritation develops.
First aid, inhalation		Supply fresh air. If required, provide artificial respiration. Keep patient warm. Consult doctor if symptoms persist. In case of unconsciousness place patient stably in side position for
First aid, skin		Immediately rinse with water. If skin irritation continues, consult a doctor.
UN/NA class	-	1993
ECOLOGICAL PROPERTIES		
Aquatic toxicity, *Daphnia magna*, 48-h LC50	mg/l	10000
USE & PERFORMANCE		
Manufacturer		Dorf Ketal Chemicals, LLC
Outstanding properties		provides good stability in solvent-borne printing inks, good adhesion promotion, and crosslinking properties resulting in improved water, solvent, and heat resistance, high reactivity during curing, low or no discoloration, no release of acetylacetone.
Recommended for products		printing ink for flexible packaging
Recommended applications		adhesion promoters and crosslinkers used as additives for flexographic and gravure printing inks.
Concentrations used	wt%	1-4

Tyzor TPT

PARAMETER	UNIT	VALUE
GENERAL INFORMATION		
Name		Tyzor TPT
CAS #	-	546-68-9
Composition		28.1% (TiO_2), tetra-isopropyl titanate
Formula		
Chemical class		titanate
Active matter	wt%	100
PHYSICAL PROPERTIES		
State	-	liquid
Color	-	clear, yellowish
Boiling point	°C	232
Melting point	°C	19.0
Density at 25°C	kg/m³	960
Kinematic viscosity at 20°C	cSt	1.477
Solubility in water at 25°C	g/l	moisture sensitive
Viscosity at 25°C	mPas	3.5
HEALTH & SAFETY		
Flash point	°C	49
Effect of exposure, swallowing (human)		Harmful if swallowed.
Effect of repeated or overexposure (human)		May cause damage to organs.
Exposure, personal protection		Safety glasses, protective clothing based on chemical resistance data, chemical-resistant gloves, general and local exhaust ventilation.
First aid, eye		Immediately flush eyes with plenty of water, occasionally lifting the upper and lower eyelids. Check for and remove any contact lenses. Rinse opened eye for several minutes under running water. Obtain medical attention if irritation develops.
USE & PERFORMANCE		
Manufacturer		Dorf Ketal Chemicals, LLC

Tyzor TPT

PARAMETER	UNIT	VALUE
Outstanding properties		improved adhesion of the ink to the substrate, enhanced curing and heat resistance, improved chemical, solvent, and water resistance, decreased drying times, lowered cure temperatures, and increased lamination bond strength.
Recommended for polymers		PE, PP, PA, cellophane, PVC, metalized plastic
Recommended for products		printing inks, coatings, paint additives.
Recommended applications		adhesion promoters between inorganic (e.g., aluminum) and organic substrates (PE, PP, PA, cellophane, PVC), en-hanced the performance of printing inks in a variety of applications, specifically flexible packaging

3.37 Zirconates
Tyzor® 212

PARAMETER	UNIT	VALUE
GENERAL INFORMATION		
Name		Tyzor® 212
Composition		zirconium chelate containing 52% of acrive ingredient in 1-propanol
Zirconium content	wt%	12
PHYSICAL PROPERTIES		
State	-	liquid
Odor	-	alcohol-like
Color		yellow
Boiling point	°C	97
Melting point	°C	<-25
Density at 20°C	kg/m³	1060
HEALTH & SAFETY		
NFPA classification	Flammability	4
	Health	2
	Instability	0
HMIS classification	Flammability	4
	Health	2
	Physical hazard	0
Carcinogenicity	No known significant effects or critical hazards.	
Flash point	°C	22
Flash point method	-	PMCC
First aid, eye	Check for and remove any contact lenses. Immediately flush eyes with plenty of water for at least 15 minutes, occasionally lifting the upper and lower eyelids. Get medical attention immediately.	
First aid, inhalation	Move exposed person to fresh air. If not breathing, if breathing is irregular or if respiratory arrest occurs, provide artificial respiration or oxygen by trained personnel. Loosen tight clothing such as a collar, tie, belt or waistband. Get medical attention immediately	
First aid, skin	In case of contact, immediately flush skin with plenty of water for at least 15 minutes while removing contaminated clothing and shoes. Wash clothing before reuse. Clean shoes thoroughly before reuse. Get medical attention immediately.	

3.37 Zirconates
Tyzor® 212

PARAMETER	UNIT	VALUE
UN #	-	1274
USE & PERFORMANCE		
Manufacturer	Dorf Ketal	
Outstanding properties	Tyzor® 212 will form bonds with hydroxyl and carboxyl functionality and work to crosslink these sites on a polymer. In oil well fracturing fluids with a pH of 8.5-11 Tyzor® 212 is used for delayed crosslinking of guar dispersions. At temperatures above 120°F, it is activated and causes a dramatic increase in viscosity. Tyzor® 212 is also used to crosslink polymers used in inks, paints, and coatings to improve the durability of the dried film and to increase the speed of cure. Additionally, Tyzor® 212 can be used to treat surfaces like metal and glass to promote adhesion.	
Recommended for products	coatings, energy, inks, oil well fracturing fluids, paints	

Tyzor® 217

PARAMETER	UNIT	VALUE
GENERAL INFORMATION		
Name	Tyzor® 217	
CAS #	-	60676-90-6
Composition	zirconium lactate	
IUPAC name	1-hydroxy-1-oxopropan-2-olate;zirconium(4+)	
Empirical formula	-	C12H20O12Zr
Molecular mass	daltons	447.50
Zirconium content	wt%	5.4
PHYSICAL PROPERTIES		
State	-	liquid
Odor	-	characteristic
Color	colorless to light yellow	
Boiling point	ºC	100
Melting point	ºC	0
Density at 20ºC	kg/m³	1200
pH	-	6.5-7.5
HEALTH & SAFETY		
NFPA classification	Flammability	0
	Health	0
	Instability	0
HMIS classification	Flammability	0
	Health	0
	Physical hazard	0
Carcinogenicity	None of the ingredients are listed by IARC and NTP	
First aid, eye	Remove contact lenses if worn, if possible. Rinse opened eye for several minutes under running water. Then consult a doctor.	
First aid, inhalation	Supply fresh air. If required, provide artificial respiration. Keep patient warm. Consult doctor if symptoms persist. In case of unconsciousness place patient stably in side position for transportation.	
First aid, skin	Immediately rinse with water. If skin irritation continues, consult a doctor	
USE & PERFORMANCE		
Manufacturer	Dorf Ketal	

Tyzor® 217

PARAMETER	UNIT	VALUE
Outstanding properties		reacts with hydroxyl groups, carboxyl groups, and other nucleophiles. It is used to treat surfaces like glass and metal to improve the adhesion of paints and coatings. It can also be added to paint, coating, and ink formulations to crosslink binders and pigments to improve cohesive strength. In addition, Tyzor® 217 is used to treat paper and textile surfaces to increase hardness and improve hydrophobicity.
Recommended for products		coatings, inks, adhesives, fracturing fluids

Tytan APZ900

PARAMETER	UNIT	VALUE
GENERAL INFORMATION		
Name		Tytan APZ900
Composition		35% zirconium tetrapropanolate, 55% proprietary hydroxy acid ester, 10% 1-propanol
Common name		zirconium, hydroxy acid ester complex
Titanium content	wt%	9.6
PHYSICAL PROPERTIES		
State	-	liquid
Odor	-	alcohol-like
Color	-	yellow
Initial boiling point	°C	97
Density at 20°C	kg/m³	1110-1120
Viscosity at 25°C	mPas	88
HEALTH & SAFETY		
Flash point	°C	22
Flash point method	-	CC
First aid, eye		Immediately irrigate with eyewash solution or clean water, holding the eyelids apart, for at least 15 minutes. Obtain medical attention.
First aid, inhalation		Remove patient form exposure. Apply artificial respiration if breathing has ceased or showed signs of failing. Obtain medical attention.
First aid, skin		Wash immediately with water followed by soap and water.
UN/NA #	-	1993
USE & PERFORMANCE		
Manufacturer		Borica Co., Ltd.

Tytan APZ900

PARAMETER	UNIT	VALUE
Outstanding properties		Tytan APZ900 is a zirconium complex with α-hydroxy acid ester that can be used as an adhesion promoter for printing, crosslinker for paints and as a catalyst for esterification. Due to its chelated structure, it shows delayed reaction at ambient temperature and thus facilitates storage stability of inks. Compared with organo-titanate adhesion promoters, it shows good compatibility with PVB, CAB and CAP inks and provides minimal discoloration during the sterilizing process in the printing of milk pouches.
Recommended for polymers		CAB, CAP, PVB
Recommended for products		paints, inks

Zircoate Zr-402

PARAMETER	UNIT	VALUE
GENERAL INFORMATION		
Name		Zircoate Zr-402
CAS #	-	1071-76-7
EC number	-	213-995-3
General description		zirconate in butanol solution
Common synonym		tetra-n-butyl zirconate
Empirical formula		(n-C4H9O)4Zr
Formula		

PARAMETER	UNIT	VALUE
Molecular mass	daltons	383.68
PHYSICAL PROPERTIES		
State	-	solution of solid
Color	-	white
Density at 20°C	kg/m³	1050
Viscosity at 20°C	mPas	100
USE & PERFORMANCE		
Manufacturer		Xanadu Technologies Limited
Outstanding properties		crosslinking agent, catalyst, and as an adhesion promoter. It increases surface hardness, improves adhesion promotion and scratches resistance, enhances coloring effects and iridescence, and improves heat and light reflection and corrosion resistance.

CPI Antony Rowe
Eastbourne, UK
March 06, 2023